Physics of Energy Sources

The Manchester Physics Series

General Editors
J.R. FORSHAW, H.F. GLEESON, F.K. LOEBINGER

School of Physics and Astronomy,
University of Manchester

Properties of Matter	B.H. Flowers and E. Mendoza
Statistical Physics *Second Edition*	F. Mandl
Electromagnetism *Second Edition*	I.S. Grant and W.R. Phillips
Statistics	R.J. Barlow
Solid State Physics *Second Edition*	J.R. Hook and H.E. Hall
Quantum Mechanics	F. Mandl
Computing for Scientists	R.J. Barlow and A.R. Barnett
The Physics of Stars *Second Edition*	A.C. Phillips
Nuclear Physics	J.S. Lilley
Introduction to Quantum Mechanics	A.C. Phillips
Dynamics and Relativity	J.R. Forshaw and A.G. Smith
Vibrations and Waves	G.C. King
Mathematics for Physicists	B.R. Martin and G. Shaw
Particle Physics *Fourth Edition*	B.R. Martin and G. Shaw
Physics of Energy Sources	G.C. King

Physics of Energy Sources

GEORGE C. KING

School of Physics and Astronomy
Manchester University

Registered Offices
John Wiley & Sons, Inc., 111 River Street, Hoboken, NJ 07030, USA
John Wiley & Sons, Ltd., The Atrium, Southern Gate, Chichester, West Sussex, PO19 8SQ, UK

Editorial Office
The Atrium, Southern Gate, Chichester, West Sussex, PO19 8SQ, UK

For details of our global editorial offices, customer services, and more information about Wiley products visit us at www.wiley.com.

Library of Congress Cataloging-in-Publication Data

Name: King, George C., author.
Title: Physics of energy sources / George C. King, University of Manchester, UK.
Other titles: Manchester physics series.
Description: First edition. | Chichester, UK ; Hoboken, New Jersey : John Wiley & Sons, Inc.,
   [2018] | Series: Manchester physics series
Identifiers: LCCN 2016050484 (print) | LCCN 2016054262 (ebook) | ISBN
   9781119961673 (hardback ; cloth) | ISBN 111996167X (hardback ; cloth) |
   ISBN 9781119961680 (pbk.) | ISBN 1119961688 (pbk.) | ISBN 9781118698440
   (pdf) | ISBN 9781118698426 (epub)
Subjects: LCSH: Power resources. | Renewable energy sources. | Physics.
Classification: LCC TJ163.2 .K53 2018 (print) | LCC TJ163.2 (ebook) | DDC
   621.3101/53–dc23
LC record available at https://lccn.loc.gov/2016050484

Cover design: Wiley
Cover images: Top circle: Gemasolar plant © SENER/TORRESOL ENERGY;
Bottom circle: Courtesy of ESA https://images.nasa.gov/#/details-PIA03149.html

Set in 11/13pt Computer Modern by Aptara Inc., New Delhi, India.
Printed and bound in Singapore by Markono Print Media Pte Ltd

10 9 8 7 6 5 4 3 2 1

'But why are such terrific efforts made just to find new particles?' asked Mr Tompkins.

'Well, this is science,' replied the professor, 'the attempt of the human mind to understand everything around us, be it giant stellar galaxies, microscopic bacteria, or these elementary particles. It is interesting and exciting, and that is why we are doing it.'

From *Mr Tompkins Tastes a Japanese Meal*, by George Gamow (*Mr Tompkins in Paperback*, Cambridge University Press (1965), p.186).

To my family: Michele, May, George and May.

Contents

Editors' preface to the Manchester Physics Series   xi

Author's preface   xiii

1 Introduction   **1**
  1.1 Energy consumption   1
  1.2 Energy sources   3
  1.3 Renewable and non-renewable energy sources   5
  1.4 The form and conversion of energy   6
      1.4.1 Thermal energy sources   7
      1.4.2 Mechanical energy sources   7
      1.4.3 Photovoltaic sources   7
      1.4.4 Energy storage   8
  Problems 1   9

2 The atomic nucleus   **11**
  2.1 The composition and properties of nuclei   12
      2.1.1 The composition of nuclei   12
      2.1.2 The size of a nucleus   14
      2.1.3 The distributions of nuclear matter and charge   19
      2.1.4 The mass of a nucleus   21
      2.1.5 The charge of a nucleus   24
      2.1.6 Nuclear binding energy   27
      2.1.7 Binding energy curve of the nuclides   30
      2.1.8 The semi-empirical mass formula   32
  2.2 Nuclear forces and energies   35
      2.2.1 Characteristics of the nuclear force   35
      2.2.2 Nuclear energies   36
      2.2.3 Quantum mechanical description of a particle
            in a potential well   39
  2.3 Radioactivity and nuclear stability   47
      2.3.1 Segré chart of the stable nuclides   48
      2.3.2 Decay laws of radioactivity   49
      2.3.3 α, β and γ decay   57
  Problems 2   67

3 Nuclear power   **71**
  3.1 How to get energy from the nucleus   71
  3.2 Nuclear reactions   73
      3.2.1 Nuclear reactions   73

	3.2.2	*Q*-value of a nuclear reaction	74
	3.2.3	Reaction cross-sections and reaction rates	76
3.3	Nuclear fission		82
	3.3.1	Liquid-drop model of nuclear fission	83
	3.3.2	Induced nuclear fission	86
	3.3.3	Fission cross-sections	87
	3.3.4	Fission reactions and products	88
	3.3.5	Energy in fission	90
	3.3.6	Moderation of fast neutrons	92
	3.3.7	Uranium enrichment	93
3.4	Controlled fission reactions		97
	3.4.1	Chain reactions	97
	3.4.2	Control of fission reactions	101
	3.4.3	Fission reactors	103
	3.4.4	Commercial nuclear reactors	105
	3.4.5	Nuclear waste	107
3.5	Nuclear fusion		109
	3.5.1	Fusion reactions	110
	3.5.2	Energy in fusion	111
	3.5.3	Coulomb barrier for nuclear fusion	113
	3.5.4	Fusion reaction rates	113
	3.5.5	Performance criteria	115
	3.5.6	Controlled thermonuclear fusion	117
	Problems 3		123
4	**Solar power**		**127**
4.1	Stellar fusion		128
	4.1.1	Star formation and evolution	128
	4.1.2	Thermonuclear fusion in the Sun: the proton–proton cycle	131
	4.1.3	Solar radiation	132
4.2	Blackbody radiation		134
	4.2.1	Laws of blackbody radiation	135
	4.2.2	Emissivity	137
	4.2.3	Birth of the photon	141
4.3	Solar radiation and its interaction with the Earth		145
	4.3.1	Characteristics of solar radiation	145
	4.3.2	Interaction of solar radiation with Earth and its atmosphere	147
	4.3.3	Penetration of solar energy into the ground	155
4.4	Geothermal energy		159
	4.4.1	Shallow geothermal energy	160
	4.4.2	Deep geothermal energy	161
4.5	Solar heaters		162
	4.5.1	Solar water heaters	162
	4.5.2	Heat transfer processes	165
	4.5.3	Solar thermal power systems	172
4.6	Heat engines: converting heat into work		174
	4.6.1	Equation of state of an ideal gas	175

		4.6.2	Internal energy, work and heat: the first law of thermodynamics	177
		4.6.3	Specific heats of gases	181
		4.6.4	Isothermal and adiabatic expansion	183
		4.6.5	Heat engines and the second law of thermodynamics	185
	Problems 4			196

5 Semiconductor solar cells — **201**

	5.1	Introduction		201
	5.2	Semiconductors		204
		5.2.1	The band structure of crystalline solids	204
		5.2.2	Intrinsic and extrinsic semiconductors	208
	5.3	The p–n junction		214
		5.3.1	The p–n junction in equilibrium	214
		5.3.2	The biased p–n junction	217
		5.3.3	The current–voltage characteristic of a p–n junction	219
		5.3.4	Electron and hole concentrations in a semiconductor	222
		5.3.5	The Fermi energy in a p–n junction	227
	5.4	Semiconductor solar cells		229
		5.4.1	Photon absorption at a p–n junction	229
		5.4.2	Power generation by a solar cell	231
		5.4.3	Maximum power delivery from a solar cell	235
		5.4.4	The Shockley–Queisser limit	238
		5.4.5	Solar cell construction	240
		5.4.6	Increasing the efficiency of solar cells and alternative solar cell materials	243
	Problems 5			248

6 Wind power — **251**

	6.1	A brief history of wind power		251
	6.2	Origin and directions of the wind		253
		6.2.1	The Coriolis force	253
	6.3	The flow of ideal fluids		256
		6.3.1	The continuity equation	257
		6.3.2	Bernoulli's equation	258
	6.4	Extraction of wind power by a turbine		263
		6.4.1	The Betz criterion	265
		6.4.2	Action of wind turbine blades	268
	6.5	Wind turbine design and operation		271
	6.6	Siting of a wind turbine		277
	Problems 6			280

7 Water power — **283**

	7.1	Hydroelectric power		284
		7.1.1	The hydroelectric plant and its principles of operation	284
		7.1.2	Flow of a viscous fluid in a pipe	286
		7.1.3	Hydroelectric turbines	288
	7.2	Wave power		291
		7.2.1	Wave motion	292

	7.2.2	Water waves	306
	7.2.3	Wave energy converters	319
7.3	Tidal power		324
	7.3.1	Origin of the tides	325
	7.3.2	Variation and enhancement of tidal range	335
	7.3.3	Harnessing tidal power	341
Problems 7			346

8   Energy storage — **349**

8.1	Types of energy storage		350
8.2	Chemical energy storage		351
	8.2.1	Biological energy storage	351
	8.2.2	Hydrogen energy storage	351
8.3	Thermal energy storage		352
8.4	Mechanical energy storage		355
	8.4.1	Pumped hydroelectric energy storage	355
	8.4.2	Compressed air energy storage	357
	8.4.3	Flywheel energy storage	361
8.5	Electrical energy storage		364
	8.5.1	Capacitors and super-capacitors	365
	8.5.2	Superconducting magnetic storage	367
	8.5.3	Rechargeable batteries	368
	8.5.4	Fuel cells	370
8.6	Distribution of electrical power		372
Problems 8			374

Solutions to problems   377

Index   397

Editors' preface to the Manchester Physics Series

The Manchester Physics Series is a set of textbooks at first degree level. It grew out of the experience at the University of Manchester, widely shared elsewhere, that many textbooks contain much more material than can be accommodated in a typical undergraduate course; and that this material is only rarely so arranged as to allow the definition of a short self-contained course. The plan for this series was to produce short books so that lecturers would find them attractive for undergraduate courses, and so that students would not be frightened off by their encyclopaedic size or price. To achieve this, we have been very selective in the choice of topics, with the emphasis on the basic physics together with some instructive, stimulating and useful applications.

Although these books were conceived as a series, each of them is self-contained and can be used independently of the others. Several of them are suitable for wider use in other sciences. Each Author's Preface gives details about the level, prerequisites, etc., of that volume.

The Manchester Physics Series has been very successful since its inception over 40 years ago, with total sales of more than a quarter of a million copies. We are extremely grateful to the many students and colleagues, at Manchester and elsewhere, for helpful criticisms and stimulating comments. Our particular thanks go to the authors for all the work they have done, for the many new ideas they have contributed, and for discussing patiently, and often accepting, the suggestions of the editors.

Finally we would like to thank our publishers, John Wiley & Sons, Ltd., for their enthusiastic and continued commitment to the Manchester Physics Series.

J. R. Forshaw
H. F. Gleeson
F. K. Loebinger
August 2014

Author's preface

We live in a technological age where energy plays a central role. Because of its importance, issues regarding the availability and cost of energy and the environmental impact are never far from the daily news. This book describes the main sources of energy that are available to us together with the underlying physics that governs them. In particular, it deals with nuclear power, solar power, wind power, wave and tidal power. The book also describes the ways in which energy can be stored for future use. Studying the physics of energy sources has various advantages. First, such a study encompasses a wide range of physics from classical physics to quantum physics. In this way it supports other undergraduate courses in the physical sciences and engineering. Secondly, energy sources represent real applications of fundamental physics, and although energy sources are being continuously developed, the underlying physics that governs them remains the same. The book is addressed mainly to science and engineering students, who require knowledge of the physical principles governing the operation of energy sources. It is based on an introductory 24-lecture course entitled "Physics of Energy Sources" given by the author at the University of Manchester. The course was attended by first- and second-year undergraduate students taking physics or a joint honours degree course with physics, but it should also be useful for students on engineering and environmental science degree courses. The book covers the topics given in the course although it amplifies the material delivered in the lectures. A basic knowledge of differentiation and integration is assumed and simple differential equations are used, while undue mathematical complication and detail are avoided.

The organisation of the book is as follows. Chapter 1 deals with energy consumption and outlines the main energy resources available to us and the physical characteristics of energy sources. The transformation of energy from one form to another is considered, together with the role of energy storage. Chapter 2 deals with the properties of the atomic nucleus, nuclear forces and radioactivity, and forms a foundation for the understanding of nuclear fission and nuclear fusion. These are dealt with in Chapter 3, which describes how we get energy from the nucleus, by both the fission of heavy nuclei and the fusion of light nuclei. Chapter 4 describes the origin and properties of solar radiation, its interaction with the Earth and how thermal energy can be harvested from sunlight. The conversion of thermal energy into mechanical energy is also discussed. Chapter 5 is devoted to semiconductor solar cells and includes a description of the band structure of semiconductors and the action of the p–n junction. Chapter 6 deals with the harnessing of wind power by wind turbines and introduces the elements of fluid mechanics. Chapter 7 describes water power in its various forms, including hydroelectric power, wave power and tidal power. This includes a discussion of wave motion and also the origin of the tides. Chapters 4, 5, 6 and 7 are thus about renewable energy sources. Finally, Chapter 8 describes various ways in which energy can be stored for future use. Fossil fuels are not dealt with explicitly. However, these fuels are usually used to produce thermal energy that drives steam turbines. In this respect, they have much in common with, for example nuclear power, where again the energy is first converted into thermal energy to drive steam turbines, and similar thermodynamic principles apply.

Worked examples are included in the text. In addition, each chapter is accompanied by a set of problems that form an important part of the book. These have been designed to deepen the understanding of the reader and develop their skill and self-confidence in the use of the physics. Hints and solutions to these problems are given at the end of the book. It is, of course, beneficial for the reader to try to solve the problems before consulting the solutions.

I am particularly indebted to Fred Loebinger who was my editor throughout the writing of the book. He read the manuscript with great care and physical insight and made numerous and valuable comments and suggestions. I am grateful to the members of the Manchester Physics Series Editorial Board – Fred Loebinger, Helen Gleeson, Jeff Forshaw and Jenny Cossham – for helpful suggestions regarding the content and organisation of the book. I am also grateful to my colleagues David Binks, Mark Dickinson and Jeff Forshaw for valuable and enlightening discussions about various topics in physics, and to Michele Siggel-King for constructing some of the figures.

George C. King
January 2017

1

Introduction

Energy is essential to our lives. Our bodies need energy to function
and to perform physical activities. And the technological age in which
we live needs a reliable energy supply for heating, lighting, commu-
nication, transport, food production, manufacturing goods, and so
on. Because of their importance, issues such as the supply and cost
of energy and the environmental impact make frequent appearances
on the daily news. In this introductory chapter we consider energy
consumption and the energy resources available to us. We consider
the general characteristics of energy sources and the transformation
of energy from one form to another to suit the end use. We also
consider the role of energy storage.

1.1 Energy consumption

We consume energy in maintaining our vital bodily functions, such as
the operation of the heart and lungs, the maintenance of body temperature, brain function and digestion of the food we eat. Roughly
speaking, in maintaining these functions we consume energy at the
rate of $\sim$100 J/s; a power of $\sim$100 W. We also expend energy when
we do physical work. Suppose, for example, that we climb stairs
and rise at the rate of 0.5 m/s in vertical height. If our mass is
75 kg, our rate of doing work is 75 kg $\times$ 9.8 m/s^2 $\times$ 0.5 m/s $=$ 368 W.
The amount of physical activity that a person does depends on
their lifestyle. Suppose, however, that, averaged over the course of a
24-hour period, we consume energy at the average rate of 125 W in
maintaining our metabolic rate and performing physical work. This
amounts to $\sim$10 MJ of energy per day. This energy comes from the
chemical energy stored in the food that we eat; a tin of baked beans,

Physics of Energy Sources, First Edition. George C. King.
© 2018 John Wiley & Sons, Ltd. Published 2018 by John Wiley & Sons, Ltd.

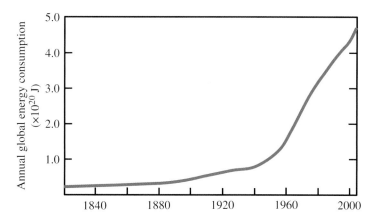

Figure 1.1 Illustration of the dramatic rise in annual global energy consumption that occurred between 1820 and 2010.

for comparison, contains ∼1.5 MJ of energy. We also need energy to heat and light our houses, to run washing machines and refrigerators, to travel to work, to use computers, to fly to a foreign country on holiday, and so on. Furthermore, energy is needed to produce the food we eat, to manufacture and transport the goods we buy, etc. Overall, the total energy consumption per person per day in the UK is ∼450 MJ. When we consider energy consumption, it is perhaps more meaningful to use the kilowatt-hour (kWh) unit of energy. This is the energy consumed by a 1 kW electric fire in 1 hour and the conversion factor is 1 kWh = 3.6 MJ. So 450 MJ/day = 125 kWh/day, which is the amount of power consumed by five 1 kW electric fires running day and night. This figure of 125 kWh per person per day is typical for a European country. In the USA, the energy consumption per person is about twice as high, while in underdeveloped countries it is considerably lower. Averaged over all countries, energy consumption is ∼60 kWh per person per day and this amounts to a total global energy consumption of ∼5×10^{20} J/year.

Global consumption of energy continues to increase because of advances in technology, growth in world population and economic growth, factors that are interrelated. Figure 1.1 illustrates the dramatic increase in annual global consumption of energy that occurred between 1820 and 2010. As an example of a technological advance, James Watt patented his steam engine in 1769 and this enabled the Earth's deposits of fossil fuels such as coal to be unlocked. This signalled a sharp increase in energy consumption, and once industrialisation occurred, the rate of consumption increased dramatically; over the course of the 20th century, global use of energy increased more than 10-fold. The world's population has also increased dramatically over the last few hundred years, rising from 1 billion in 1800 to 7.4 billion in 2016. Indeed the curves for global energy consumption and global population follow each other quite closely. Presently, global population is increasing at a rate of just over 1% per year. The rate of economic growth is different for different countries. However, averaged over all countries, economic growth also increases at about 1% per year. Taking the various factors into account, it is predicted that the growth in global energy consumption over the next 30 years will be $\sim$2% per year.

A complementary aspect of energy consumption is the efficiency with which energy is used. No source of energy is cheap or occurs without some form of environmental disruption, and it is important that energy is used as efficiently as possible. One particular advance can be seen in the use of electric light bulbs. It is estimated that lighting consumes about 20% of the world's electricity. Traditional incandescent light bulbs with a wire filament are only about 5% efficient, while new types of lighting are much more efficient. LED lighting, for example is about 20% efficient.

1.2 Energy sources

The main sources of energy available to us are:

- fossil fuels

- solar energy

- biofuels

- wind energy

- nuclear energy

- waves and tidal energy

- hydroelectric energy

- geothermal energy.

Most of the energy available to us comes directly or indirectly from the Sun. The Sun gets its energy from nuclear fusion reactions that heat its core to a temperature of $\sim$10^7 K. Energy is transported to the Sun's surface and maintains the surface at a temperature of $\sim$6000 K. The hot surface acts as a blackbody radiator emitting electromagnetic radiation and it is this radiation or sunlight that delivers solar energy to the Earth. The total solar power that falls on the Earth is enormous, $\sim$1.7 $\times$ 10^{17} W, which is about 25 MW for every person in the world.

Sunlight provides us with energy in various ways. *Photosynthesis* is the process by which plants and other organisms use sunlight to transform water, carbon dioxide, and minerals into oxygen and organic compounds. Fossil fuels that we burn, including oil, coal and natural gas, were formed over millions of years by the action of heat and pressure on the fossils of dead plants. *Bioenergy* comes from biofuels that are produced directly or indirectly from organic matter, including plant material and animal waste; an example is rapeseed oil, which produces oil for fuel. Wood also fits into this category and, indeed, burning wood is by far the oldest source of energy used by humankind. Hydroelectric power, wind power and wave power can also be traced back to the Sun. Solar energy heats water on the Earth's surface, causing it to evaporate. The water vapour condenses into clouds and falls as precipitation. This fills the reservoirs of hydroelectric plants, and the potential energy of the stored water provides a supply of energy. The Sun's warming of the Earth's surface produces winds that circulate the globe and which can be used to drive wind turbines. The winds also produce ocean waves whose kinetic energy can be harvested. More directly, solar energy can be captured by solar water heaters or alternatively by photovoltaic devices, which convert sunlight into electrical energy directly. The Sun even plays a role in the formation of the tides, which result from the motions of the Moon, Sun and Earth. The rising and falling tides contain potential and kinetic energies that can be harvested.

We also get energy from human-induced nuclear reactions. So far, nuclear power has exploited fission reactions of heavy, radioactive elements such as uranium. However, as we will see, nuclear fusion of light elements such as deuterium and tritium has great potential as an energy source of the future. Finally, the Earth itself is a source of energy called geothermal energy. This is stored as thermal energy beneath the Earth's surface. It results from the processes involved in the formation of the Earth and from the decay of radioactive elements within its crust and appears, for example, as hot water springs in various regions of the world.

The annual consumption of energy with respect to energy source varies from country to country and from year to year. However, to get an impression of energy consumption by energy source, Figure 1.2 shows the data for the USA in 2014. We see that 81% of energy consumption came from fossil fuels, while nuclear energy and renewable sources provided the remainder.

The energy sources listed above are called *primary* energy sources. Electricity, on the other hand, is described as a *secondary* energy source, as it derives from the conversion of energy from a primary source. Electricity has significant advantages as an energy carrier. It can be conveniently transported and distributed via a national grid, and for many energy needs it is easier to use than the primary energy source itself. The other important secondary energy source is hydrogen gas, which can be burnt or used in *fuel cells*.

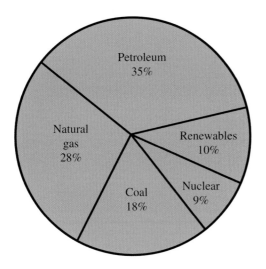

Figure 1.2 Annual energy consumption for the USA in 2014, by energy source – 81% of energy consumption came from fossil fuels, while nuclear energy and renewable sources provided the remainder.

1.3   Renewable and non-renewable energy sources

Energy sources can be classified as either renewable or non-renewable. We define a renewable source as one in which the energy comes from a natural and persistent flow of energy that occurs in the environment. Hydroelectric energy, solar energy, wind energy, wave energy, tidal energy and geothermal energy are renewable sources and so is bioenergy, so long as the trees and crops are replaced. Non-renewable sources are finite stores of energy, such as coal and oil, and nuclear fuels such as uranium. These non-renewable sources are not sustainable in the longer term. The distinction between renewable and non-renewable energy sources is illustrated in Figure 1.3. Closely associated with renewable energy sources is *sustainability*.

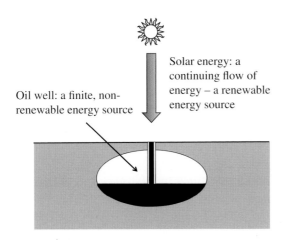

Figure 1.3 Illustration of the distinction between renewable and non-renewable energy sources, using the examples of solar energy and energy from the fossil fuel oil. Solar energy flows continuously from the Sun, while reserves of oil are finite.

Sustainable development can be broadly defined as living, producing and consuming in a manner that meets the needs of the present without compromising the ability of future generations to meet their needs. Renewable sources are much more compatible with sustainable development than non-renewable sources.

Fossil fuels, although non-renewable, have the advantage that their energy densities are high, i.e. they yield a large amount of energy per unit mass or per unit volume. For example, a litre of oil contains 35 MJ of energy. Moreover, the output of a power plant using fossil fuels is controllable. However, the burning of fossil fuels produces substantial amounts of pollution and increases the concentration of CO_2 in the atmosphere, which enhances the greenhouse effect. Nuclear fuels have an even higher energy density. In fact, their energy density is $\sim 10^6$ times greater than that of a fossil fuel and, again, the output of a nuclear reactor can be controlled. But, of course, nuclear power presents its own challenges, including the long-term storage of spent nuclear fuel. In general, renewable energy sources produce less atmospheric pollution than fossil fuels and do not emit CO_2 gas directly, the exception being the burning of biofuels. Furthermore, because they extract their energy from natural flows of energy that are already compatible with the environment, they produce minimum *thermal* pollution. They may also offer the possibility of a country becoming self-sufficient in energy. A disadvantage of some renewable sources is that they produce energy intermittently; solar cells need the Sun and wind turbines need the wind. Hence, the energy they deliver cannot be controlled in the same way as, say, a nuclear power station. However, this disadvantage is mitigated by the use of energy storage. Renewable energy is also usually more expensive than that obtained from fossil fuels, and renewable energy plants may have a significant impact on the local environment. For example, a hydroelectric plant can greatly affect the local ecology and may also cause the displacement of local inhabitants. At present, renewable sources contribute a much smaller fraction of global energy than do fossil fuels, although this fraction is expected to increase significantly in the future; presently, for example, renewable energy accounts for roughly a fifth of global electricity production.

1.4   The form and conversion of energy

According to the kind of energy they deliver, we can broadly divide sources into the following categories: thermal energy sources, mechanical energy sources and photovoltaic sources.

1.4.1   Thermal energy sources

Fossil fuels are a store of chemical energy that is a form of potential energy associated with the chemical bonds of the molecules of the fuel. Burning the fuel breaks these bonds and releases energy, mostly in the form of thermal energy. Nuclear fission reactions release potential energy that is stored in the nuclei that undergo fission and this energy becomes converted into thermal energy in the core of the reactor. In both cases, the thermal energy is converted into mechanical energy by a steam turbine, which is a type of *heat engine*. A fundamental aspect of the conversion of thermal energy into mechanical energy is that it is governed by the laws of thermodynamics, and these limit the efficiency of the conversion process, as we shall see in Chapter 4. For example, the efficiency of a conventional, coal-fired power plant for converting thermal energy into mechanical energy may be ~35%. Alternatively, thermal energy can be used to heat buildings directly, thus avoiding thermodynamic limitations. Here, the thermal energy is transported as steam through large-diameter insulated pipes, and such *district heating* is common in some countries.

1.4.2   Mechanical energy sources

These sources deliver mechanical energy directly, as in the case of a wind turbine. The wind causes the blades of a turbine to rotate and the rotation of the turbine shaft delivers mechanical energy directly. Hence, the thermodynamic limitations of thermal to mechanical energy conversion are avoided. Nevertheless, methods of extracting mechanical energy from a particular source also have inherent limitations to their efficiency. In the case of a wind turbine, we will see that the maximum efficiency of a turbine for extracting energy from the wind is 59%.

1.4.3   Photovoltaic sources

Photovoltaic solar cells have the advantage that they convert sunlight into electrical energy directly so again the thermodynamic limitations of thermal to mechanical energy conversion are avoided. However, as we shall see, there are a number of factors that limit the efficiency of solar cells. In practice, the efficiency of a commercial solar cell for converting solar energy into electrical energy is ~20%.

The efficiency with which a particular source of energy can be transformed from one form to another is described as the *quality* of the source. Waste hot water from a manufacturing process at 60°C would be described as low quality. This is because at this relatively low temperature the efficiency of a heat engine to convert the thermal energy of the water into mechanical energy is very low, less than 12%. On the other hand, electricity has high quality. For example, it can be converted into mechanical energy by an electric motor with very high efficiency, $\sim 95\%$.

1.4.4 Energy storage

Energy has to be provided when it is needed. Some energy sources, such as nuclear power stations and hydroelectric plants can provide a continuous supply of energy. On the other hand, sources such as wind turbines and solar cells produce energy intermittently; these sources may not generate enough energy when it is needed or, alternatively, they may generate excess energy. Energy storage systems allow the excess energy to be stored and used at a later time. And with increasing use of intermittent renewable sources, energy storage is becoming increasingly important. It is also the case that demand for energy varies substantially throughout the seasons and throughout the day; it tends to peak in the morning and afternoon, and fall to a minimum during the night. Power stations should ideally be operated at a fairly constant output level and close to where they operate most efficiently. But it does not make economic sense, and is a waste of energy, to have a power supply system whose capacity exceeds peak demand. Stored energy can supply the extra energy when required. So energy storage is able to even out variations in both supply and demand.

The chapters that follow deal with nuclear power, solar power, wind power and water power. Chapter 2 deals with the properties of the atomic nucleus, nuclear forces and radioactivity, and forms a foundation for the understanding of nuclear fission and nuclear fusion. These are dealt with in Chapter 3, which describes how we get energy from the nucleus, by both the fission of heavy nuclei and the fusion of light nuclei. Chapter 4 describes the origin and characteristics of solar radiation, its interaction with the Earth and how thermal energy can be harvested from sunlight. We also discuss the conversion of thermal energy into mechanical energy. Chapter 5 is devoted to semiconductor solar cells and their underlying principles of operation including the action of the *p–n junction*. Chapter 6 deals with the harnessing of wind power by wind turbines. Chapter 7 describes water power and the various ways in which the energy is harvested, including hydroelectric, wave and tidal power. Finally, Chapter 8 describes various ways in which energy can be stored for future use. Fossil fuels are not dealt with explicitly. However, fossil fuels are used to produce thermal energy that, in turn, produces steam to drive a turbine. In this respect they have much in common with, for example, nuclear power, where again the energy is first converted into thermal energy to drive steam turbines, and similar thermodynamic principles apply.

Problems 1

1.1 Show that $1\,\text{kWh} = 3.6\,\text{MJ}$.

1.2 Estimate the amount of energy (in kWh) that is used for lighting in an average house in a year.

1.3 (a) Estimate the amount of energy in kWh/passenger for a round trip between London and New York, which are separated by a flying distance of 5600 km. Assume that the plane uses 12 L/km and that the plane has 400 passengers on board and that the fuel has an energy density of 36 MJ/L. (b) Compare this value with the amount of energy required to commute by car 5 days a week for 48 weeks. Take the distance travelled each day to be 30 km, the fuel consumption to be 15 km/L, the energy density of the fuel to also be 36 MJ/L and assume that there are no passengers in the car.

1.4 A particular chocolate bar contains 230 food calories. Through what vertical distance could this amount of energy, in principle, lift a 1 tonne motor car? Note that 1 food calorie = 1 kcal = 4.2 kJ.

2

The atomic nucleus

Radioactivity was discovered by Antoine Henri Becquerel in 1896. He found that the mineral crystal he was investigating caused a photographic plate to become blackened. It was an accidental discovery because he had been looking for the emission of X-rays from the crystal; X-rays had also recently been discovered. However, the crystal happened to contain some uranium, which produced the nuclear radiation. There followed a period of intense research on the nature of this radiation and the materials that emitted it. Marie and Pierre Curie isolated and identified the radioactive element radon, while Ernest Rutherford found three distinct forms of nuclear radiation, which he characterised by their ability to penetrate matter and ionise air. The first type of radiation, which penetrated the least, but caused the most ionisation, he called alpha (α) rays. The second type, with intermediate penetration and ionisation, he called beta (β) rays, and the third type, which produced the least ionisation but penetrated the most, he called gamma (γ) rays. It was subsequently discovered that α-rays are composed of helium nuclei, β-rays are composed of electrons and γ-rays are composed of high-energy photons. Becquerel, the Curies and Rutherford all received Nobel prizes for their discoveries. A turning point in the understanding of the atomic nucleus came in a series of pioneering experiments by Rutherford and his collaborators at the University of Manchester. They directed a beam of α particles at a thin gold foil and observed how the α particles were deflected by the foil. Based upon their observations, Rutherford postulated a model of the atomic nucleus that remains familiar today: a model in which all the positive charge of the atom and all the mass is concentrated in an extremely small region called the nucleus. It could be said that nuclear physics began with Rutherford's discovery of the atomic nucleus.

Physics of Energy Sources, First Edition. George C. King.
© 2018 John Wiley & Sons, Ltd. Published 2018 by John Wiley & Sons, Ltd.

In parallel with these experiments on the atomic nucleus there were the revolutionary developments of quantum mechanics and relativity. These considerably aided the understanding of the results of the experiments, and indeed these experiments provided convincing evidence for the validity of the new theories. In 1905 Einstein published his equation describing the equivalence of mass and energy:

$$E = mc^2. \tag{2.1}$$

Einstein's equation shows that huge amounts of energy are released when mass–energy conversion takes place, as happens in nuclear fission and nuclear fusion. The fission of heavy nuclei, such as uranium, is a major source of power today, while the fusion of light nuclei is the energy source that powers the stars, including our Sun, and has the potential to play a key role in providing the energy needs of the future. Nuclear fission and nuclear fusion are discussed in Chapter 3.

In this chapter we describe the composition of nuclei and their basic properties, including their size, mass and electric charge. The characteristics of the forces that bind the constituents of a nucleus together are also described together with the resulting nuclear *binding energies*. We will see how the binding energy is intimately related to the energy that we can obtain from nuclear fission and fusion through the equivalence of mass and energy. We will also see that binding energy plays a central role in the stability of nuclei and that unstable nuclei decay to more stable nuclei. The various ways in which radioactive decay can occur will also be described.

2.1 The composition and properties of nuclei

2.1.1 The composition of nuclei

At the centre of an atom is a positively charged nucleus. The nucleus is very small compared with the overall size of the atom; an atomic diameter is $\sim 10^{-10}$ m, while the diameter of a nucleus is $\sim 10^{-14}$ m, a factor of $\sim 10^4$ smaller. The nucleus contains just two kinds of particles, protons and neutrons. Protons and neutrons are much heavier than electrons, by a factor of approximately 2000, and so nearly all the atomic mass is concentrated in the nucleus. A nucleus is characterised by the number of protons and neutrons it contains. The number of protons is called its *atomic number Z*. Since a proton has a charge $+e$, where e is the magnitude of the electronic charge, the nuclear charge is equal to $+Ze$. The number of atomic electrons must be equal to the number of protons in the nucleus to maintain charge neutrality, and so an electrically neutral atom must therefore have

Z electrons. The number of neutrons a nucleus contains is called its *neutron number*, N. A neutron has a mass that is very close to that of a proton, but it is electrically neutral. The proton and neutron are collectively known as *nucleons*, as they are both found in the nucleus. The total number of protons and neutrons is called the *nucleon number*, A, which is more usually called the *mass number* or *atomic mass number* of the nucleus. (This follows, as nuclear masses are measured on a scale in which the proton and the neutron have masses that are close to one fundamental unit of mass. A is then the *integer* nearest to the ratio between the nuclear mass and the fundamental mass unit.) A particular species of nucleus is called a *nuclide* and is specified by its values of A, Z and N as:

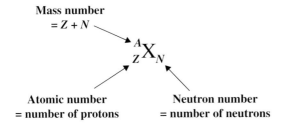

where X is the chemical symbol. An example is $^{58}_{28}\text{Ni}_{30}$, pronounced nickel-58. Note that we do not need to write both the chemical symbol and the atomic number because the chemical symbol tells us the value of Z; for example, every nickel atom has $Z = 28$. It is also usually not necessary to include N, as $N = A - Z$. Thus we can simply and conveniently write ^{A}X; for example ^{58}Ni. (Including Z and N is useful when we are trying to balance Z and N in a nuclear reaction or decay process.)

Naturally occurring samples of most elements contain atoms with the same atomic number Z but with different values of mass number A. Nuclides with the same Z but different N are called *isotopes*. Typically an element may have two or three stable isotopes, although some, like gold, have just one while iodine has nine. A familiar example is chlorine ($Z = 17$). About 76% of naturally occurring chlorine nuclei have $N = 18$, while 24% have $N = 20$. These fractions are called the *natural abundances* of the respective isotopes. The chemical properties of an element are determined by its atomic electrons. As different isotopes of the same element have the same number of electrons, they have the same chemical properties. On the other hand, different isotopes have slightly different physical properties – in particular, properties that depend on mass. For example, the $^{235}_{92}\text{U}$ and $^{238}_{92}\text{U}$ isotopes of uranium can be physically separated because of the slightly different diffusion coefficients of the gases $^{235}_{92}\text{UF}_6$ and $^{238}_{92}\text{UF}_6$. This has important application in enriching naturally occurring uranium for nuclear fission reactors.

Figure 2.1 Schematic diagram of Rutherford's apparatus for observing the scattering of α particles by a thin gold foil. α particles from the source are collimated into a narrow beam and directed at the foil. The angle θ through which an α particle is deflected at the foil is measured by observing the point at which the particle strikes the fluorescent screen.

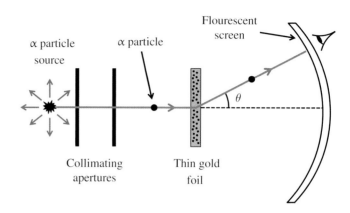

2.1.2   The size of a nucleus

A measure of nucleus size was first obtained by Rutherford and his collaborators, Hans Geiger (inventor of the Geiger counter) and Ernest Marsden, the latter still being an undergraduate student at the time. Rutherford had already established that α particles are doubly ionised helium atoms, He^{++}, where both electrons have been removed from the atom. He then used α particles as a *probe* of the nucleus. In these experiments, beams of α particles were fired at thin metal foils and the way these particles were deflected or *scattered* by the foils was investigated. (Firing energetic projectiles at sub-atomic particles remains, of course, a principal means of investigating their nature.) These investigations culminated in Rutherford's model of the structure of the atom: *all the positive charge of the atom, and consequently all the mass, is concentrated in an extremely small region called the nucleus.*

The Rutherford scattering experiment

A schematic diagram of the apparatus for Rutherford's scattering experiment is shown in Figure 2.1. The radioactive source emits α particles of well-defined kinetic energy E. This energy is typically about 5 MeV[1] and is measured in a separate experiment by observing the motion of the α particles in crossed electric and magnetic fields. The α particles are collimated into a narrow beam using a combination of apertures and are directed onto a thin gold foil. The foil is so thin that most of the α particles pass through it with little energy loss. However, the charged α particles suffer deflections

---

[1] The electron volt (eV) is defined as the energy that an electron gains when it falls through a potential of 1 volt. It is equal to 1.602×10^{-19} J. The eV is a convenient energy unit for atomic energies, while MeV $= 10^6$ eV is a convenient unit for nuclear energies.

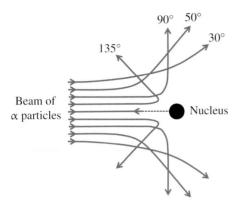

Figure 2.2 The scattering of α particles by a gold nucleus. The scattering angle increases as the α particles approach closer to the nucleus, and these particles can be scattered through very large angles, up to 180°.

because of their electrostatic interactions with the positive and negative charges in the gold atoms, although the electrons are so light that they do not deflect the α particles appreciably. The scattered α particles are detected on a flourescent screen that produces a tiny flash of light when an α particle strikes it. In this way it is possible to measure the *angular distribution* of the scattered α particles, i.e. the number of detected α particles as a function of scattering angle θ.

In practice, an α-particle beam is wide compared with the dimensions of a nucleus and this results in a range of scattering angles, as illustrated in Figure 2.2. To Rutherford's astonishment it was observed that a significant number of the α particles were deflected through very large angles, up to 180°. We emphasise that large angular deflections of an α particle can only occur if the positive charge of the atom is concentrated into a region of extremely small size. We illustrate this with an *order of magnitude* calculation for the deflection angle θ that an α particle will suffer when it comes close to a nucleus. This is illustrated schematically in Figure 2.3, where the distance b is called the *impact parameter*. If there were no electrostatic interaction between the incident α particle and the nucleus, the α particle would miss the nucleus by distance b. As the α particle passes by the nucleus, which has positive charge $+Ze$, it experiences

Figure 2.3 The deflection angle, θ, that an α particle experiences as it passes close to a gold nucleus. The distance b is called the impact parameter. The deflection angle is given by the ratio $\Delta p/p$, where p is the momentum of the incident α particle and Δp is the sideways momentum that results from the electrostatic repulsion.

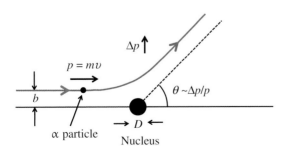

an impulse Δp due to electrostatic repulsion. For a deflection that occurs when the α particle is close to the nucleus we can write $\Delta p \sim F\Delta t$, where F is the repulsive electrostatic force and Δt is a measure of the time during which the α particle experiences the force. Δt is given approximately by the time it takes the α particle to transverse the nuclear region, which is D/v, where D is the characteristic size of the nucleus and v is the velocity of the α particle. In a close encounter, the distance of nearest approach is $\sim D$. At this distance the electrostatic force is given by

$$F = \frac{2eZe}{4\pi\varepsilon_0 D^2},$$

(2.2)

where ε_0 is the permittivity of free space. The deflection angle θ is given by the ratio $\Delta p/p$, where p is the momentum of the incident α particle and Δp is the sideways momentum that results from the electrostatic interaction. Hence we have

$$\theta \sim \frac{F\Delta t}{p} = \frac{2Ze^2}{4\pi\varepsilon_0 D^2}\frac{D/v}{mv} = \frac{Ze^2}{4\pi\varepsilon_0 DE},$$

(2.3)

where m is the mass of the α particle and $E = \frac{1}{2}mv^2$ is its kinetic energy. We see that the deflection angle is inversely proportional to nuclear size D. For gold ($Z = 79$) and for an α particle with an energy of 5 MeV, we obtain

$$\theta \sim \frac{2 \times 10^{-14}}{D} \text{ rad/m}.$$

For a large angle of deflection, D must be $\sim 10^{-14}$ m.

Worked example

An α particle of kinetic energy 5 MeV is deflected through an angle of $180°$ when it is incident on a thin foil of gold. Determine the distance of closest approach of the α particle to the nucleus and hence obtain an upper limit to the radius of the nucleus.

Solution

The largest deflection angle, i.e. $180°$, occurs when an α particle is in a head-on collision with the nucleus, i.e. $b = 0$. In such a collision, all the kinetic energy of the α particle is converted into potential energy, when it momentarily comes to rest at its distance of closest approach. We can then write:

$$E = \frac{1}{2}mv^2 = \frac{1}{4\pi\varepsilon_0}\frac{2Ze^2}{R},$$

where m and v are the mass and velocity of the α particle of charge $+2e$, Z is the atomic number of the gold nucleus and R is the distance of closest approach.

Hence,

$$R = \frac{2Ze^2}{(4\pi\varepsilon_0)(mv^2/2)}$$

$$= \frac{2 \times 79 \times (1.60 \times 10^{-19})^2}{(4 \times \pi \times 8.9 \times 10^{-12})(5 \times 10^6 \times 1.60 \times 10^{-19})} = 45 \times 10^{-15}\,\text{m}$$

$$= 45 \text{ femtometres (fm)},$$

where $1\,\text{fm} = 10^{-15}$ m. This value is an upper limit to the radius of a gold nucleus, as the value is limited by the energy of the α particle. We note that an α particle of energy 5 MeV has a velocity of $1.5 \times 10^7\,\text{m/s}$, which is much less than the velocity of light and this allowed us to use a non-relativistic calculation. We have also neglected the recoil of the nucleus, which is a good approximation as the nucleus is much more massive than the α particle.

Rutherford made a detailed analysis of the angular distribution to be expected for α particles that are deflected through large angles from a positively charged nucleus of small dimension. He assumed that the α particle does not penetrate the nuclear region so that the α particle and the nucleus act like point charges as far as the electrostatic force is concerned. He also assumed that the mass of the nucleus is so much greater than the mass of the α particle that the nucleus remains fixed in space during the scattering process. His analysis was based on a classical treatment of the electrostatic force between charged particles under the conservation of energy and angular momentum about the target nucleus. This analysis leads to an expression for $\mathcal{N}(\theta)$, which is the number of detected α particles detected *per unit area* on the flourescent screen at a deflection angle θ:

$$\mathcal{N}(\theta) = \left(\frac{1}{4\pi\varepsilon_0}\right)^2 \left(\frac{Ze^2}{2E}\right)^2 \frac{Int}{r^2} \frac{1}{\sin^4(\theta/2)}, \tag{2.4}$$

where I is the number of incident α particles, n is the number of nuclei per unit volume, t is the thickness of the foil and r is the distance of the screen from the foil.[2] This expression predicts how $\mathcal{N}(\theta)$ depends on the deflection angle θ, the atomic number Z of the atoms in the foil, and the kinetic energy E of the incident α particles. The critical features are the inverse-square dependence on E and the

---

[2] For a derivation of Rutherford's scattering formula see, for example, J. Lilley, *Nuclear Physics*, John Wiley & Sons Ltd, 2001.

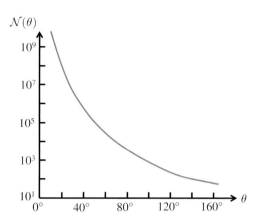

Figure 2.4 The $[\sin^4(\theta/2)]^{-1}$ angular dependence of $\mathcal{N}(\theta)$, the number of α particles detected per unit area on the fluorescent screen. Note the logarithmic scale for $\mathcal{N}(\theta)$. Although $\mathcal{N}(\theta)$ falls rapidly with increasing scattering angle θ, a significant number of α particles are detected at very large angles.

strong dependence on θ. The dependence of $\mathcal{N}(\theta)$ on θ is shown in Figure 2.4, where the logarithmic scale of the vertical axis may be noted. Even though the number of detected particles is predicted to fall very sharply with increasing θ, the number at large angles remains appreciable, unlike the predictions of rival theories. Geiger and Marsden confirmed the predictions of Rutherford's expression for heavy nuclei like gold and, in particular, the angular behaviour of the scattered α particles, and consequently the Rutherford model of the atom was adopted. By comparing the observed angular behaviour of the α particles with the predictions of his expression, Rutherford was able to obtain a better value for the nuclear radius, which he found to be approximately 10 fm.

Rutherford was fortunate in two ways. The scattering of an α particle by a nucleus is properly described by quantum mechanics. However, for this particular case, the quantum mechanical treatment of the scattering process gives exactly the same result as the classical treatment. Moreover, a 5 MeV α particle does not have enough kinetic energy to penetrate the nucleus of a high-Z nuclide like gold and so Rutherford's assumption of two point charges repelled by electrostatic repulsion was valid.

Experiments using foils of metals with low Z, like aluminium, were also investigated. For these light metals, it was found that Rutherford's expression no longer fitted the experimental data at large angles. This is because the electrostatic repulsion between the α particle and the low-Z nucleus is reduced to the extent that the α particles could penetrate the nucleus at their distances of closest approach. In this case, the nucleus and the α particles can no longer be treated as point charges. Moreover, the α particle comes under the influence of the nuclear force, which operates within the nucleus. This divergence from Rutherford's expression can also be observed by detecting the number of α particles scattered through a particular angle, say $60°$, and varying the energy of the incident α particle. The results of such an experiment using a lead foil are shown in

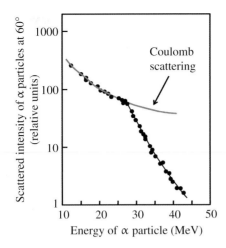

Figure 2.5 Experimental data for the scattering of α particles through a fixed angle of 60° by a lead target shows the breakdown of the Rutherford scattering formula. At an α-particle energy of about 27.5 eV, the data abruptly depart from scattering due to purely electrostatic repulsion. Above this energy the α particles get close enough to the nucleus to come under the influence of the strong but short-range nuclear force.

Figure 2.5. At lower energies of the α particles, the scattering data are in agreement with the predictions of Rutherford's expression, indicated by the blue curve. However, when the energy of the incident α particle is increased to the point where it can penetrate the nucleus, Rutherford's model breaks down. The α particle and the nucleus can no longer be treated as point charges and the α particle comes under the influence of the nuclear force. All models of physical systems have limitations. However, seeing under what circumstances they break down usually leads to new physical insight.

2.1.3   The distributions of nuclear matter and charge

Since Rutherford's first scattering experiments, atomic nuclei have been extensively investigated by firing high-energy particles at them. In addition to α particles, these experiments have used neutrons and electrons. Electrons are not affected by the nuclear force that exists between protons and neutrons and so they probe the *distribution of nuclear charge*. By contrast, neutrons have no charge, but interact with protons and other neutrons through the nuclear force and so they probe the *distribution of nuclear matter*.

From quantum mechanics we know that particles have a wave-like character. This was demonstrated by Clinton Davisson and Lester Germer in 1927. They fired a beam of electrons at the surface of a metal, which happened to be a well-crystallised piece of nickel. They found that the electrons were diffracted by the crystal structure, just as a beam of X-rays is diffracted by a crystal. Any particle with momentum p has an associated wavelength λ that is given by the *de Broglie* relationship:

$$\lambda = \frac{h}{p}, \tag{2.5}$$

where h is Planck's constant. If we want to 'see' an object, we need to use radiation that has a wavelength that is smaller, or at least of the same size, as the dimensions of the object. For example, the electrons in an electron microscope have a de Broglie wavelength that is $\sim 10^5$ times smaller than the wavelength of visible light. Consequently, they are used to see objects that are too small to be seen with a conventional microscope. So it is with nuclei; the de Broglie wavelength of the incident probe particles must be small compared with the nuclear dimensions. We have seen that nuclear diameters are ~ 10 fm, and so we require the de Broglie wavelength of the incident particles to be less than about this value. This means that neutrons must have an energy that is greater than about 2 MeV, while electrons must have an energy greater than about 100 MeV. This is a relativistic energy for an electron because its rest mass is 0.51 MeV.

The typical form of the charge distribution of a nucleus, obtained from electron scattering experiments, is illustrated schematically in Figure 2.6. In this figure, the charge density, $\rho(r)$, is plotted as a function of radial distance, r, from the centre of the nucleus. The charge distribution can be characterised by two parameters: the *mean radius R*, where the density is half its central value and the *skin thickness* over which $\rho(r)$ drops from 90% to 10% of its maximum value. When a range of nuclides was investigated, it was found that

- the nuclear charge density is roughly constant in the central region of a nuclide and is nearly the same for all nuclides;

- the mean nuclear radius R steadily increases with mass number A according to:

$$R = R_0 A^{1/3}, \tag{2.6}$$

where R_0 is a constant equal to 1.2 fm.

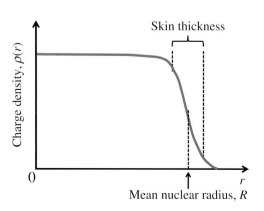

Figure 2.6 Typical form of the distribution of charge density, $\rho(r)$, for a nucleus as a function of radial distance, r, from the centre of the nucleus. R represents the mean nuclear radius. The skin thickness is the distance over which $\rho(r)$ falls from 90% to 10% of its maximum value. The charge density is roughly uniform over the central region of a nucleus.

Complimentary experiments using neutron scattering show that the mass distribution of a nucleus is very similar to its charge distribution; the radii of nuclides deduced from charge and mass distributions give values that are the same within about 0.1 fm. The physical explanation of this consistency between mass and charge distributions is that the combination of nuclear and electrostatic forces results in a roughly constant mix of protons and neutrons in the nucleus. That nuclear matter has a constant 'density' is consistent with the nuclear radius being proportional to $A^{1/3}$, as its volume is then proportional to A and its mass is also proportional to A. It also follows that the number of nucleons per unit volume is roughly constant, which, as we shall see, gives valuable physical insights into the nature of the nuclear force. Experimental measurements show that the density of nuclear matter is $\sim 2 \times 10^{17}\,\mathrm{kg/m^3}$, which is 2×10^{14} times larger than the density of water.

In contrast to nuclei, where nuclear radius steadily increases with mass number, all atoms have roughly the same size. This constancy in atomic radius arises because, as the number of electrons in an atom increases, so also does the number of protons in the nucleus. This increases the nuclear charge, and the attractive electrostatic force acting on the electrons correspondingly increases.

2.1.4 The mass of a nucleus

A good deal of valuable information can be obtained about nuclei from the accurate determination of their masses, as we shall see. However, it is in general very difficult to measure the mass of a bare nucleus, because we would have to strip away all the atomic electrons. Instead we measure the mass of the atom and deduce the nuclear mass from the atomic mass. Or rather, we measure the mass of the atomic ion where just one electron has been removed; it is straightforward to account for the mass of the missing electron. The mass of the atomic ion is measured by exploiting its behaviour in electric and magnetic fields using the technique of *mass spectroscopy*.

A schematic diagram of a conventional mass spectrometer is shown in Figure 2.7. It consists of an *atomic ion source*, a *velocity selector*, a *momentum selector* to separate the atomic ions according to their mass and an *ion detector*. The ion source may consist of a vapour of the atom of interest, which is bombarded with an electron beam, and this ionises the atoms. Alternatively, it may be a discharge tube in which the electrodes are coated with the atoms of interest. Either way, the ions are produced with a broad range of thermal energies and consequently they have a broad range of kinetic energies. This would smear out any peaks in the mass spectrum so that individual mass peaks could not be resolved. The purpose of the

Figure 2.7 Schematic diagram
of a mass spectrometer. The
collimating apertures define a
narrow beam of ions from the
source. The velocity selector
filters the ions according to
their velocity and the
velocity-selected ions are
injected into the momentum
selector. The momentum
selector disperses the ions
according to their mass and a
mass spectrum is recorded on
the ion detector. The velocity
and momentum selectors are
immersed in magnetic fields
B_1 and B_2, respectively, as
indicated by the $\otimes$ symbols.
In the velocity selector there
is also an electric field E that
is perpendicular to the
magnetic field B_1.

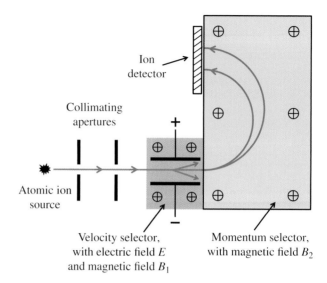

Figure 2.7 Schematic diagram of a mass spectrometer. The collimating apertures define a narrow beam of ions from the source. The velocity selector filters the ions according to their velocity and the velocity-selected ions are injected into the momentum selector. The momentum selector disperses the ions according to their mass and a mass spectrum is recorded on the ion detector. The velocity and momentum selectors are immersed in magnetic fields B_1 and B_2, respectively, as indicated by the $\otimes$ symbols. In the velocity selector there is also an electric field E that is perpendicular to the magnetic field B_1.

velocity selector is to filter the ions according to their kinetic energy so that the ion beam emerging from the selector has a very narrow spread of velocities. In the velocity selector there is a pair of parallel electrostatic plates that have opposite polarity. These produce an electric field E that is perpendicular to the direction of the ion beam. These plates are immersed in a magnetic field B_1 whose direction is perpendicular to the electric field direction and also to the ion beam direction. The electric and magnetic fields act in opposite directions and ions pass through the selector for which $qE = qvB_1$, where q is the charge of the ion and v is its velocity, which is therefore given by

$$v = \frac{E}{B_1}. \tag{2.7}$$

The velocity-selected ions are then injected into the momentum selector, which has a magnetic field B_2 that is again perpendicular to the ion beam direction. The ions pass through this magnetic field and follow a circular path according to

$$B_2qv = \frac{Mv^2}{r}, \tag{2.8}$$

where M is the mass of the ion. The radius r depends on the ion momentum, i.e.

$$r = \frac{Mv}{qB_2} \tag{2.9}$$

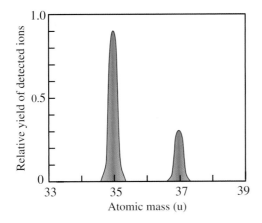

Figure 2.8 A mass spectrum of a naturally occurring sample of chlorine. The relative areas of the peaks due to the ^{35}Cl and ^{37}Cl ions are in the ratio 3:1, reflecting the natural abundances of the two isotopes.

Usually the magnetic fields of the velocity and momentum selectors are the same, i.e. $B_1 = B_2 = B$, and, substituting for v from Equation (2.7), we obtain

$$M = \frac{rqB^2}{E}. \tag{2.10}$$

Ions of different mass have different values of r and are dispersed across the detector as illustrated in Figure 2.7. By analogy, a glass prism disperses white light according to wavelength. The ion detector was traditionally a photographic plate, but now an electronic detector is used. Note that r is directly proportional to M, which means that the mass scale of the spectrometer is linear. Figure 2.8 illustrates a mass spectrum of chlorine, showing the peaks of the two isotopes ^{35}Cl and ^{37}Cl. The relative abundances of these two isotopes are 76% and 24%, respectively, which can be deduced from the relative areas of the peaks in the spectrum.

The standard unit of mass for nuclear and atomic physics is the *atomic mass unit* (u), which is defined such that the mass of a ^{12}C *atom* is exactly 12 u. In terms of the kg:

$$1\,\text{u} = 1.660539 \times 10^{-27}\,\text{kg}. \tag{2.11}$$

Using Einstein's equation $E = mc^2$, we can also express mass in MeV/c^2, where

$$1\,\text{u} \equiv 931.5\,\text{MeV}/c^2. \tag{2.12}$$

When we discuss the production of energy from nuclear reactions, we will see that it is mass difference that determines the amount of energy that is produced through the equivalence of mass and energy, $E = mc^2$, Equation (2.1). An important aspect of mass spectroscopy is that it can determine the masses of the initial nuclei taking part in a nuclear reaction and those of the product nuclei. From their mass difference, the energy produced in the reaction can be predicted. This requires the masses to be determined to $\sim$1 part in 10^7 or 10^8. To achieve this accuracy, the spectrometer is usually calibrated for one particular atomic mass and other masses are measured with respect to the known mass. This is because it is much easier to measure a mass *difference* with high precision than the *absolute* value of the mass from absolute measurements of quantities on the right-hand side of Equation (2.10). The mass difference is deduced from the separation of the two mass peaks on the ion detector, one due to the calibration mass and the other due to the unknown mass.

Worked example

Singly charged ^{238}U ions are injected into a mass spectrometer and are observed to strike the detector at a radius, r, of 25 cm. If an ion beam containing a mixture of singly charged ^{238}U and ^{235}U ions were injected into the mass spectrometer, what would be the separation of the positions at which the two ion species strike the detector?

Solution

From Equation (2.10) we obtain

$$\frac{E}{qB^2} = \frac{r}{M} = \frac{0.25}{M_{238}} \text{ m/unit mass.}$$

Because the mass difference δM between ^{238}U and ^{235}U is very small compared with the mass, M_{238}, of ^{238}U, we make the approximation

$$\delta r \approx \frac{\mathrm{d}r}{\mathrm{d}M}\delta M = \frac{E}{qB^2}\delta M = \frac{0.25}{M_{238}}\delta M = \frac{0.25 \times 3}{238} = 0.0032 \text{ m.}$$

The separation will be twice δr and equal to 6.4 mm.

New experimental techniques have made it possible to confine atomic ions in *traps* consisting of combined electrostatic and magnetic fields. The ions oscillate in these fields and can be contained in the trap for extended periods of time, rather like a chemical sample in a test tube. By measuring the *cyclotron frequency* of the ion in the magnetic field, the mass of the ion can be determined. As this is a measurement of frequency, the accuracy is extremely high: ~ 1 part in 10^{11} for stable nuclides.

2.1.5   The charge of a nucleus

The determination of nuclear charge was first achieved by Henry Moseley in 1913. He did this by observing the effects of the nuclear charge on *atomic* energy levels. In particular, he studied in detail the X-ray spectra of the atoms. These spectra contain sharp peaks called

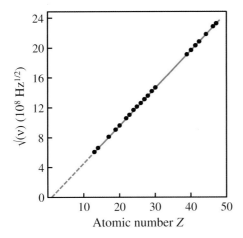

Figure 2.9 A plot of the square root of the frequency ν of the K_α characteristic X-ray line of an atom versus its atomic number Z. The linear dependence of $\sqrt{\nu}$ on Z is in agreement with Moseley's law.

characteristic X-rays that occur at different wavelengths for different elements. Moseley measured the wavelengths of these characteristic X-rays by diffracting them with a crystal. Fortunately, the technique of crystal diffraction had just been developed by W.L. Bragg and his father W.H. Bragg. Moseley found that the wavelengths and frequencies of the characteristic X-ray lines vary smoothly from one element to the next. Furthermore, he found that the measured frequencies of the most intense line in the spectra, known as the K_α line, could be fitted within experimental accuracy by the empirical formula

$$\nu = C(Z-1)^2, \tag{2.13}$$

where ν is the X-ray frequency, Z is the nuclear charge and C is a constant with the value 2.48×10^{15} Hz. Equation (2.13) is called *Moseley's law*. It gives a linear relationship between the square root of the frequency $\sqrt{\nu}$ and $(Z-1)$, which is illustrated in Figure 2.9. This graph has an intercept at $Z = 1$, confirming Moseley's law. To find the value of Z for an element we simply measure the frequency of its K_α characteristic line and insert its value into Equation (2.13).

It is important to understand the physics behind the results of any experiment. Moseley interpreted his results using an extension of the Bohr model of the atom, which was published in the same year. In Bohr's model of the hydrogen atom, the single electron can exist in orbitals around the proton nucleus that have well defined energies. Bohr obtained the following expression for the energy E_n of the nth orbital:

$$E_n = -\frac{mZ^2 e^4}{\varepsilon_0^2 8h^2} \frac{1}{n^2}, \tag{2.14}$$

where m, e, ε_0 and h are fundamental constants, Z is the charge on the nucleus and n is called the *principal quantum number* of the orbital. n can take on values 1, 2, 3, ... ($Z = 1$ for hydrogen, but we will retain the symbol Z for reasons that will soon become apparent).

The minus sign indicates that the electron is bound to the proton, where E_n gives the binding energy; energy E_n is required to eject the electron from the nth orbital. Inserting the values of the fundamental constants, we find

$$E_n = -13.6Z^2 \frac{1}{n^2} \text{ eV}. \tag{2.15}$$

When an electron makes a transition from initial orbital n_i to lower-lying orbital n_f, energy is released in the form of a photon whose energy E_{ph} is $E_i - E_f$, which is given by

$$E_{\text{ph}} = -13.6Z^2 \frac{1}{n_i^2} - \left(-13.6Z^2 \frac{1}{n_f^2} \right)$$
$$= 13.6Z^2 \left(\frac{1}{n_f^2} - \frac{1}{n_i^2} \right) \text{ eV}. \tag{2.16}$$

For example, when an electron in the $n = 4$ orbital of hydrogen falls to the $n = 3$ orbital, a photon is released with an energy of 1.89 eV. This corresponds to a wavelength that is given by $\lambda = hc/E$, and which is equal to 564 nm. This lies in the visible region of the electromagnetic spectrum.

In multi-electron atoms, the electrons exist in *shells* that surround the nucleus. Again these shells are characterised by their principal quantum number n. Roughly speaking, electrons in the same shell are at about the same distance from the nucleus. X-rays correspond to transitions between the *inner* shells of an atom. They are produced, for example, when a metal target such as copper is bombarded with a beam of very energetic ($\sim$10–100 keV) electrons. An incident electron may knock out an electron from the innermost shell of the target atom, with $n = 1$, as shown in Figure 2.10(a). The ejected electron leaves behind a vacancy or hole in that shell. This vacancy may then be filled by an atomic electron from another shell, say the $n = 2$ shell, as shown in Figure 2.10(b). This is accompanied by the emission of an X-ray, whose energy is equal to the energy difference between the energies of initial ($n = 2$) and final ($n = 1$) shells of the atom. These

Figure 2.10 The process of X-ray emission from an atom. (a) An incident electron knocks out an atomic electron from the $n = 1$ shell, leaving a vacancy in that shell. (b) The resulting vacancy in the $n = 1$ shell is filled by an atomic electron from the $n = 2$ shell and this is accompanied by the emission of an X-ray.

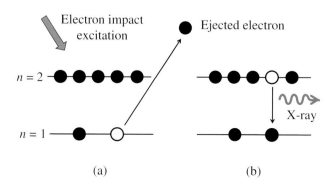

transitions are exactly analogous to those between the outer shells of atoms. However, the inner-shell electrons are much closer to the nucleus and, consequently, are much more tightly bound. As a result, the energy differences between the inner shells are correspondingly large and the X-rays have short wavelengths, $\sim$0.1 nm.

The $n = 1$ shell of an atom contains two electrons, and when one is knocked out of the shell one electron remains. Thus an electron that falls from the $n = 2$ shell into the vacancy in the $n = 1$ shell sees an *effective nuclear charge*, Z_{eff}, equal to $+ (Z - 1)e$, due to the Z protons in the nucleus and the remaining electron in the $n = 1$ shell; the $n = 1$ electron *screens* the $n = 2$ electron from the nucleus. Moseley modified the Bohr model to take account of the effective nuclear charge that an electron experiences. In terms of Moseley's modified formula, we write the energy level E_n of an atom as

$$E_n = -13.6\frac{Z_{\text{eff}}^2}{n^2} \text{ eV.} \tag{2.17}$$

The energy of a resulting X-ray is then

$$E_{\text{X-ray}} = 13.6 \times Z_{\text{eff}}^2 \left(\frac{1}{n_f^2} - \frac{1}{n_i^2} \right) \text{ eV.} \tag{2.18}$$

Taking $Z_{\text{eff}} = (Z - 1)$ with n_i and n_f equal to 2 and 1, respectively, we obtain

$$E_{\text{X-ray}} = 10.2 \times (Z - 1)^2 \text{ eV.} \tag{2.19}$$

The frequency, ν, of the emitted X-ray is then

$$\nu = \frac{E_{\text{X-ray}}}{h} = \frac{10.2(Z - 1)^2}{4.136 \times 10^{-15}} = 2.47 \times 10^{15}(Z - 1)^2 \text{ Hz.} \tag{2.20}$$

The form of this equation is in agreement with Moseley's Equation (2.15). Moreover, the value of the constant in Equation (2.20) is in good agreement with the value of the constant C obtained by Moseley.

Moseley's work provided for the first time a way to determine the atomic number Z of an element. As a result, the correct sequence of elements in the periodic table could be established. Previously the ordering had been in terms of their atomic mass, which had led to some inaccurate ordering. In addition, Moseley found gaps that indicated the existence of elements that were not known at the time. These elements were subsequently discovered, some many years later.

2.1.6 Nuclear binding energy

The neutrons and protons in a nucleus are held together by strong attractive forces, as we will describe in Section 2.2. Therefore, work must be done to separate these nucleons from each other until they are a large distance apart, i.e. energy must be supplied to the nucleus to separate it into its individual nucleons. The energy that must be added to a nucleus to separate it into its individual neutrons and protons is called the *binding energy, B, of the nucleus*. Binding energy is not an energy that resides in a nucleus, like kinetic energy. Rather, it is the energy equivalent of mass through Einstein's mass–energy relationship

$$E = mc^2. \tag{2.1}$$

We thus expect the mass of the nucleus to be less than the sum of the masses of the constituent nucleons, and this is indeed the case. The difference between the mass of the nucleus and the constituent nucleons is called the *mass defect* Δm. In terms of the mass defect, the binding energy is equal to Δmc^2.

We recall that it is atomic masses that are measured in mass spectroscopy. If the mass of an atom is $_Z^A M$, the mass of its nucleus is $(_Z^A M - Zm_e)$, where m_e is the mass of the electron. Hence, the mass difference, Δm, between a nucleus and its constituent nucleons is

$$Zm_p + Nm_n - \left(_Z^A M - Zm_e\right) = Z\left(m_p + m_e\right) + Nm_n - _Z^A M \tag{2.21}$$

where m_p and m_n are the masses of the proton and the neutron, respectively. But $(m_p + m_e)$ is just the mass of a neutral hydrogen atom. Hence, we have

$$\Delta m = \left(ZM_H + Nm_n - _Z^A M\right) \tag{2.22}$$

and finally, the binding energy B of the nucleus is given by

$$B = \left(ZM_H + Nm_n - _Z^A M\right) c^2, \tag{2.23}$$

where M_H is the mass of the *neutral* hydrogen atom.

The more tightly bound the nucleons are in a nucleus, the more energy it takes to separate them. For a given value of $A (= Z + N)$, there may be different combinations of Z and N. The combination that gives the largest total binding energy will be the most stable nuclide. It follows from Equation (2.23) that the most stable nuclide will also be the one with the lowest mass, $_Z^A M$.

The electrons in an atom also have binding energy. For example, the electron in the ground state of hydrogen has a binding energy of 13.6 eV, i.e. this amount of energy must be added to the atom to remove the electron. We are able to neglect the binding energies of the atomic electrons when deducing the binding energy of a nucleus for two main reasons. First, electron binding energies are negligible compared with the rest-mass energies that appear in Equation (2.23).

For example, an atom with an atomic number A of, say, 100 has a rest mass energy of $\sim 10^{11}$ eV, while the most tightly bound electrons, those in the inner shells of an atom, have energies of $\sim 10^4$ eV. Secondly, electronic binding energies tend to cancel out in calculations of differences in nuclear binding energy between two different nuclides.

As an example of nuclear binding energy, we consider deuterium $^2_1$D, which is an isotope of hydrogen, with $Z = 1$. The deuterium nucleus contains a single neutron and a single proton. From precise mass spectroscopic measurements we know that the masses of deuterium and hydrogen atoms are, respectively, 2.014 102 and 1.007 825 u, while the mass of the neutron is 1.008 665 u. Hence, the mass defect is

$$\Delta m = (1.007\,825 + 1.008\,665 - 2.014\,102)\,\mathrm{u} = 0.002\,388\,\mathrm{u}$$
$$= (0.002\,388 \times 1.661 \times 10^{-27})\,\mathrm{kg} = 3.965 \times 10^{-30}\,\mathrm{kg},$$

and the binding energy B is

$$3.965 \times 10^{-30} \times (2.998 \times 10^8)^2\,\mathrm{J}$$
$$= \frac{3.965 \times 10^{-30} \times (2.998 \times 10^8)^2}{1.602 \times 10^{-19}}\,\mathrm{eV} = 2.224\,\mathrm{MeV}.$$

We have given all the steps of this calculation for the sake of clarity. However, we note that it is more convenient to use the conversion factor

$$c^2 = 931.5\,\mathrm{MeV/u}, \tag{2.24}$$

so using Equation (2.23) directly, we have

$$B = (1.007\,825 + 1.008\,665 - 2.014\,102)\mathrm{u} \times 931.5\,\mathrm{MeV/u}$$
$$= 2.224\,\mathrm{MeV}.$$

We can also determine the binding energy by bringing a proton (^{1}H) and a neutron together to form deuterium and measuring the energy that is emitted in the reaction, which is in the form of a γ-ray:

$$^1\mathrm{H} + {}^1\mathrm{n} \to {}^2\mathrm{D} + \gamma. \tag{2.25}$$

The observed energy of the γ-ray, less a small correction due to the recoil energy, is found to be 2.225 MeV in close agreement with the value deduced from the mass spectroscopic measurements. Alternatively, we can have the reverse reaction called *photo-dissociation*, in which we split deuterium into a proton (^{1}H) and a neutron by the absorption of a γ-ray:

$$^2\mathrm{D} + \gamma \to {}^1\mathrm{H} + {}^1\mathrm{n}, \tag{2.26}$$

as illustrated schematically in Figure 2.11. The minimum energy that the γ-ray must have to do this is equal to the binding energy, again

Figure 2.11 The disintegration of a deuterium nucleus into a proton (^{1}H) and a neutron by the absorption of a γ-ray.

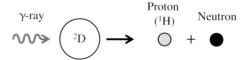

corrected for the recoil of the final products. The measured value for the minimum energy is 2.224 MeV, again in close agreement with the value deduced from the mass spectroscopic measurements.

Worked example

Calculate the binding energy of the $^{62}_{28}$Ni nuclide. Compare its mass defect to the total mass of the atom, which is 61.928 346 u.

Solution

Using Equation (2.23) we find the binding energy to be

$$B = (28 \times 1.007\,825 + 34 \times 1.008\,665 - 61.928\,346)u = 0.585\,364\,u$$
$$= 0.585\,364 \times 931.5\,\mathrm{MeV} = 545.3\,\mathrm{MeV}.$$

This is the energy that would be required to completely pull apart a $^{62}_{28}$Ni nucleus and separate its 62 nucleons. The binding energy *per nucleon* is 8.795 MeV. The ratio of the mass defect to the atomic mass is 0.585 364/61.928 346, i.e. the mass defect is about 1% of the mass of the atom.

2.1.7   Binding energy curve of the nuclides

We could repeat the exercise of the previous section to deduce the nuclear binding energy B for all the elements in the periodic table from their accurately measured atomic mass. The deduced values of B could then be plotted versus atomic mass number A. However, we obtain more physical insight into the nature of the nucleus by plotting B/A versus A, where B/A is the *binding energy per nucleon*. The resulting curve is presented in Figure 2.12 for the stable nuclides. Several features of this curve are immediately apparent:

• The binding energy per nucleon is roughly constant at $\sim$8 MeV except for nuclides lighter than ^{12}C. The relative flatness of the curve shows that B is approximately proportional to A. This is a very significant feature as it indicates that each nucleon in the nucleus is attracted only to its nearest neighbours. If each nucleon attracted all the other $(A - 1)$ nucleons, there would be a total of $A(A - 1)$

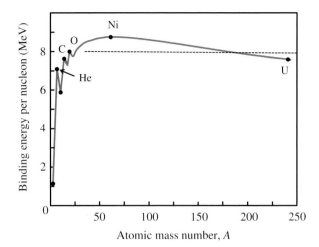

Figure 2.12 A plot of binding energy per nucleon B/A versus atomic mass number A for the stable nuclides. Except for the region of low mass number, the binding energy curve has a roughly constant value of ~ 8 MeV, indicating that the total binding energy of a nuclide is approximately proportional to A. Peaks occur at ^{4}He, ^{12}C and ^{16}O, as these nuclides have particularly high values of binding energy. Also indicated are the nuclides Ni (nickel), which occurs close to the maximum of the curve, and U (uranium), which occurs at the high end of the atomic mass range.

bonds and B would instead be proportional to $A(A-1) \sim A^2$, not to A. In turn, this demonstrates that the nuclear force must operate over a very short range. This is analogous to the molecules in a drop of water, where the molecules interact only with their nearest neighbours.

• The curve has a broad maximum at $A \sim 60$. This feature is of great physical and practical importance. It indicates that we can release energy by splitting a heavy nucleus into two less massive and more tightly bound nuclei that lie closer to the maximum in the curve (nuclear fission). Alternatively, we can release energy by fusing two light nuclei together to produce a more tightly bound nucleus, which again lies closer to the curve maximum (nuclear fusion).

• The curve gradually reduces in height at high A. This is due to the electrostatic repulsion of the protons. In contrast to the short-range nuclear attractive force, the electrostatic force is long range, falling as $1/r^2$, where r is the proton–proton separation. Consequently, each proton repels *all* the other protons. This repulsion reduces the binding energy. Eventually, for very large A, the electrostatic repulsion dominates the nuclear attraction and nuclides with A greater than about 210 are unstable and radioactively decay.

• There is a pronounced spike in the curve at $A = 4$. This shows that the binding energy of the helium nucleus, i.e. the α particle, is particularly large. There are also peaks, although less pronounced, at $A = 12$ and 16. These spikes are an indication that neutrons and protons are arranged in shells, just as the electrons in an atom are arranged in shells. These more tightly bound nuclides have *closed shells* of nucleons. This is again analogous to the situation for atoms that have closed electronic shells like the rare gases, which have relatively high ionisation energies.

2.1.8   The semi-empirical mass formula

There are some similarities between nucleons in a nucleus and molecules in a drop of water, as we have already hinted. In particular, the density of water is constant and independent of the size of the water drop. Moreover, the amount of energy it takes to vapourise a drop of water is directly proportional to the volume of the drop, i.e. the amount of energy required per molecule is constant. Similarly, the density of nuclear matter is approximately the same for all nuclides and the binding energy per nucleon is fairly constant, except for the lightest nuclides. These considerations led to the *liquid drop model* of the nucleus, where individual nucleons are considered to be analogues of molecules in a liquid, held together by short-range interactions and surface tension effects. Molecules are held together by *covalent bonding*, which operates over a short range so that molecules bind only to their nearest neighbours, just as nucleons bind only to their nearest neighbours. We have also hinted at indications that nuclei have a *shell-like* structure where the nucleons occupy shells, just as electrons occupy shells in atoms. Such indications lead to the *shell model* of the nucleus. In this model, each nucleon moves independently with an average potential energy due to the nuclear force of all the other nucleons and the electrostatic force of the other protons (see also Section 2.2.3).

Based upon these considerations, we can attempt to derive a formula for the binding energy $B(Z, A)$ of a nuclide in terms of its Z and A values. This formula has the following five contributions.

Volume term   The value of B/A is fairly constant in the binding energy curve across most of the range of the stable nuclides, i.e. B depends roughly linearly on A. Hence, the most obvious, and also the most important, term in a formula for $B(Z, A)$ is a term $a_{\mathrm{vol}} A$ that depends linearly on A and where a_{vol} is a constant. As the volume of a nucleus is approximately proportional to A, this term is called the volume term.

Surface term   The nucleons at the surface of a nucleus are less tightly bound than those in the interior because they have no neighbours outside the surface. This leads to a decrease in binding energy. As the radius of a nucleus is proportional to $A^{1/3}$, the surface area is proportional to $A^{2/3}$. This results in the surface term $-a_{\mathrm{surf}} A^{2/3}$, where a_{surf} is a constant and the minus sign indicates that this term reduces the binding energy. This surface tension effect helps to explain why nucleii are nearly spherical in shape, the shape that water droplets seek to attain.

Coulomb term   This term arises from the electrostatic repulsion of the protons that tends to make the nucleus less tightly bound. Electrostatic repulsion is a long-range force and so each proton interacts with all the other $(Z-1)$ protons. As there are Z protons, the total repulsive Coulomb energy is proportional to $Z(Z-1) \sim Z^2$. The Coulomb energy also depends on the inverse of the radius of the nucleus. We thus obtain the Coulomb term $-a_{\text{Coul}}Z^2/A^{1/3}$, where a_{Coul} is a constant and, again, the minus sign indicates that this term reduces the binding energy. [We can compare this result with the electrical potential energy of a uniformly charged sphere of radius R and total charge Q, which is equal to $3/5\,(Q^2/4\pi\varepsilon_0 R)$.]

(These first three terms would also arise for the case of a charged liquid drop in accordance with the liquid drop model of the nucleus. The following two terms are accounted for in terms of the way in which nucleons occupy the shell-like structure of a nucleus.)

Asymmetry term   The chart of stable nuclides shows that there is a preference for light nuclei to have $Z \approx N$, i.e. $Z \approx A/2$, (see also Section 2.3.1), although this tendency is reduced when A is large and electrostatic repulsion of the protons increases. A suitable term in $B(Z,A)$ to take this into account is $-a_{\text{sym}}(A-2Z)^2/A$ and is called the *asymmetry term*, where a_{sym} is a constant. This term is zero when $Z = A/2$ but when $Z \neq A/2$, it appears as a negative term, reducing the binding energy. Also, the asymmetry term reduces in importance as A increases.

Pairing term   The nuclear force favours *pairing* of like nucleons; the nuclide is more tightly bound if both Z and N are even, but less tightly bound if both Z and N are odd. This is accounted for by a pairing term δ in $B(Z,A)$. δ is positive if Z and N are both even, negative if both Z and N are odd, and zero if either Z or N is odd (A odd). δ can be expressed as $a_{\text{pair}}A^{-3/4}$.

Combining the five terms, we obtain the formula $B(Z,A)$ for the binding energy of a nuclide:

$$B(Z,A) = a_{\text{vol}}A - a_{\text{surf}}A^{2/3} - \frac{a_{\text{Coul}}Z(Z-1)}{A^{1/3}} - \frac{a_{\text{sym}}(A-2Z)^2}{A} \pm \delta.$$

$$(2.27)$$

In Section 2.1.6. we saw that we could deduce the binding energy, B, of a nuclide from a knowledge of its atomic mass, $_Z^A M$:

$$B = \left(ZM_{\text{H}} + Nm_{\text{n}} - {}_Z^A M\right)c^2,$$

$$(2.23)$$

When we substitute $B(Z, A)$ from Equation (2.27) into Equation (2.23), we obtain

$$_Z^A M = ZM_{\mathrm{H}} + Nm_{\mathrm{n}} - B(Z, A)/c^2. \qquad (2.28)$$

This is the *semi-empirical mass formula*. It enables accurate values of an atomic mass to be predicted and it applies to both stable and, very usefully, unstable nuclides. This makes it of great practical importance because it can predict, for example, the mass difference in a particular nuclear reaction and hence the amount of energy that can be obtained from that reaction. The formula is described as semi-empirical because it is based on both experimental observations and theoretical considerations.

By dividing Equation (2.27) by mass number A, we obtain a formula for $B(Z, A)/A$, the *binding energy per nucleon*. Furthermore, we can fit the resulting formula to the binding energy curve in Figure 2.12, by adjusting the values of the constants $a_{\mathrm{vol}}, a_{\mathrm{surf}}, a_{\mathrm{Coul}}, a_{\mathrm{sym}}$, and a_{pair} to give the best agreement between $B(Z, A)/A$ and the binding energy curve. One set of values is:

$$a_{\mathrm{vol}} = 15.5\,\mathrm{MeV}; \quad a_{\mathrm{surf}} = 16.8\,\mathrm{MeV}; \quad a_{\mathrm{Coul}} = 0.72\,\mathrm{MeV};$$
$$a_{\mathrm{sym}} = 23\,\mathrm{MeV}; \quad \text{and} \quad a_{\mathrm{pair}} = 34\,\mathrm{MeV}.$$

Worked example

Use the semi-empirical mass formula to find the total binding energy of the $_{28}^{62}\mathrm{Ni}$ nucleus and the mass of the nickel atom.

Solution

$A = 62$, $Z = 28$ and $N = 34$, and evaluating the five terms in Equation (2.27) gives:

1. $a_{\mathrm{vol}}A = 15.5 \times 62 = 961$ MeV;

2. $-a_{\mathrm{surf}}A^{2/3} = -16.8 \times 62^{2/3} = -263$ MeV;

3. $-\dfrac{a_{\mathrm{Coul}}Z(Z-1)}{A^{1/3}} = -\dfrac{0.72 \times 28 \times 27}{62^{1/3}} = -138$ MeV;

4. $-\dfrac{a_{\mathrm{sym}}(A - 2Z)^2}{A} = -\dfrac{23 \times [62 - (2 \times 28)]^2}{62} = -13.4$ MeV;

5. $+a_{\mathrm{pair}}A^{-3/4} = 34 \times 62^{-3/4} = 1.53$ MeV.

The total binding energy is the sum of these five terms, which is equal to 548.1 MeV. This compares with the correct value of

545.3 MeV that was obtained from precise spectroscopic mass measurements, a difference of 0.5%. Using Equation (2.28), the mass of the nickel atom is

$$(28 \times 1.007\ 825) + (34 \times 1.008\ 665) - \frac{548.1}{931.5} = 61.93\ \text{u}.$$

This compares with the correct value of 61.928 346 u, and is only different from it by 0.003%.

2.2 Nuclear forces and energies

2.2.1 Characteristics of the nuclear force

The neutrons and protons in a nucleus are bound together by the so-called *nuclear force*. This force is understood to be a secondary effect of the *strong force* that binds quarks together to form the neutrons and protons themselves. This more powerful strong force is *mediated* by particles called *gluons*, which act to hold quarks together. Quarks, gluons and their dynamics are mostly confined within nucleons, but residual influences extend slightly beyond nucleon boundaries to give rise to the nuclear force. (The strong force is considered to be one of the four fundamental forces of nature, the others being the gravitational force, the electromagnetic force and the *weak nuclear force*, which we will meet when we discuss β decay.) The nuclear force that binds the neutrons and protons together has the following properties:

• It has a strongly attractive component, which acts only over a very small range, ∼ few fm.

• It has a repulsive component at very short distances, less than ∼0.5 fm, which keeps the nucleons apart.

• Within its range of operation, the nuclear force is much stronger than the electrostatic force. For example, the nuclear force acting between two protons at a separation of 2 fm is about 100 times stronger than the electrostatic force. Consequently, the nuclear force can overcome the electrostatic repulsion of the protons in the nucleus.

• It is charge-independent, i.e. the nuclear force between protons is the same as the nuclear force between neutrons and the same as the nuclear force between protons and neutrons.

• There is no nuclear force between electrons and either protons or neutrons.

• The nuclear force favours *pairing* of protons and of neutrons with opposite *spins*. Hence the α particle with two protons and two neutrons is an exceptionally stable nucleus, as we have already seen.

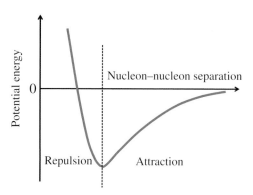

Figure 2.13 The potential energy of two nucleons as a function of their separation, resulting from the strong nuclear force between them. A minimum in the potential energy occurs at an equilibrium separation where the repulsive component of the nuclear force is balanced by the attractive component.

The nuclear force between two nucleons results in a potential energy curve of the form illustrated in Figure 2.13, which shows the potential energy as a function of the separation of the two nucleons. The curve applies to the case of two neutrons or a proton and a neutron, for which there is no electrostatic repulsion. At large separation, the potential energy is zero. As the two nucleons approach each other, at around a few fm, they experience the strongly attractive nuclear force that leads to a decrease in potential energy. At short distances, below about 0.5 fm, the nuclear force becomes repulsive and the potential energy increases sharply. A minimum in the potential energy occurs at a separation where the attractive and repulsive forces are equal and opposite. This is a position of equilibrium and so the two nucleons tend to maintain this separation.

2.2.2   Nuclear energies

Figure 2.13 illustrates the potential energy between *two* nucleons as a function of their separation. We now consider the energies of nucleons that exist within a nucleus. We saw above that there is an equilibrium separation between nucleons and so the nucleons within a nucleus tend to maintain an average separation with respect to their neighbours. Furthermore, as the nuclear force has a very short range, nucleons interact only with their nearest neighbours. Consequently, a nucleon has an approximately constant energy due to the attraction of its neighbours. We can consider the nucleons in a nucleus as particles that exist within a *potential well*. Such a potential well for the neutrons in the nucleus is shown in Figure 2.14(a). In this figure, the horizontal axis represents the radial distance, r, from the centre of the nucleus and the vertical axis represents the potential energy, $V(r)$, of the neutrons. The bottom of the well is flat as the neutrons have an approximately constant energy of attraction and so their potential energy is approximately constant in the nuclear interior. Because the neutrons are bound to the nucleus, the bottom of the well lies below the zero of potential energy. The well does not have

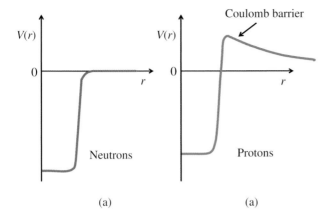

(a)   (a)

Figure 2.14 (a) The potential energy function, $V(r)$, for neutrons that are bound in a nucleus, where the potential energy V is plotted as function of radial distance r from the centre of the nucleus. In the interior of the nucleus, the potential energy is approximately constant, as depicted by the flat bottom of the potential well. (b) $V(r)$ for protons that are bound in a nucleus, again plotted as function of r. In addition to the nuclear force, protons also experience electrostatic repulsive due to other protons. The addition of this electrostatic repulsion to the nuclear attraction reduces the depth of the potential well for protons. Further away from the nucleus, the protons feel just the long-range electrostatic force, and the combination of the attractive nuclear force and the repulsive electrostatic force results in the formation of a Coulomb barrier.

sharp sides because the nuclear force does not fall off abruptly and also because the distribution of neutrons near the surface does not fall off abruptly (see Figure 2.6). The width of the well is essentially the nuclear radius. The depth of the well depends on the number of nucleons in the nucleus but is ~50 MeV. When it is away from the nucleus, a neutron does not experience any force and its potential energy is zero. If it approaches the nuclear surface, it experiences the attractive nuclear force and 'falls' into the nuclear interior with a gain in kinetic energy corresponding to a decrease in its potential energy.

The protons in a nucleus experience the same nuclear force as the neutrons. Hence, they have a component in their potential energy function due the nuclear force that is the same as for the neutrons. In addition, however, protons also experience the repulsive electrostatic force due to other protons. When we add this electrostatic repulsion to the attraction due to the nuclear force, we get the total potential energy function $V(r)$ for the protons that is shown in Figure 2.14(b). Within the nucleus, the electrostatic repulsion reduces the depth of the potential well. Further away from the nucleus, the protons feel just the long-range electrostatic force. The combination of the attractive nuclear force and the repulsive electrostatic force results in the formation of a *potential barrier* as shown in Figure 2.14(b). This is called the *Coulomb barrier*. It plays a crucially important role in maintaining the stability of the Universe, as it prevents low-energy nuclei coming into contact with each other and forming a heavier nucleus, even though that heavier nucleus would be more stable. As the positively charged nuclei approach each other, they are repelled by the long-range electrostatic force before they can come under the influence of the much stronger nuclear force. Indeed, the Coulomb barrier hinders the generation of energy by nuclear fusion, as we will see in Chapter 3. It also plays a role in the mechanism of the radioactive decay of nuclei by α particle emission, as we will discuss in Section 2.3.3.

We can get a rough estimate of the kinetic energy of a nucleon in a nucleus from an important principle of quantum mechanics; namely the *uncertainty principle*. One statement of this principle is that it is not possible to know the position and the momentum of a particle completely. This is represented by the expression

$$\Delta x \Delta p \geq \frac{\hbar}{2}, \tag{2.29}$$

where Δx is the uncertainty in the particle's position, Δp is the uncertainty in its momentum, and $\hbar$ is $h/2\pi$, where h is Planck's constant. It follows that the momentum p of the particle must be at least as large as $\hbar/2\Delta x$. Hence, an estimate of the lower limit for the kinetic energy $E = p^2/2m$ of a nucleon is given by

$$E \sim \frac{\hbar^2}{8m(\Delta x)^2}.$$

Taking a nuclear dimension of 1 fm for Δx we obtain a value for E of ~ 5 MeV. When we repeat this exercise for an electron in an atom, and take an atomic dimension of 0.1 nm, we obtain the kinetic energy of the electron to be ~ 1 eV, a factor $\sim 10^6$ smaller. We can begin to see that nuclear reactions will produce much greater amounts of energy than are produced in chemical reactions, which involve atomic electrons. The uncertainty principle also explains why electrons do not reside in the nucleus. Electrons are around 2000 times less massive than nucleons. If an electron were to be confined to a region of dimension ~ 1 fm, and taking relativistic effects into account, the electron would have a kinetic energy of ~ 100 MeV. This would require a potential barrier of this magnitude to bind the electron and such a barrier does not exist.

When a particle is confined in a potential well, the confinement leads naturally to discrete energy levels that the particle can occupy. This is one of the important results of quantum mechanics, as we describe in Section 2.2.3. It is the basis of the *shell model* of the nucleus, where we consider that neutrons and protons exist inside nuclei in certain allowed energy levels within a potential well. The situation is analogous to that for atoms, which similarly have discrete energy levels, except that the nuclear energies are the order of a million times larger, as noted earlier. Atomic energy levels are occupied by the atomic electrons. Electrons are *spin*-$^1/_2$ particles and consequently they fill the available atomic energy levels according to the *Pauli exclusion principle*. This states that no two spin-$^1/_2$ particles can have the same set of quantum numbers. Consequently, each energy level can hold only a certain number of electrons; additional electrons have to go into levels that are at higher energy. Neutrons and protons are also spin-$^1/_2$ particles and, again, fill the available energy levels of a nucleus according to the Pauli exclusion principle.

We illustrate this arrangement of nucleons in a nucleus with the example of the light nuclide ^{12}C. Because this nuclide has relatively few protons, we can neglect electrostatic repulsion and consider only the potential energy due to the nuclear force. The resulting potential wells are then the same for both neutrons and protons. The potential wells and the first few possible energy levels for ^{12}C are illustrated schematically in Figure 2.15; the separations of the energy levels are not to scale. The potential wells for the neutrons and protons are conventionally drawn back to back as shown. ^{12}C has six protons and six neutrons and these populate the energy levels according to the Pauli exclusion principle, with two nucleons in the lowest energy level and four nucleons in the next level. Protons and neutrons are not identical particles and so the Pauli exclusion principle is not violated when they occupy the same energy levels, as shown.

It is found that nuclei with either Z or $N = 2, 8, 20, 28, 50, 82$ and 126 have a larger value of binding energy per nucleon than predicted by the semi-empirical mass formula. These numbers are the so-called *magic numbers*. Their existence provides compelling evidence for the shell model of the nucleus, where these values of Z correspond to full shells of nucleons, either neutrons or protons. An analogous situation occurs for atoms. For particular values of Z, corresponding to the rare gases, helium, neon and argon, the ionisation energy of the respective atom is relatively large. Again these atoms have full shells of electrons.

2.2.3 Quantum mechanical description of a particle in a potential well

Calculating the exact potential energy function, $V(r)$, of a nucleus and the resulting energies of the individual nucleons is a formidable task and must be solved with sophisticated methods of quantum mechanics. However, we can gain valuable physical insight into the existence of nuclear energy levels by considering the much simpler case of a particle confined to a one-dimensional potential well. (Indeed, a particle in a one-dimensional well is an important example in quantum mechanics and is applied to many physical situations.) Moreover, we can extend our discussion to see what happens when a particle strikes a potential barrier of finite height; a case that is of importance in α-particle decay.

We are familiar with the case of a taut string that is held between two fixed points. When the string is plucked, standing waves are formed on the string. In particular, only standing waves of particular wavelengths can exist. The condition is $n\lambda/2 = L$, where λ is the wavelength, L is the length of the string and $n = 1, 2, 3, \ldots$ (see also Section 4.2.3). Only these values of λ are allowed because of the *boundary conditions* at the fixed ends. As we have already noted,

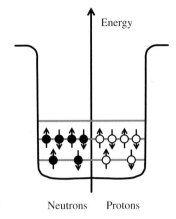

Figure 2.15 Schematic diagram of the first few energy levels in the ^{12}C nucleus, showing the arrangement of the six neutrons and six protons in the levels, which are not drawn to scale. The potential wells for the neutrons and protons are conventionally drawn back to back, as shown. Because ^{12}C has relatively few protons, the electrostatic energy due to proton repulsion can be neglected and only the potential energy due to the nuclear force needs to be considered.

particles on an atomic or subatomic scale have wave-like properties. Accordingly, the state of such a particle is described by a *wave function*, just as a disturbance on a taut string is described as a wave. In general the wave function is a function of the three spatial coordinates, x, y and z, and of time t, i.e. $\psi(x, y, z, t)$. The physical meaning of the wave function can be interpreted in the following way. The square of the wave function of a particle at a particular point gives the probability of finding the particle at that point. The particle is most likely to be found where $|\psi|^2$ is large (we write the square of the absolute value of ψ^2 as ψ may be a complex quantity). In some cases, the value of ψ is independent of time, in which case the wave function depends only on the spatial coordinates, i.e. $\psi = \psi(x, y, z)$. This occurs, for example, when the particle is in a state of definite energy, which does not change with time. We can determine the wave function of a particle and its associated energy using the *Schrödinger equation*, named after Erwin Schrödinger who introduced it. The Schrödinger equation plays the role in quantum mechanics that Newton's laws play in classical mechanics. For a particle of mass m and energy E that moves in one dimension along the x-axis, the time-independent Schrödinger equation is

$$-\frac{\hbar^2}{2m}\frac{\mathrm{d}^2\psi(x)}{\mathrm{d}x^2} + V(x)\psi(x) = E\psi(x), \tag{2.30}$$

where $V(x)$ is the potential energy and $\hbar$ is as defined earlier. Schrödinger's equation is really a statement that the kinetic energy of the particle [represented by the term $-(\hbar^2/2m)(\mathrm{d}^2\psi(x)/\mathrm{d}x^2)$] plus the potential energy [represented by the term $V(x)\psi(x)$] is equal to the total energy of the particle [represented by the term $E\psi(x)$], cf. the classical expression:

$$\frac{p^2}{2m} + V(x) = E. \tag{2.31}$$

We can readily apply Schrödinger's equation to the example of a particle in a one-dimensional infinite potential well.

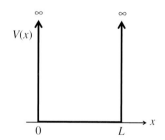

Figure 2.16 The one-dimensional infinite potential well. The potential energy $V(x)$ is zero between $x = 0$ and $x = L$, but rises abruptly to infinity at $x = 0$ and $x = L$, so that any particle in the well will be confined to the region $0 < x < L$.

A particle in a one-dimensional infinite potential well

The one-dimensional infinite potential well is shown in Figure 2.16. The potential energy $V(x)$ is zero between $x = 0$ and $x = L$, and infinite elsewhere. We consider a particle of mass m and total energy E that can move within the well along the x-axis. As $V(x)$ rises abruptly to infinity at $x = 0$ and $x = L$, the particle is confined to the region $0 < x < L$ and outside this region $\psi(x)$ must be zero, i.e. the particle is confined between two infinitely high rigid walls.

Between $x = 0$ and $x = L$, $V(x) = 0$ and from Equation (2.30), the wave function must satisfy

$$-\frac{\hbar^2}{2m}\frac{d^2\psi(x)}{dx^2} = E\psi(x) \qquad (2.32)$$

The general solution to this equation is

$$\psi(x) = A\sin kx + B\cos kx, \qquad (2.33)$$

where

$$k = \sqrt{2mE/\hbar^2}. \qquad (2.34)$$

As $\psi(x) = 0$ at $x = 0$, we have $B = 0$, giving

$$\psi(x) = A\sin kx. \qquad (2.35)$$

As $\psi(x) = 0$ at $x = L$, we have

$$k_n = n\pi/L : n = 1, 2, 3, ... \qquad (2.36)$$

k can have an infinite number of values given by Equation (2.36) and for each value of k there is the corresponding wave function:

$$\psi_n(x) = A\sin\frac{n\pi}{L}x. \qquad (2.37)$$

[This result is completely analogous to the case of standing waves on a taut string, because the differential equations describing the two situations are of the same form. If in Equation (2.32) we substitute $E = p^2/2m$, with $\lambda = h/p$, the de Broglie wavelength of the particle, we obtain

$$\frac{d^2\psi(x)}{dx^2} = -\left(\frac{2\pi}{\lambda}\right)^2\psi(x).$$

The differential equation describing standing waves on a taut string is

$$\frac{\partial^2 f_n(x)}{\partial x^2} = -\left(\frac{2\pi}{\lambda}\right)^2 f_n(x),$$

where λ is the wavelength and $f_n(x)$ describes the variation of the amplitude A of the wave along the length L of the string. $f_n(x)$ has the form $f_n(x) = A_n\sin(n\pi x/L)$.]

Substituting for $k = \sqrt{2mE/\hbar^2}$ in Equation (2.35) and substituting the resulting $\psi_n(x)$ in Equation (2.32), we obtain the possible energy levels E_n for the particle in the well:

$$E_n = \frac{n^2\pi^2\hbar^2}{2mL^2} = \frac{n^2h^2}{8mL^2}. \qquad (2.38)$$

We see that the particle can only have certain discrete values of total energy E_n, which are given by Equation (2.38). For the infinite

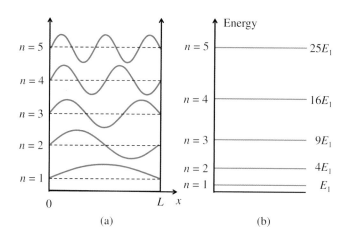

Figure 2.17 (a) The wave functions for a particle confined in an infinite potential well, for quantum number $n = 1$ to 5. These functions are identical to those for standing waves on a string. (b) The corresponding, quantised energies of the particle in the well, which scale as n^2.

potential well there is an *infinite* number of energy levels, each having its own quantum number n and a corresponding wave function given by Equation (2.35). Figure 2.17(a) shows the wave functions for $n = 1, 2, 3, 4$ and 5. Again these functions are identical to those for a standing wave on a string. Figure 2.17(b) shows the energy level diagram for the system. The energy levels are quantised and are proportional to n^2. We can use Equation (2.38) to make an estimate of the energy of a nucleon in a nucleus. Taking $n = 1$ and $L = 5$ fm, and using the known values of the fundamental constants, we obtain $E_1 \sim 8$ MeV.

A particle in a one-dimensional finite potential well

The potential well confining the nucleons in a nucleus is not infinite in extent (see Figure 2.14), and so we next consider the case of a particle that is bound in a potential well where one of the walls is not infinitely high. This well is illustrated in Figure 2.18 and resembles the form of the potential wells of Figure 2.14. In this case, the potential $V(x)$ is given by

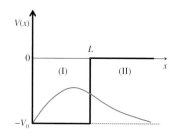

Figure 2.18 A potential well in which one of the walls is not completely rigid; the potential energy does not become infinite at $x = L$. Also shown is the wave function that corresponds to the ground state of a particle in the well.

$$V(x) = \begin{cases} \infty, & x < 0 \\ -V_0, & 0 \leq x \leq L \\ 0, & x > L \end{cases}$$

When the *total* energy E of the particle (kinetic energy plus potential energy) is negative, having a value between $-V_0$ and 0, the particle is trapped in the well. The solution of the Schrödinger equation follows the treatment above, but now there are two separate regions, (I) and (II), to consider, as indicated on Figure 2.18, with two separate solutions of the wave equation. As usual, these solutions must comply with the boundary conditions. At $x = 0$, $\psi(x) = 0$, as in the previous example. At $x = L$, both $\psi(x)$ and $d\psi(x)/dx$ must be continuous: if $\psi(x)$ or $d\psi(x)/dx$ were discontinuous at any point then $d^2\psi(x)/dx^2$ would be infinite, which is unphysical.

In region (I); $0 \leq x \leq L$, we have $V(x) = -V_0$, and

$$-\frac{\hbar^2}{2m}\frac{\mathrm{d}^2\psi(x)}{\mathrm{d}x^2} - V_0\psi(x) = E\psi(x). \tag{2.39}$$

Taking into account the boundary condition at $x = 0$ we obtain the solution

$$\psi(x) = A\sin k_1 x, \tag{2.40}$$

where $k_1 = \sqrt{2m(E + V_0)/\hbar^2}$. Again, the wave function has a sinusoidal form.

In region (II), $x > L$, the potential energy is zero and we have

$$-\frac{\hbar^2}{2m}\frac{\mathrm{d}^2\psi(x)}{\mathrm{d}x^2} = E\psi(x). \tag{2.41}$$

E is a negative quantity, as the particle is bound and so the solution of Equation (2.41) is

$$\psi(x) = B\exp(k_2 x) + C\exp(-k_2 x), \tag{2.42}$$

where $k_2 = \sqrt{-2mE/\hbar^2}$. To ensure that the wave function is finite for $x \to \infty$, B must be zero. Hence, the wave function in region (II) is given by

$$\psi(x) = C\exp(-k_2 x). \tag{2.43}$$

This quantum mechanical result is dramatically different from the classical result. Classically, the particle is trapped between the walls and cannot exist in the region $x > L$, as its energy E lies below the top of the wall. Quantum mechanically, however, it can penetrate into the classically forbidden region. This ability to penetrate a classically forbidden region is one of the most important properties of a quantum particle and arises because of its wave nature.

Applying the boundary conditions on $\psi(x)$ and $\mathrm{d}\psi(x)/\mathrm{d}x$ at $x = L$, we obtain

$$k_2 = -k_1 \cot k_1 L. \tag{2.44}$$

k_1 and k_2 are not independent parameters. They are defined by

$$E = -\frac{\hbar^2 k_2^2}{2m} \quad \text{and} \quad E = \frac{\hbar^2 k_1^2}{2m} - V_0, \tag{2.45}$$

giving

$$k_1^2 + k_2^2 = \frac{2mV_0}{\hbar^2}. \tag{2.46}$$

We thus have two simultaneous equations for k_1 and k_2. Given the value of V_0, these equations can be solved graphically (or numerically) to obtain the values of k_1 and k_2, and hence the energies of the possible bound states. The form of the ground-state wave function for a particle in the well is shown in Figure 2.18 as the blue curve.

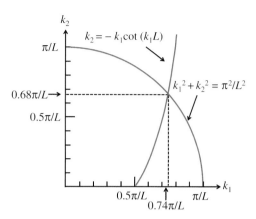

Figure 2.19 Plots of k_2 versus k_1 to determine the possible energies of a particle in a finite well potential such as that shown in Figure 2.18, with a well depth $V_0 = \hbar^2\pi^2/2mL^2$. In this case, there is just one bound state.

Worked example

Suppose the potential well shown in Figure 2.18 has a depth V_0 that is equal to $\hbar^2\pi^2/2mL^2$. Determine the possible energies of the particle in the well.

Solution

From Equation (2.46), we have

$$k_1^2 + k_2^2 = \left(\frac{2m}{\hbar^2}\right)\left(\frac{\hbar^2\pi^2}{2mL^2}\right) = \left(\frac{\pi^2}{L^2}\right).$$

In Figure 2.19, k_2 is plotted versus k_1, in units of π/L, resulting in a quadrant of a circle of radius π/L. We also have $k_2 = -k_1 \cot k_1 L$. Again, plotting k_2 versus k_1, in units of π/L, we obtain the second curve in Figure 2.19. The two curves cross for $k_2 = 0.68\pi/L$, $k_1 = 0.74\pi/L$. Then the energy of the particle, given by either $E = -\hbar^2 k_2^2/2m$ or $E = \hbar^2 k_1^2/2m - V_0$, is equal to $-0.46\,\hbar^2\pi^2/2mL^2$. Inspection of Figure 2.19 shows that there is only one intersection of the two graphs and therefore only one possible energy level. Moreover, if the radius $(k_1^2 + k_2^2)^{1/2}$ were less than $0.5\pi/L$, i.e. $V_0 < \hbar^2\pi^2/8mL^2$, there would be no bound states.

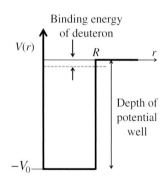

Figure 2.20
A three-dimensional finite potential well to model the deuteron nucleus. This model gives just one bound state of the nucleus with a binding energy of approximately 2 MeV, which is in agreement with the experimentally measured value.

The deuteron

The proton and the neutron in the deuteron nucleus are bound together by the nuclear force. Because it is a bound system the nucleons move in an attractive potential, representing their mutual attraction, and their total energy must be negative. We can model the deuteron nucleus by a *three-dimensional* form of the finite potential well. This well is illustrated in Figure 2.20, where the horizontal

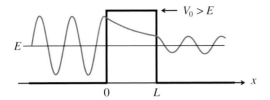

$V_0 > E$

Figure 2.21 A particle of energy E approaches a barrier of height V_0, where $E < V_0$. Quantum mechanically, the particle can tunnel through the barrier. The wave function of the particle exhibits an oscillatory behaviour on either side of the barrier and has an exponentially decreasing amplitude within the barrier.

axis represents the radial distance r. In a similar manner to the one-dimensional case above, we obtain the relation

$$k_2 = -k_1 \cot k_1 R, \tag{2.47}$$

where

$$k_1 = \sqrt{2m(E + V_0)/\hbar^2}, \quad k_2 = \sqrt{-2mE/\hbar^2} \tag{2.48}$$

and R is taken to be the diameter of the deuteron nucleus. Equation (2.47) is exactly analogous to Equation (2.44). Note that Equations (2.47) and (2.48) connect the depth V_0 of the potential well to the diameter R of the nucleus. Knowing the value of R from scattering experiments, we can use the model to predict the value of V_0, which is found to be approximately 35 MeV. The model gives just one bound energy level in this potential well, as in the worked example above. Moreover, this level lies just below the top of the well with a binding energy of about 2 MeV, in agreement with the known values of the binding energy that we saw in Section 2.1.6. It is fortunate that deuterium does have a bound state, i.e. that it is a stable nucleus, because deuterium is an essential step in the fusion of nuclei in stars, including our own.

Barrier penetration

The other case that is of interest to us is where a particle is incident upon a potential barrier. In particular, we consider the case where the total energy of the particle E is less than the height of the barrier. This case is illustrated in Figure 2.21, where the particle approaches the barrier from the left. The potential $V(x)$ is given by

$$V(x) = \begin{cases} 0, & x < 0 \\ V_0, & 0 \leq x \leq L \\ 0, & x > L. \end{cases}$$

Classically the particle cannot surmount the barrier; it would simply be reflected backwards. However, quantum mechanically the particle has a finite probability of being found to the right of the barrier, i.e. *tunnelling* through the barrier. We can make an analogy here with barrier penetration by a wave in light optics. If a beam of light is incident upon a glass/air interface at an angle greater than the critical angle, the light beam is completely reflected from the interface. However, there exists a wave disturbance that penetrates a few wavelengths into the air space. If a second block of glass is brought

Figure 2.22 When light is incident upon the glass/air interface of a prism at an angle greater than the critical angle, it is totally internally reflected. However, because of its wave nature, the light can penetrate an air gap and pass into a second prism when the gap is of the order of the wavelength of the light. The critical angle for glass of refractive index 1.5 is 42°.

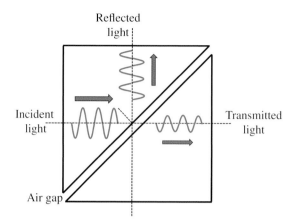

sufficiently close to the glass/air interface, light will pass across the gap into the second glass block, as illustrated in Figure 2.22. The penetration of a particle through a potential barrier is analogous to this because of the wave nature of the particle.

To determine the wave function in the three regions of the potential energy function shown in Figure 2.21, we proceed as before by solving Schrödinger's equation for each region and matching the wave function and their derivatives at $x = 0$ and $x = L$. This is fairly involved mathematically.[3] What we find is that in the regions $x < 0$ and $x > L$, the wave functions exhibit an oscillatory behaviour (cf. Equation 2.33) while in the region $0 \leq x \leq L$, the wave function has an exponential form (cf. Equation 2.42). The resulting wave functions are illustrated in Figure 2.21. The probability T that the particle tunnels through the barrier is proportional to the square of the ratio of the amplitudes of the wave functions to the left and to the right of the barrier. When the probability T is much smaller than 1, it is found that

$$T \approx \left[\frac{16E\,(V_0 - E)}{V_0^2} \right] e^{-2\beta}, \qquad (2.49a)$$

where

$$\beta = \frac{L\sqrt{2m\,(V_0 - E)}}{\hbar}. \qquad (2.49b)$$

There is a finite probability of the incident particle tunnelling through the barrier. Moreover, the probability depends exponentially on the energy difference through the term $\sqrt{m\,(V_0 - E)}$ and on the barrier width L. This means that the probability is very sensitive to these parameters. Because of the negative sign of the exponent, the

---

[3] This problem is discussed in detail in, for example, A.C. Phillips, *Introduction to Quantum Physics*, John Wiley & Sons Ltd, 2003.

probability decreases very rapidly with increasing energy difference $(V_0 - E)$ and increasing barrier width L.

Barrier penetration is of great importance for nuclear fusion. A fusion reaction can occur when two nuclei tunnel through the barrier due to their mutual electrostatic repulsion and approach each other sufficiently closely that the attractive nuclear force causes them to fuse together.

2.3 Radioactivity and nuclear stability

Natural radioactivity was discovered before the existence of the nucleus had been established. In all, there are about 2500 known nuclides, but less than 10% of these are stable against radioactive decay; the majority are unstable and decay to other nuclides emitting particles and high-energy electromagnetic radiation. The essential idea is that *an unstable nucleus will decay to a nucleus that is more stable, by which we mean more tightly bound.* The timescales for radioactive decay range from a tiny fraction of a second to billions of years. Of the radioactive nuclides, over 60 can be found in nature and indeed we are surrounded by natural radioactivity. It comes from the rocks and soil that make up the planet, from the building materials of our homes and even from the food we eat. For example, potassium is present in many of our foods and the ^{40}K isotope of potassium constitutes the major radioactive nuclide in our bodies. Some of the naturally occurring radioactive nuclides in nature, such as ^{238}U, originated in the interior of stars and still exist because their *half-lives* are comparable to the age of the Earth (the age of the Earth is $\sim 5 \times 10^9$ years, while ^{238}U has a half-life of 4.5×10^9 years). Others exist because long-lived nuclides like ^{238}U decay in a chain of separate steps and produce a range of radioactive nuclides in the process. Although these product nuclides may have short half-lives, they are being continually created from the long-lived *parent nuclide*. Natural radioactivity also arises from radioactive isotopes that are produced by high-energy cosmic rays that rain down upon the Earth. For example, cosmic rays interact with molecules in the atmosphere to produce the ^{14}C isotope of carbon, which is used in the technique of *radiocarbon dating.*

Radioactive nuclides are also produced artificially in particle accelerators and nuclear reactors. For example, radioactive nuclides for *nuclear medicine* are produced by bombarding stable nuclides with energetic particles in a cyclotron. In one application the radioactive isotope ^{131}I is used to identify and treat cancerous nodules in the thyroid gland. A minute amount of ^{131}I is injected into the patient. The speed with which the isotope is concentrated in the thyroid is a measure of how well the thyroid is working. Furthermore, employing sophisticated imaging techniques, the γ radiation from the decay of ^{131}I can be used to obtain an image of the thyroid, which can show up abnormalities in, for example, thyroid size. If cancerous nodules are discovered, they can then be destroyed by a larger dose of ^{131}I. The half-life of ^{131}I is about 8 days, so that long-term effects of nuclear radiation are avoided. In the case of nuclear reactors, some of the radioactive nuclides they produce have half-lives of many thousands of years. These must be isolated from biological systems for periods of the order 10^5–10^6 years and one of the challenges of nuclear energy is to store these radioactive products securely over such long timescales.

2.3.1   Segré chart of the stable nuclides

We can organise nuclides on a so-called Segré chart, named after the nuclear physicist Emilio Segré. Each nuclide is indicated by a point on the chart where its neutron number N (vertical axis) is plotted against its proton number Z (horizontal axis). The Segré chart for the *stable* nuclides is shown in Figure 2.23. The points corresponding to the stable nuclides follow a curve called the *line of stability* and define a rather narrow region of stability. Also included in the figure is the straight line $N = Z$.

For Z less than about 16, we see that $Z \approx N$. This is because, for light nuclei, the greatest stability is achieved when the number of neutrons and protons is approximately equal. For example, ^{4}He, ^{12}C and ^{16}O all have the same number of neutrons and protons. For larger values of Z, N grows faster than Z and the ratio N/Z eventually reaches a value of about 1.6. This is because of the increasing influence of the electrostatic repulsion of the protons. We saw in Section 2.1.8 that nuclear binding energy is approximately proportional to $A(= Z + N)$, while electrostatic repulsion energy is approximately proportional to Z^2, and of course a quadratic term

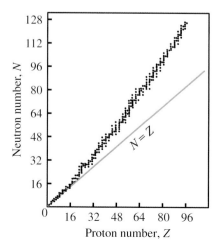

Figure 2.23 The Segré chart for the stable nuclides. Each nuclide is indicated by a data point, where its neutron number N is plotted versus its atomic number Z. Also shown is the straight line $N = Z$.

increases faster than a linear term. The nuclear instability due to the electrostatic repulsion is minimised by having more neutrons than protons. The chart also reveals a tendency for stable nuclei to have even Z and even N. This tendency arises because two nucleons of the same species, either neutrons or protons, can couple together to form an especially strong bond. This makes a particularly large contribution to the nuclear binding energy, as we discussed in Section 2.1.8. ^{208}Pb is the heaviest stable nuclide, although ^{209}Bi has a half-life of 2×10^{19} years and can be considered to be stable for all practical purposes.

The Segré chart also reveals the various isotopes of the elements, which have the same Z but different values of N. The isotopes of a particular element have the same number of electrons and hence the same chemical properties. All these isotopes therefore fit into the same box in the periodic table of the elements. The nuclear properties of the isotopes of a given element are, however, generally quite different from one another.

In addition to the stable nuclides, there are the several thousand unstable nuclides. These have values of N and Z that lie outside the region of stability. They decay to stable nuclides, typically by α or β decay as described in Section 2.2.3.

2.3.2 Decay laws of radioactivity

Rutherford discovered that the number of radioactive nuclei in a sample decreases *exponentially* with time. This result is of practical importance. For example, we can use it to predict when a radioactive product from a nuclear reactor will have decayed sufficiently for it to be considered safe. The exponential time dependence is also of fundamental importance in that it indicates that *radioactivity is a statistical process*. We cannot predict how long a *particular* nucleus will survive before it decays. In a sample of uranium, for example, a particular nucleus may decay in the next second or it may not decay for a further billion years. On the other hand, we can say that about half the nuclei in the sample will decay within the half-life of the nuclide. For example, 2 mg of ^{238}U contains approximately 5×10^{18} nuclei. Knowing its half-life, we can say that about 25 of these nuclei will decay each second. By analogy, when we toss a coin, we cannot predict whether the coin will turn up heads or tails. However, we can say that if we toss the coin 100 times, the number of heads will be close to 50.

The random nature of radioactivity was demonstrated by Rutherford and Geiger by observing the fluctuations of the number of α particles emitted by a particular source in a given time interval. For example, suppose we count the number of α particles that are emitted by a certain radioactive source in a fixed period of time, say 10 seconds, and we do this many times. We would find that the number of detected α particles would vary; sometimes it would be five counts say, sometimes four counts and sometimes seven counts, etc. The number would be random, although it would be within a limited range. The radioactive sample used by Rutherford and Geiger had a very long lifetime so that the nuclear *activity* could be assumed to be constant during the experiment. They plotted the number of times that n decays were recorded in the fixed period of time and they found that the distribution of n values was in agreement with the *Poisson distribution* for random events. The Poisson probability formula is

$$P(n) = \frac{m^n \mathrm{e}^{-m}}{n!}, \tag{2.50}$$

where $P(n)$ is the probability for getting n counts in the counting period, m is the mean or average count rate and $n!$ denotes the factorial function. (Happily, the mnemonic to remember the Poisson formula is the word 'mnemonic', i.e. m to the n times e to the minus m over n factorial.)

Worked example

A long-lived radioactive source is found to produce 1900 decays in a period of 2000 seconds. Plot the Poisson distribution $P(n)$ for the number, n, of decays that would be counted in a period of 4.0 seconds, for n between 0 and 6.

Solution

The *average* number of decays that would be expected to occur in 4.0 s is

$$\frac{1900}{2000} \times 4 = 3.80.$$

Using Equation (2.50), we obtain:

$$P(0) = \frac{\mathrm{e}^{-3.8} \times 3.8^0}{1} = 0.022, \text{ since } 0! \text{ is defined as equal to } 1;$$

$$P(1) = \frac{\mathrm{e}^{-3.8} \times 3.8^1}{1} = 0.085;$$

$$P(2) = \frac{\mathrm{e}^{-3.8} \times 3.8^2}{2 \times 1} = 0.160.$$

Similarly, $P(3) = 0.205$, $P(4) = 0.194$, $P(5) = 0.148$, $P(6) = 0.094$. The sum of these probabilities is equal to 0.908. If we also added the probabilities for $n > 6$, the total sum would be 1, as it should be. The resultant Poisson distribution is plotted in Figure 2.24.

Although we cannot say which particular nuclei in a sample will decay, we can say that if a sample contains N radioactive nuclei,

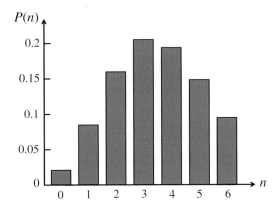

Figure 2.24 An example of a Poisson distribution where $P(n)$ is the probability of n counts in a particular counting period. In this example the mean number of counts is 3.80.

then the rate $(= -dN/dt)$ at which nuclei will decay is proportional to N, as observed experimentally. The negative sign indicates the decrease in the number of radioactive nuclei. Hence we can write

$$-\frac{(dN/dt)}{N} = \lambda, \tag{2.51}$$

where λ is called the *decay constant*. The left-hand side of this equation is the probability per unit time for the decay of a nucleus. That this probability is constant, regardless of the age of the nucleus, is the basic assumption of the statistical theory of radioactive decay. It readily follows that *the probability that a given nucleus will decay in time dt is equal to* λdt.

The observed, exponential decrease in the number of radioactive nuclei with time follows directly from the assumption of a constant probability per unit time λ for the decay of a nucleus. Thus, integrating Equation (2.51) from $N = N_0$ at $t = 0$ to N at $t = t$, gives

$$\int_{N_0}^{N} \frac{dN}{N} = -\lambda \int_0^t dt. \tag{2.52}$$

The solution of this equation is

$$N(t) = N_0 e^{-\lambda t}. \tag{2.53}$$

This is the familiar law of radioactive decay, and is the same for all forms of radioactive decay. Of course, the total number of nuclei in the sample does not change. The radioactive nuclei are simply converted into other, more stable nuclei. The half-life of a nuclide $t_{1/2}$ is defined as the time it takes for half the nuclei to decay. Then,

$$\frac{N_0}{2} = N_0 e^{-\lambda t_{1/2}}, \tag{2.54}$$

giving

$$t_{1/2} = \frac{\ln 2}{\lambda} = \frac{0.693}{\lambda}. \tag{2.55}$$

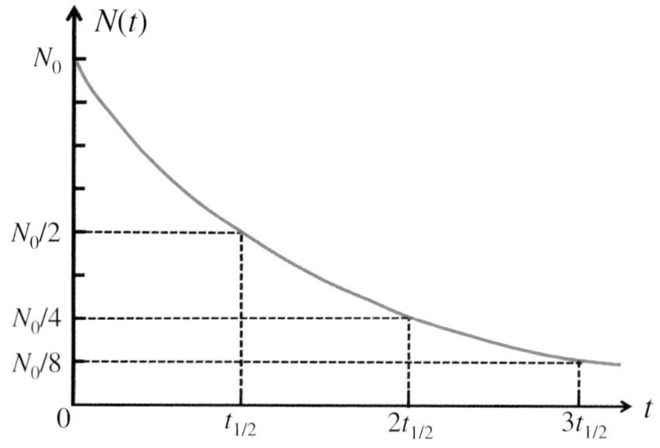

Figure 2.25 The exponential decay of the number $N(t)$ of radioactive nuclei versus time t. $N(t)$ reduces by a factor of 2 for every half-life $t_{1/2}$.

The number, $N(t)$, of radioactive nuclei reduces by a factor of 2 for every half-life as illustrated by Figure 2.25, and after n half-lives the number remaining is $(1/2)^n \, N_0$.

Note that the number reduces by a factor of 2 every half-life, regardless of the instant of time we chose for $t = 0$. Clearly a nuclide will still be radioactive after one half-life, and indeed will remain so after several half-lives, although the decay rate will decrease.

In practice it is very difficult to measure the number of radioactive nuclei present in a given sample. It is much easier to measure the number of radioactive decays per unit time. This decay rate is called the *activity* $\mathcal{A}$ and is simply the absolute value of dN/dt:

$$\mathcal{A} = \left| \frac{dN}{dt} \right|. \tag{2.56}$$

Hence, substituting for dN/dt from Equation (2.51):

$$\begin{aligned} \mathcal{A}(t) &= \lambda N(t) \\ &= \lambda N_0 e^{-\lambda t}, \end{aligned} \tag{2.57}$$

giving

$$\mathcal{A}(t) = \mathcal{A}_0 e^{-\lambda t}, \tag{2.58}$$

where $\mathcal{A}_0$ is the initial activity at $t = 0$. We see that the activity also decays exponentially with time. When we plot the natural logarithm of $\mathcal{A}(t)$ versus t, we obtain a straight line with a slope of $-\lambda$, as shown Figure 2.26. Hence, if we have experimental data for $\mathcal{A}(t)$ we can plot the data in this way to obtain the decay constant λ of the nuclide.

The SI unit for activity is the becquerel (Bq), named after Antoine Henri Becquerel, where

$$1 \text{ becquerel} = 1 \text{ decay/s}.$$

However, a unit that is often used in practice is the curie (Ci), which is approximately the activity of 1 g of ^{226}Ra. It is now defined as

$$1 \text{ Ci} = 3.70 \times 10^{10} \text{ decays/s}.$$

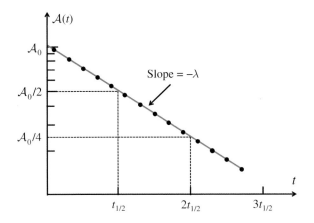

Figure 2.26 A
semi-logarithmic plot of
activity $\mathcal{A}(t)$ versus time t.
The slope of the straight line
is equal to $-\lambda$, where λ is the
decay constant of the
radioactive nuclide. $\mathcal{A}(t)$
reduces by a factor of 2 for
every half-life $t_{1/2}$.

The curie is a very large unit and laboratory sources are typically measured in millicuries (mCi) or microcuries (μCi). Note that the activity $\mathcal{A}$ of a source says nothing about the kind of particle or radiation emitted by the decaying nucleus or the energy of the particles or radiation.

Another useful parameter is the *mean lifetime* τ, which is defined as the average time that a nucleus is likely to survive before it decays. We can obtain the relationship between the mean lifetime and the decay constant λ as follows. The decay rate is equal to $|dN/dt|$. Hence, the number of nuclei that decay between t and $t + dt$ is equal to

$$|dN/dt| \times dt = \mathcal{A}(t)dt.$$

Figure 2.27 is a linear plot of $\mathcal{A}(t)$ versus t. The shaded area of the plot is $\mathcal{A}(t)dt$ and corresponds to the number of nuclei with lifetimes between t and $t + dt$. The sum of the lifetimes of these nuclei makes

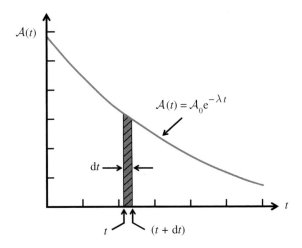

Figure 2.27 A linear plot of
activity $\mathcal{A}(t)$ versus time t.
The shaded area of width dt
corresponds to the number of
nuclei that have lifetimes
between t and $(t + dt)$.

a contribution $\mathcal{A}(t)\mathrm{d}t \times t$ to the total sum of all lifetimes. Thus the total sum of the lifetimes of all the nuclei is

$$\int_0^\infty t\mathcal{A}(t)\,\mathrm{d}t. \tag{2.59}$$

Substituting for $\mathcal{A}(t)$ from Equation (2.57) and integrating the resultant integral, we obtain

$$\int_0^\infty \lambda t N_0 \mathrm{e}^{-\lambda t}\mathrm{d}t = \frac{N_0}{\lambda}. \tag{2.60}$$

As the total number of nuclei is N_0, the mean lifetime τ is given by

$$\tau = \frac{1}{\lambda}. \tag{2.61}$$

We see that the mean lifetime τ is equal to the reciprocal of the decay constant λ. The mean lifetime is often simply called the *lifetime* of the nuclide. From Equations (2.55) and (2.61) we have

$$\tau = \frac{t_{1/2}}{\ln 2}. \tag{2.62}$$

Worked example

A quantity of 2.0 kg of natural uranium is found to emit α radiation characteristic of ^{235}U at a rate of 1.15×10^6 decays/s. Given that the abundance of ^{235}U in a naturally occurring sample of uranium is 0.72%, deduce the half-life of ^{235}U.

Solution

2000 g of natural uranium contains

$$\frac{2000 \times 0.0072 \times 6.02 \times 10^{23}}{235} = 3.69 \times 10^{22} \text{ atoms of } ^{235}\text{U},$$

where 6.02×10^{23} is Avagadro's number N_A.

$$\therefore \lambda = \frac{\mathcal{A}}{N} = \frac{1.15 \times 10^6}{3.69 \times 10^{22}} = 3.11 \times 10^{17}\,\mathrm{s}^{-1},$$

$$\Rightarrow t_{1/2} = \frac{\ln 2}{\lambda} = 2.22 \times 10^{16}\,\mathrm{s} = 7.04 \times 10^8 \text{ years.}$$

Worked example

Given that the probability a particular nucleus will decay between times t and $t + \mathrm{d}t$ is given by $P(t)\,\mathrm{d}t$, show that $P(t)\,\mathrm{d}t = \mathrm{e}^{-\lambda t}\,\mathrm{d}t$.

Solution

Let $P_s(t)$ be the probability the nucleus survives time t. The probability $P_s(t + dt)$ the nucleus survives time $t + dt$ is equal to

$$\begin{pmatrix} \text{probability that nucleus} \\ \text{survives time } t \end{pmatrix} \times \begin{pmatrix} \text{probability that nucleus} \\ \text{does not decay in time } dt \end{pmatrix}.$$

Therefore, $P_s(t + dt) = P_s(t) \times (1 - \lambda dt)$, as the probability that the nucleus does decay in time $dt = \lambda dt$. This gives

$$\frac{P_s(t + dt) - P_s(t)}{dt} = \frac{d}{dt} P_s(t) = -\lambda P_s(t),$$

which has the solution $\ln[P_s(t)] = -\lambda t + A$, where A is the constant of integration.

As $P_s(t) = 1$ for $t = 0$, $A = 0$. Hence, $P_s(t) = e^{-\lambda t}$.

The probability $P(t)dt$ that a particular nucleus has lifetime between times t and $t + dt$ is equal to

$$\begin{pmatrix} \text{probability that nucleus} \\ \text{survives time } t \end{pmatrix} \times \begin{pmatrix} \text{probability that nucleus} \\ \text{does decay in time } dt \end{pmatrix},$$

giving

$$P(t)dt = P_s(t)\lambda dt = \lambda e^{-\lambda t} dt.$$

The function $P(t)$ is an example of a *probability distribution*. In this case it has the units of probability per unit time. Given the probability distribution for the variable t, we can find the mean value of any function $f(t)$ of t, denoted as $\langle f(t) \rangle$. It is given by

$$\langle f(t) \rangle = \frac{\int f(t)P(t)dt}{\int P(t)dt},$$

where the limits of the integrals cover the entire range of t. For example, the mean lifetime of the nucleus is given by

$$\langle t \rangle = \frac{\int_0^\infty \lambda t e^{-\lambda t} dt}{\int_0^\infty \lambda e^{-\lambda t} dt} = \frac{1}{\lambda},$$

in agreement with our previous result.

Decay chains

A common situation occurs for high mass number radioactive nuclides where the decay of a nucleus produces another radioactive nucleus. And indeed there may be a *chain* of radioactive decays, e.g. $^{238}\text{U} \rightarrow ^{234}\text{Th} \rightarrow ^{234}\text{U} \rightarrow \ldots \rightarrow ^{206}\text{Pb}$. The original nucleus is called the *parent* nucleus and the succeeding generations are called its *daughter, grand-daughter*, etc. Suppose we have a parent A that decays to a radioactive daughter B that in turn decays to a stable grand-daughter C. At time $t = 0$, we have $N_A = N_0, N_B = N_C = 0$. We are interested in how N_A and N_B vary with time. We can say at once that the total number of nuclei will remain constant:

$$N_A + N_B + N_C = \text{constant} = N_0. \tag{2.63}$$

For nucleus A we have

$$N_A = N_0 e^{-\lambda_A t}, \tag{2.64}$$

with $dN_A = -\lambda_A N_A dt$.
For nucleus B we have

$$dN_B = +\lambda_A N_A \, dt - \lambda_B N_B dt. \tag{2.65}$$

This corresponds to the number of nuclei B produced in the decay of nucleus A minus the number decaying to nucleus C. This leads to the differential equation

$$\lambda_B N_B + \frac{dN_B}{dt} = \lambda_A N_0 e^{-\lambda_A t}. \tag{2.66}$$

Using the result

$$\frac{d}{dt}\left(N_B e^{\lambda_B t}\right) = e^{\lambda_B t}\left(\lambda_B N_B + \frac{dN_B}{dt}\right), \tag{2.67}$$

we obtain

$$\frac{d}{dt}\left(N_B e^{\lambda_B t}\right) = \lambda_A N_0 e^{(\lambda_B - \lambda_A)t}. \tag{2.68}$$

We integrate this equation:

$$\int d\left(N_B e^{\lambda_B t}\right) = \int \lambda_A N_0 e^{(\lambda_B - \lambda_A)t} dt$$

to give

$$\left(N_B e^{\lambda_B t}\right) = \frac{\lambda_A N_0 e^{(\lambda_B - \lambda_A)t}}{(\lambda_B - \lambda_A)} + \text{constant}.$$

Using the boundary condition, $N_B = 0$ at $t = 0$ and simplifying we finally obtain an expression for the number of N_B nuclei:

$$N_B = N_0 \frac{\lambda_A}{(\lambda_B - \lambda_A)}(e^{-\lambda_A t} - e^{-\lambda_B t}). \tag{2.69}$$

The quantity that is measured in practice is the activity $\mathcal{A}_B$, which is equal to $\lambda_B N_B$:

$$\mathcal{A}_B = N_0 \frac{\lambda_A \lambda_B}{(\lambda_B - \lambda_A)}(e^{-\lambda_A t} - e^{-\lambda_B t}). \tag{2.70}$$

The activity of the parent is $\mathcal{A}_A = \lambda_A N_A = \lambda_A N_0 e^{-\lambda_A t}$. Hence, the ratio of activities is

$$\frac{\mathcal{A}_B}{\mathcal{A}_A} = \frac{N_0 \lambda_B}{N_0 (\lambda_B - \lambda_A)} (e^{-\lambda_A t} - e^{-\lambda_B t}). \qquad (2.71)$$

If the parent is long-lived compared with the daughter, i.e. $\lambda_A \ll \lambda_B$, the parent decays at an essentially constant rate, and

$$\frac{\mathcal{A}_B}{\mathcal{A}_A} \approx 1 - e^{-\lambda_B t} \approx 1 - e^{-t/\tau_B}. \qquad (2.72)$$

After a few mean lives τ_B of the shorter-lived daughter, the exponential term becomes negligible and the activity of the daughter becomes equal to the activity of the parent: $\mathcal{A}_B = \mathcal{A}_A$. This condition is called *secular equilibrium*.

This analysis can be readily extended to the case where there are many steps in the decay chain, where the daughter and succeeding generations have lifetimes that are short compared with the parent lifetime. It explains why radioactive nuclides with very short lifetimes are found in nature: the short-lifetime radioactive nuclei are in equilibrium in decay families with long lifetime parents.

2.3.3 α, β and γ decay

When an unstable nucleus decays to become a more stable nucleus, it may emit an α particle or a β particle. Furthermore, this may result in a daughter nucleus that is in an excited state, which then decays towards its ground state by emitting a γ-ray. Early on in the study of radioactivity it was discovered that α particles are bare helium nuclei, He^{++}, that β particles are electrons (or positrons) and that γ-rays are highly energetic photons. Figure 2.28 illustrates the effect of a magnetic field on each of them, assuming they have equal energy. Being relatively light, negatively charged β particles suffer large deflections in the magnetic field. The much heavier and positively charged α particles are deflected by a much less amount and in the opposite direction. γ-rays are uncharged and are not deflected at

Figure 2.28 The behaviour of α particles, β^- particles and γ-rays in a magnetic field, all having the same energy. The relatively light, negatively charged β^- particles suffer large deflections in the magnetic field. The much heavier and positively charged α particles are deflected by a much smaller amount and in the opposite direction. The γ-rays are uncharged and are not deflected at all.

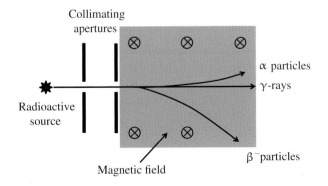

all. These emissions that occur in radioactive decay are collectively called *nuclear radiation.*

α, β and γ decay are the principal types of radioactive decay. However, it might be the case that a heavy radioactive nucleus decays to an unstable nucleus that has a large excess of neutrons and therefore lies very far from the region of nuclear stability. In that case, the unstable product nucleus will usually decay by emitting some of the excess neutrons. Importantly, the time that elapses before these neutrons are emitted may be relatively long, ~ several seconds. We will see that these so-called *delayed neutrons* are crucial to the control of a nuclear fission reactor.

α decay

In α decay, a high-Z parent nucleus X decays to a daughter nucleus X$'$ by emitting an α particle:

$$_Z^A X_N \rightarrow {}_{Z-2}^{A-4} X'_{N-2} + \alpha. \tag{2.73}$$

This lowers both the nuclear charge and the nuclear mass. It also liberates energy because the decay products are more tightly bound than the parent nucleus. And since the α particle has a relatively large binding energy, there is a correspondingly large energy release. This energy release, called the *Q-value*, is the energy equivalent of the mass defect of the reaction:

$$Q = \left(M_X - M_{X'} - M_{He} \right) c^2, \tag{2.74}$$

where, taking the mass of the electrons into account, M_X, $M_{X'}$ and M_{He} are the *atomic* masses of the parent, the daughter and the helium atom. Clearly, it follows that α decay can only occur when the atomic mass of the parent is greater than the sum of the atomic masses of the daughter and of helium. The released energy is shared as kinetic energy between the daughter nucleus and the α particle and, as the α particle has the much smaller mass, it carries away most of the released energy.

Worked example

^{241}Am, an isotope of americium, α decays to ^{237}Np and is sometimes used as a laboratory source of α particles. Show that α decay is indeed energetically possible and determine the kinetic energy of the α particle if the ^{241}Am nucleus is initially stationary. The atomic masses of ^{241}Am, ^{237}Np and ^{4}He are 241.056 824, 237.048 168 and 4.002 603 u respectively.

Solution

The mass defect of the reaction $^{241}_{95}\text{Am} \rightarrow {}^{237}_{93}\text{Np} + \alpha$ is

$$241.056\ 824\ \text{u} - (237.048\ 168 + 4.002\ 603)\ \text{u} = 0.006\ 057\ \text{u}.$$

This is a positive quantity with an equivalent energy of

$$0.006\ 057 \times 931.50 = 5.64\ \text{MeV},$$

so that α decay is indeed energetically possible. Both momentum and energy must be conserved in the reaction. As the ^{241}Am nucleus was initially stationary, we have

$$M_{\text{D}} V_{\text{D}} = m_\alpha v_\alpha;$$
$$\frac{1}{2} M_{\text{D}} V_{\text{D}}^2 + \frac{1}{2} m_\alpha v_\alpha^2 = Q,$$

where M_{D} and m_α are the masses of the daughter nucleus and the α particle, respectively, and V_{D} and v_α are their velocities. Combining these equations we readily obtain

$$\frac{1}{2} m_\alpha v_\alpha^2 = \frac{M_{\text{D}}}{m_\alpha} \left(\frac{1}{2} M_{\text{D}} V_{\text{D}}^2 \right),$$

which shows that the α particle takes away almost all of the released energy as kinetic energy. Substituting for $\frac{1}{2} M_{\text{D}} V_{\text{D}}^2$, we obtain

$$\frac{1}{2} m_\alpha v_\alpha^2 = \frac{Q}{(m_\alpha/M_{\text{D}} + 1)}.$$

Taking M_{D} to be 237 u and m_α to be 4 u, we find that the kinetic energy of the α particle is

$$\frac{5.64}{(4/237 + 1)} = 5.55\ \text{MeV}.$$

This is a typical value for the energy of an emitted α particle and in this example, accounts for 98% of the released energy.

A striking feature of α-particle decay is that the lifetimes of the nuclides that α decay vary *enormously*, while the kinetic energies of the emitted α particles lie within a rather narrow range; lifetimes vary from a tiny fraction of a second to $\sim 10^{10}$ years, while the kinetic energies lie between about 4 and 9 MeV. For example, ^{232}Th has a half-life of 1.4×10^{10} years and emits an α particle with a kinetic energy of 4.08 MeV, while ^{218}Th has a half-life of 1.0×10^{-7} s and emits an α particle with a kinetic energy of 9.85 MeV. A factor of ~ 2 in α-particle energy results in a factor of $\sim 10^{24}$ in half-life. These characteristics of α decay were explained by G. Gamow and independently by E. Condon and R Gurney in 1928 using quantum wave mechanics. They explained the paradox of α-particle emission

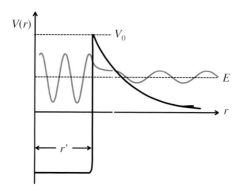

Figure 2.29 Potential energy function, $V(r)$, for an α particle trapped inside a daughter nucleus, as a function of radial distance r. The radius r' can be taken as the sum of the radius of the daughter nucleus and of the α particle. Outside the nucleus, $V(r)$ is the potential due to the $1/r$ Coulomb repulsion between the α particle and the nucleus. Inside the nucleus, $V(r)$ is represented by a square well, which reflects the potential energy due to the nuclear force. The horizontal line E represents the kinetic energy of the α particle at a large distance from the nucleus. The figure also shows schematically the wave function of the α particle. It is oscillatory inside the nucleus, decreases exponentially in amplitude within the barrier and becomes oscillatory again outside the nucleus. Quantum mechanically, the α particle has a finite probability of tunnelling through the Coulomb barrier.

in terms of *barrier penetration*, and in fact their successful application of quantum wave mechanics was one of its earliest and most convincing verifications.

Barrier penetration model of α decay   In the model of Gamow *et al.*, it is assumed that the α particle, which as we have seen is a particularly tightly bound system, is pre-formed inside the parent nucleus. We then have the picture of the α particle moving in a spherical region that is determined by the *daughter* nucleus, which has a nuclear charge of $(Z - 2)$. Figure 2.29 represents the potential energy function $V(r)$ for the α particle as a function of its radial distance r from the centre of the daughter nucleus. Outside the nucleus, $V(r)$ is the electrostatic potential:

$$V(r) = \frac{2e(Z-2)e}{4\pi\varepsilon_0 r}, \tag{2.75}$$

where $2e$ is the charge on the α particle. Inside the nucleus $V(r)$ is represented by a square well, which reflects the potential energy due to the nuclear force. This combination of electrostatic and nuclear potentials gives rise to a Coulomb barrier (see also Section 2.2.2). The horizontal line E represents the kinetic energy of the α particle at a large distance from the nucleus. If we assume that the daughter and α particle are uniformly charged spheres, we can estimate the electrostatic potential energy V_0 when they are just touching, i.e. the height of the Coulomb barrier. This is given by

$$V_0 = \frac{2e(Z-2)e}{4\pi\varepsilon_0 r'}, \tag{2.76}$$

where r' is the sum of the radii of the daughter nucleus and the α particle. We can estimate these radii using the relationship $R = R_0 A^{1/3}$, with $R_0 = 1.2 \times 10^{-15}$ m. Taking the example of $^{241}_{95}$Am, we estimate r' to be

$$1.2 \times (4^{1/3} + 237^{1/3}) \times 10^{-15} = 9.32 \times 10^{-15} \text{ m}.$$

Taking this value for r':

$$V_0 = \frac{2 \times 93 \times (1.6 \times 10^{-19})^2}{4\pi \times 8.85 \times 10^{-12} \times 9.32 \times 10^{-15}} \mathrm{J} = 4.6 \times 10^{-12} \ \mathrm{J}$$
$$= 29 \ \mathrm{MeV}.$$

This value for the height of the Coulomb barrier is typical for nuclides that decay by α-particle emission. Clearly, it is much larger than the energies of emitted α particles, which are ~ 5 MeV. Classically, therefore, α decay should not be possible. Quantum mechanically, however, there is a finite probability for the transmission of an incident particle through a potential barrier, as we saw in Section 2.2.3. For the case of a rectangular barrier, the probability T that the particle tunnels through the barrier is given by

$$T \approx \left[\frac{16E\,(V_0 - E)}{V_0^2} \right] \mathrm{e}^{-2\beta}, \tag{2.49a}$$

where

$$\beta = \frac{L\sqrt{2m\,(V_0 - E)}}{\hbar}, \tag{2.49b}$$

although in the case of α decay, the exponential term dominates the behaviour of T so that it is sufficient to take

$$T \approx \mathrm{e}^{-2\beta}.$$

The Coulomb barrier, however, is not rectangular in shape, as can be seen in Figure 2.29. Gamow *et al.* therefore approximated the shape of the Coulomb barrier by a sequence of infinitesimally narrow rectangular barriers of height $V(r) = 2(Z - 2)e^2/4\pi\varepsilon_0 r$ and width $\mathrm{d}r$. This leads to the following expression for T:

$$T \approx \exp\left\{ -\int_r \left(\frac{2\sqrt{2m\,(V(r) - E)}}{\hbar} \right) \mathrm{d}r \right\}, \tag{2.77}$$

which gives the probability that an α particle will tunnel through the Coulomb barrier when it is incident upon it. The energy E is the measured kinetic energy of the emitted particle because far from the nucleus its potential energy is zero. Figure 2.29 shows schematically the wave function of the α particle. It is oscillatory inside the nucleus, decreases exponentially in amplitude within the barrier and becomes oscillatory again outside the nucleus.

To determine the α decay rate, Gamow *et al.* assumed that the α particle bounces back and forth inside the daughter nucleus, hitting the barrier at the nuclear radius. Each time the α particle strikes the barrier there is a small probability, T, that it tunnels through the barrier and appears outside the nucleus. The number of times per second, N, that it strikes the barrier can be estimated from

$$N \sim \frac{v}{2R},\tag{2.78}$$

where v is the velocity of the α particle and $2R$ the nuclear diameter. The value of v is assumed to be comparable to the velocity of the α particle after the emission. Then the decay rate of the nucleus will be $\sim N \times T$. For example, ^{238}U decays to ^{234}Th with the emission of an α particle of kinetic energy 4.20 MeV with a half-life of 1.42×10^{17} s. Taking the diameter of ^{234}Th to be $2 \times 234^{1/3} \times 1.2 = 14.8$ fm and the α particle to have a velocity in the nucleus of 1.42×10^7 m/s, an α particle strikes the Coulomb barrier $\approx 1 \times 10^{21}$ times/s. Hence it must make, on average, $\sim 10^{38}$ attempts before it finally escapes. So, although finite, the probability of tunnelling, T, is extremely small. Using their model, Gamow *et al.* obtained α-decay rates that were in good agreement with those measured experimentally.

The height, V_0, of the Coulomb barrier and the nuclear radius, R, do not change significantly for nuclei in the limited range of the periodic table in which α-emitting nuclei are found. The enormous variation in α-decay rate from one nuclide to the next arises from the exponential dependence of the transmission probability T on $[V(r) - E]$ and on L. As the energy E of the emitted α particle increases, the term $[V(r) - E]$ decreases and the width of the barrier decreases as the α particle reaches higher up the barrier.

β decay

For a given value of mass number A there can be various possible combinations of atomic number Z and neutron number N. Some combinations result in more stable nuclides than others. Nuclides that have a combination of Z and N that is not the most stable may change these Z and N values, whilst retaining the same value of A by β decay. In a β-decay process, neutrons may be converted into protons or vice versa, bringing an unstable nucleus closer to the line of nuclear stability (see Figure 2.23). β decay can occur in three possible ways, each of which involves another charged particle to conserve electric charge:

$$
\begin{aligned}
\text{n} &\rightarrow \text{p} + \text{e}^- + \bar{\nu} & \beta^- \text{ decay}\\
\text{p} &\rightarrow \text{n} + \text{e}^+ + \nu & \beta^+ \text{ decay}\\
\text{p} + \text{e}^- &\rightarrow \text{n} + \nu & \text{electron capture}
\end{aligned}
$$

The first process is known as β-*minus* decay, which involves the creation and emission of an electron and the conversion of a neutron into a proton. The second process is called β-*plus* decay, which involves the creation and emission of a positron and the conversion of a proton into a neutron. In the third process, a proton is again converted into a neutron but now an *atomic* electron from an inner-shell orbital is captured by the nucleus. In all three processes, a further particle, called a *neutrino ν* (or *antineutrino $\bar{\nu}$*), is also emitted.

The force that governs β decay is the *weak nuclear* force. This is much weaker by a factor of $\sim 10^{12}$ than the nuclear force that binds nucleons together in a nucleus. It also has a much shorter range. Speaking in very general terms, we can say that if a force is weak, it takes a 'long time' to produce an effect. So the lifetimes for β decay tend to be long. For example, the half-life of ^{14}C which β^- decays is 5730 years.

β^- decay   The simplest example of β^- decay occurs for a free neutron:

$$\mathrm{n} \rightarrow \mathrm{p} + \mathrm{e}^- + \bar{\nu}. \tag{2.79}$$

The decay is energetically favourable because the mass of a proton is less than that of a neutron, by 0.001 388 4 u. Indeed, neutrons are unstable against β decay, with a half-life of 10.6 min. Free neutrons are produced in various nuclear reactions as we will see in Chapter 3. Such free neutrons are widely used to study materials, because the de Broglie wavelength of a *thermal* neutron, with a thermal energy kT of 1/40 eV, is about the same as the spacing of the atomic planes in a crystal. Usually, however, the neutron is bound in a nucleus and we have:

$$_Z^A\mathrm{X}_N \rightarrow _{Z+1}^A\mathrm{X}'_{N-1} + \mathrm{e}^- + \bar{\nu}. \tag{2.80}$$

Note that the Z value increases by 1 and the N value reduces by 1 but the atomic number, $A = Z + N$, remains the same.

Electron emission can take place if the mass $_Z^A m$ of the initial *nucleus* is greater than the mass $_{Z+1}^A m$ of the final *nucleus* plus the rest mass m_e of the electron:

$$_Z^A m - _{Z+1}^A m - m_\mathrm{e} > 0. \tag{2.81}$$

In terms of atomic masses $_Z^A M$, and neglecting the binding energies of the atomic electrons

$$_Z^A m = _Z^A M - Z m_\mathrm{e}; \quad _{Z+1}^A m = _{Z+1}^A M - (Z+1) m_\mathrm{e}.$$

Hence, β^- decay is allowed if

$$\left(_Z^A M - Z m_\mathrm{e}\right) - \left(_{Z+1}^A M - (Z+1) m_\mathrm{e} + m_\mathrm{e}\right) > 0,$$

which reduces to the simple result:

$$_Z^A M - _{Z+1}^A M > 0. \tag{2.82}$$

β^- decay can occur so long as the initial *atomic* mass is greater than the final *atomic* mass; the mass of the electron added to the atom is compensated by the mass of the electron emitted by the nucleus. The mass excess times c^2 is the energy E made available in the decay:

$$E = \left[{}_Z^A M - {}_{Z+1}^{A} M \right] c^2. \tag{2.83}$$

An important example of β^- decay is

$${}_6^{14}C \rightarrow {}_7^{14}N + e^- + \bar{\nu},$$

which is used in *radioactive carbon dating*. As noted previously, the action of cosmic rays on atmospheric molecules produces ^{14}C, and for every 10^{12} atoms of stable ^{12}C there are about 1.5 atoms of ^{14}C. Both of these isotopes are taken up in this proportion by living organisms such as plants and trees when they absorb carbon dioxide. When the organism dies, it stops taking in carbon. The ^{14}C decays by β^- emission to ^{14}N with a half-life of 5730 years and hence the exponential decay of the ^{14}C serves as a clock to measure the time from when the organism died. The technique of radioactive dating thus involves determining the relative amounts of the unstable ^{14}C isotope and the stable ^{12}C isotope in the remains of the organic matter. For example, if the number of ^{14}C atoms was 1.5 per 10^{12} atoms of carbon at the moment an organism died, then that number would fall to 0.75 per 10^{12} atoms of ^{12}C after 5730 years. This half-life of ^{14}C is appropriate for measuring time intervals in the range of historical interest. The dating the Dead Sea scrolls is just one of many examples of the use of radioactive carbon dating. For much longer timescales, such as the age of rocks and that of planet Earth, radioactive nuclides with much longer half-lives are used. For example, the radioactive isotope ^{40}K decays to ^{40}Ar by electron capture with a half-life of 1.2×10^9 years, and the age of a rock, for example, can be determined by measuring the ratio of ^{40}K to ^{40}Ar in the rock.

In the first experimental investigations of β^- decay, the investigators were not aware of the emitted antineutrino and this led to some puzzling observations. The electrons were found to have a *continuous* distribution of energy, as illustrated in Figure 2.30. This would not be possible if β^-decay were a two-body process, as this would require the ejected electron to have a definite energy, which was not observed. Furthermore, angular momentum could not be conserved if the neutron and the proton were the only reaction products, as the neutron, proton and electron are all spin-$^1/_2$ fermions. Since the neutron spin is $^1/_2$, the resulting total proton-electron spin can only be either 0 or 1. The answer to these puzzling results came in 1931 when Wolfgang Pauli postulated the existence of a third body to take away the excess energy and conserve linear and angular momentum. For conservation of charge, this body must be electrically neutral,

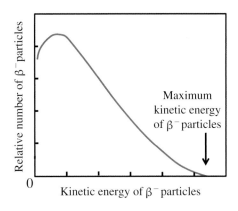

Figure 2.30 The continuous distribution of kinetic energy for β^- particles emitted in the β^- decay of a radioactive nuclide. The vertical arrow indicates the maximum kinetic energy that a β^- particle can have.

and from conservation of angular momentum, it must be a spin-$^1/_2$ particle. Moreover, the maximum energy of the ejected electron in the β^- decay was found to be equal to the decay energy E within the experimental uncertainty. This implied that the mass of the third particle must be zero or nearly zero. The third particle that Pauli postulated is the antineutrino $\bar{\nu}$. Now it is believed that the neutrino has a finite but exceedingly small mass. The fact that the antineutrino was not observed in the experiments implies, correctly, that it is not affected by nuclear or electrostatic forces. Indeed, antineutrinos (and neutrinos) hardly interact with matter at all. Many millions of neutrinos pass straight through our bodies every second, while a neutrino can readily pass through the universe without interacting. It is not surprising therefore that antineutrinos are very hard to detect, and indeed it was nearly 30 years after Pauli's postulation of their existence that they were detected experimentally.

β^+ decay   We represent a β^+ decay process as

$$_{Z}^{A}X_N \rightarrow \, _{Z-1}^{A}X'_{N+1} + e^+ + \nu. \qquad (2.84)$$

The number of protons Z reduces by 1 and the number of neutrons N increases by 1, but the atomic number A remains the same. For β^+ decay to occur, the proton must be bound in a nucleus, as the decay of a free proton is energetically forbidden. In a similar way to the above, it can be readily shown that for β^+ decay to occur, the original atomic mass must be greater than the final atomic mass by at least *two* electron masses: $_{Z}^{A}M > \, _{Z-1}^{A}M + 2m_e$.

An important aspect of β^+ decay is that it provides a source of positrons, which are required in a range of applications. For example they are used in *positron emission tomography (PET)*, which is an imaging technique that produces a three-dimensional image or picture of functional processes in the body. The positrons collide with electrons in the body, producing annihilation radiation that is detected by a camera.

Electron capture   An alternative to β^+ decay is electron capture where a nucleus captures one of its atomic electrons, with the result that a proton in the nucleus is transformed into a neutron and a neutrino is emitted:

$$
{}_{Z}^{A}X_N + e^- \rightarrow {}_{Z-1}^{A}X'_{N+1} + \nu. \tag{2.85}
$$

The captured electron usually comes from the innermost ($n = 1$) orbit of the atom as this orbit is closest to the nucleus and the *overlap* of the electronic wave function with the volume of the nucleus is greatest. The vacancy is soon filled by an electron from a higher lying orbit, which results in the emission of an X-ray. The wavelength of the X-ray is characteristic of the *daughter* nucleus. As the electron is provided by the atom, electron capture can occur if the original atomic mass is greater than the final atomic mass: ${}_{Z}^{A}M > {}_{Z-1}^{A}M$. Hence, there is a range of unstable nuclides where the difference in atomic masses is such that electron capture is allowed but β^+ decay is energetically forbidden.

γ decay

Most α and β decays, and also most nuclear reactions, leave the final nucleus in an excited state. These excited states decay rapidly towards the ground state via the emission of one or more γ-rays, typically within a timescale of $<10^{-9}$ s. The process is analogous to the way atoms in excited states decay back to their ground state, but the energies and wavelengths involved are very different. Since energy spacings in nuclei are $\sim$MeV, γ-rays have wavelengths of $\sim10^{-12}$ m, about 10^6 times shorter than visible light. Unlike α or β decay, neither the mass number A nor the atomic number Z changes during γ decay.

An important example of γ decay occurs in the decay of the radioactive isotope of caesium, ^{137}Cs, which is a fission product in the radioactive decay of ^{235}U. The ^{137}Cs isotope β decays to ^{137}Ba and leaves the ^{137}Ba nucleus in an excited state, denoted as ^{137}Ba*:

$$
{}^{137}\text{Cs} \rightarrow {}^{137}\text{Ba}^* + e^- + \bar{\nu},
$$

Figure 2.31 A ^{137}Cs nucleus may decay to a ^{137}Ba nucleus with the emission of a β^- particle, leaving the ^{137}Ba nucleus in an excited state, labelled ^{137}Ba*. The excited ^{137}Ba nucleus then decays to its ground state with the emission of a γ-ray. The energy of the γ-ray is equal to the energy difference (0.66 MeV) between the excited and ground states of the ^{137}Ba nucleus.

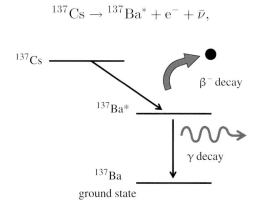

as illustrated in Figure 2.31. This excited state lies 0.66 MeV above the ground state of ^{137}Ba and decays to it with the emission of a γ-ray of this energy:

$$^{137}\text{Ba}^* \rightarrow {}^{137}\text{Ba} + \gamma$$

Through these reactions, the ^{137}Cs isotope provides an important laboratory source of γ-rays and finds application, for example, in radiation therapy in medicine. But note that the γ-rays are emitted by the ^{137}Ba nucleus and not by the ^{137}Cs nucleus.

Problems 2

2.1   When an excited state of the ^{137}Ba nucleus decays to the ground state of ^{137}Ba, it emits a γ-ray of energy 0.66 MeV. Calculate the wavelength and frequency of the γ-ray.

2.2   (a) Determine the nuclear radius and estimate the nuclear density of the following nuclides: $^{12}_{6}$C, $^{56}_{26}$Fe and $^{209}_{83}$Bi. Compare these densities with the typical density of a neutron star. (b) Show that the density of nuclear matter is roughly constant and independent of mass number A. (c) Estimate the charge density for the above nuclides. Comment on your results.

2.3   Calculate the total binding energy of the $^{120}_{50}$Sn$_{70}$ nuclide. Compare its mass defect with the total mass of the atom, which is 119.902 197 u. What is the binding energy per nucleon?

2.4   (a) An α particle is taken apart in the following steps: (1) a proton is first removed; (2) then a neutron is removed; and (3) the remaining proton and neutron are separated. Determine the energy required for each step. Compare your results with the total binding energy of an α particle. (b) Determine the amount of energy it takes to remove a neutron from an α particle. Comment on your result with respect to step (1) in part (a). The atomic and neutron masses are:

$$
\begin{array}{llll}
^4\text{He} & 4.002\ 603\ \text{u} & {}^3\text{He} & 3.016\ 029\ \text{u} \\
^3\text{H} & 3.016\ 049\ \text{u} & {}^2\text{H} & 2.014\ 102\ \text{u} \\
^1\text{H} & 1.007\ 825\ \text{u} & \text{n} & 1.008\ 665\ \text{u}
\end{array}
$$

2.5   The following table gives the binding energy per nucleon B/A for a range of nuclides. Plot the values of B/A versus A for these nuclides and on this plot make a sketch of the binding energy curve. Use your curve to estimate the value of B/A for ^{119}Pd. Suppose that a nucleus of ^{238}U were to fission into two ^{119}Pd nuclei. Estimate the energy that would be released in this process.

Nucleus	^{2}D	^{4}He	^{7}Li	^{9}Be	^{12}C	^{16}O	^{40}Ar	^{56}Fe	^{98}Mo	^{127}I	^{181}Ta	^{238}U
B/A(MeV)	1.11	7.07	5.61	6.46	7.68	7.98	8.60	8.79	8.64	8.45	8.02	7.57

2.6   For a given value of mass number A there may be various values of atomic number Z. Use the semi-empirical mass formula to show that the value of Z that gives the most tightly bound nucleus, i.e. the one with the smallest mass, is given by

$$Z_{\min} = \frac{[m_n - M_H] + a_{Coul}A^{-1/3} + 4a_{sym}}{2a_{Coul}A^{-1/3} + 8a_{sym}A^{-1}}.$$

What is the most stable nuclide with $A = 121$?

2.7 (a) Determine the Coulomb force (in N) between the two protons in an α particle assuming they are a distance 2 fm apart. (b) By what factor is the Coulomb force greater than the gravitational force between the two protons? (c) The energy required to separate all four nucleons in an α particle is about 28 MeV. Show that the nuclear force between two nucleons is $\sim$520 N by considering the range of the nuclear force.

2.8 Given that 1.0 g of ^{226}Ra has a nominal activity of 1 Ci, find the half-life of ^{226}Ra. Marie and Pierre Curie are said to have amassed 200g of ^{226}Ra by 1900. How much of that ^{226}Ra remained in 2000?

2.9 The table below shows measurements of the activity $A(t)$ of a sample of ^{128}I. Plot the data as a semi-log graph of $A(t)$ versus time t and deduce the decay constant λ and the half-life $t_{1/2}$ of ^{128}I.

Time, t (min)	0	25	50	75	100	125	150	175	200
$A(t)$ (counts/s)	950	475	238	119	59	30	15	7.4	4.0

2.10 The total number of detected α particles from a sample of ^{238}U is found to be 25 056 in a counting period of exactly 24 hours. Suppose that we do an experiment where we measure the number n of α particles that are counted in a period of 10 s and we do this a large number of times. Construct the Poisson distribution to which you could compare the experimental data. Take values of n from 0 to 4. What is the probability that more than four counts are recorded in a period of 10 s? What is the probability that at least one count is recorded in a period of 10 s? What does the fact that the detected counts follow a Poisson distribution tell us about the nature of radioactive decay?

2.11 A radioactive nuclide A decays to a daughter nuclide B that in turn decays to a stable nuclide C. Suppose that, initially, the number of A nuclei is 1000, the half-life of nuclide A is 10 min and the half-life of nuclide B is 1 min. Write a spreadsheet program to plot the populations N_A, N_B and N_C against time t, on the same graph. On another graph, plot the activities $A(t)$ of nuclides A and B against t. Investigate other combinations of half-lives.

2.12 (a) Show that carbon in the atmosphere has an activity of 15.6 decays/min/g. The relative abundances of stable ^{12}C and radioactive ^{14}C in the atmosphere are in the ratio 1:1.35 $\times$ 10^{-12} and the half-life of ^{14}C is 5730 years. (b) A 250 g sample of charcoal from an ancient fire is found to produce a β^- decay rate of 500 decays/min. How old is the charcoal?

2.13 (a) What nuclide is produced in the following radioactive decays? (i) α decay of $^{239}_{94}$Pu$_{145}$, (ii) β^- decay of $^{24}_{11}$Na$_{13}$, and (iii) β^+ decay of $^{15}_{8}$O$_7$. (b) What particle (alpha, electron or positron) is emitted in the following radioactive decays: (i) $^{27}_{14}$Si $\rightarrow$ $^{27}_{13}$Al (ii) $^{238}_{92}$U $\rightarrow$ $^{234}_{90}$Th; and (iii) $^{74}_{33}$As $\rightarrow$ $^{74}_{34}$Se?

2.14 (a) A stationary nucleus decays by α particle emission. Show that the kinetic energy of the recoiling daughter nucleus is equal to $4Q/A$ to a good approximation, where Q is the energy release and A is the mass number of the daughter. Obtain the corresponding expression for the kinetic energy of the α particle. (b) A potential hazard in homes in some geographical areas is the radioactive gas radon-222, which is produced by the α decay of radium-226. Determine the velocities of the emitted α particle and the recoiling daughter nucleus for this decay. The atomic masses of ^{226}Ra, ^{224}Rn and ^{4}He are 226.025 406, 222.017 571 and 4.002 603 u, respectively.

2.15 $^{223}_{88}$Ra may α decay to $^{219}_{86}$Ra or there is a very small probability that it may decay to $^{209}_{82}$Pb by emitting a $^{14}_{6}$C nucleus. Show that both processes are energetically possible and compare the Coulomb barriers for the two processes. Comment on your results. The atomic masses of $^{223}_{88}$Ra, $^{219}_{86}$Ra, $^{4}_{2}$He and $^{14}_{6}$C are 223.018 501, 219.009 485, 4 002 603 and 14.003 242 u, respectively.

2.16 (a) Show that β^- decay of $^{60}_{27}$Co to $^{60}_{28}$Ni is energetically possible and find the maximum energy in MeV of the neutrino produced in the decay. The atomic masses of $^{60}_{27}$Co and $^{60}_{28}$Ni are 59.933 820 and 59.930 788 u, respectively. (b) Is it allowed energetically for a free proton to decay to a free neutron?

2.17 Show that $^{57}_{27}$Co cannot β^+ decay, but can decay by electron capture. The atomic masses of $^{57}_{27}$Co and $^{57}_{26}$Fe are 56.936 294 and 56.935 396 u, respectively. The mass of an electron is 5.4858×10^{-4} u.

2.18 Plutonium-238 is an emitter of α particles of energy 5.59 MeV. This isotope has been used as a power source in spacecraft, where the energy of the emitted α particles is absorbed in a container surrounding the plutonium. Assuming that 10% of this emitted energy can be converted into electrical energy, what mass of plutonium is required to generate an electrical power of 1.5 W? The half-life of plutonium-238 is 87.7 years.

3

Nuclear power

We saw in Chapter 2 that enormous amounts of energy are released when matter is converted into energy. Such mass to energy conversion is exploited in nuclear fission and nuclear fusion reactors. Nuclear fission reactors already provide a substantial proportion of the world's energy requirements. For example, France derives over 75% of its electricity from nuclear energy. Nuclear fusion, on the other hand, remains a long-term goal and is not expected to become a major source of energy for several decades or more. Nevertheless, nuclear fusion has important advantages as an energy source, and consequently research into fusion is being pursued vigorously in many countries. In this chapter we describe how to get energy from the nucleus. We describe the basic physical processes involved in both nuclear fission and fusion and we describe the main features of fission and fusion reactors.

3.1 How to get energy from the nucleus

We can release energy by splitting a heavy nucleus into two less massive nuclei. This is the process of nuclear fission. An example is the following reaction:

$$\text{n} + {}^{235}\text{U} \rightarrow {}^{94}\text{Sr} + {}^{140}\text{Xe} + 2\text{n}, \tag{3.1}$$

where an incident neutron induces the splitting or fission of a uranium-235 nucleus. This reaction releases an energy of ~ 200 MeV. Alternatively, we can release energy by fusing together two light nuclei to form a more massive nucleus, which is the process of nuclear fusion. An example is the following reaction:

$$\text{}^{2}\text{D} + {}^{3}\text{T} \rightarrow {}^{4}\text{He} + \text{n}, \tag{3.2}$$

Physics of Energy Sources, First Edition. George C. King.
© 2018 John Wiley & Sons, Ltd. Published 2018 by John Wiley & Sons, Ltd.

in which deuterium (^{2}D) and tritium (^{3}T) nuclei fuse together, releasing 17.6 MeV of energy. For both cases, we can readily understand how energy is released in terms of the conversion of mass into energy – *if the total mass of the products of a fission or fusion reaction is less than the total mass of the interacting particles, then this mass difference is converted into energy.*

We can also understand how energy is released in terms of the binding energies of the nuclei involved; the two viewpoints are completely equivalent. The essential idea is that *going to a more tightly bound nucleus (or nuclei) releases energy.* Consider the following analogy. When we drop a stone into a deep well, potential energy is converted into kinetic energy of the stone. When the stone strikes the ground, that energy is released as thermal and sound energy. And the deeper the well to which the stone becomes bound, the greater the energy release. So if we can somehow rearrange the nucleons in the participating nuclei so that we obtain more tightly bound nuclei, energy will be released. Hence the *binding energy curve* of the nuclides provides us with a powerful basis for understanding the processes of nuclear fission and fusion. Indeed, because it underlies the origin of nuclear energy, the binding energy curve has been described as one of the most important curves in physics.

Figure 3.1 shows schematically the binding energy curve of the nuclides (see also Figure 2.12). We see that this curve reaches a maximum at about $A = 56$; for nuclides above or below this value of A, the binding energy per nucleon B/A is less. Hence, if we split a heavy nucleus, with a high value of A, into two nuclei that have a value of A closer to $A = 56$, energy will be released. This process is indicated by the fission arrow on Figure 3.1. Alternatively, if we fuse together two nuclei with low values of A to produce one with A closer to $A = 56$, energy will again be released. This process is indicated by the fusion arrow on Figure 3.1.

Figure 3.1 Schematic diagram of the binding energy curve of the nuclides. In a fusion reaction, two light nuclei fuse together to form a heavier nucleus that lies closer to the maximum of the curve at $A \sim 56$. In a fission reaction, a heavy nucleus splits into two lighter nuclei that again lie closer to the maximum of the curve.

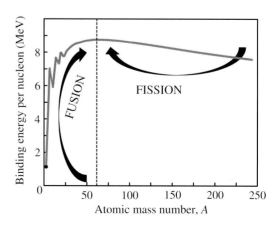

3.2 Nuclear reactions

3.2.1 Nuclear reactions

In Chapter 2 we considered the *spontaneous* decay of radioactive nuclei with the emission of α or β particles, sometimes followed by the emission of a γ-ray. By spontaneous we mean the decay of the nucleus is not initiated by any external agency; it is a natural process of the nucleus, just like the spontaneous decay of an atom in an excited state. Now we consider nuclear reactions that result from the bombardment of a nucleus by a particle or by another nucleus. Such reactions lie at the heart of nuclear energy production as exemplified by the fission and fusion reactions (Equations 3.1 and 3.2).

Many different kinds of reactions may occur during a collision between two nuclear particles. In elastic and inelastic scattering reactions, the incoming and outgoing particles are the same and may be written as

$$a + X \rightarrow Y + b,$$

where a is the incident particle, X is the target nucleus and Y and b are the products of the reaction. Rutherford scattering of α particles by gold nuclei is an example of elastic scattering and we will see that elastic collisions play a crucial role in the slowing down of very energetic neutrons in the *moderator* of a nuclear reactor. In the case of *inelastic scattering*, the incident particle excites the target nucleus to one of its excited states, denoted as X^*:

$$a + X \rightarrow X^* + b,$$
$$\qquad\quad \hookrightarrow X + \gamma.$$

The excited nucleus may then decay back to its ground state with the emission of a γ-ray within a very short time scale, typically $\sim 10^{-12} - 10^{-9}$ s. Many other kinds of reaction may occur where one or both of the interacting nuclei are changed in some way. A common reaction is one in which an incident particle is absorbed by a nucleus. Of particular interest to us, with regard to nuclear fission, is where a nucleus absorbs an incident neutron to form a *compound nucleus*. In the case of a heavy target nucleus like uranium, the compound nucleus may subsequently split into two less massive nuclei with the release of a large amount of energy, as we will discuss in Section 3.3.

Nuclear reactions are subject to the usual conservation laws for charge, momentum and energy, of course. In addition, however, the total number of nucleons must be conserved in a nuclear reaction, although neutrons may be converted into protons and vice versa.

Rutherford and his collaborators were the first to initiate nuclear reactions in the laboratory when they fired α particles at various nuclei. An example of their experiments is the transmutation of a nitrogen nucleus to an oxygen nucleus:

$$\alpha + {}^{14}\text{N} \rightarrow {}^{17}\text{O} + \text{p}.$$

(The transmutation of one element into another was the dream of the alchemists, although they wanted to turn lead into gold!) The α particles available to Rutherford were those emitted by radioactive nuclides and so the range of incident-particle energy was limited. Later in 1930, John Cockcroft and Ernest Walton built the first *particle accelerator* that was capable of inducing nuclear reactions and which enabled the energy of the incident particle to be varied. They accelerated protons to the energy of 500 keV and observed the reaction

$$\text{p} + {}^{7}\text{Li} \rightarrow {}^{4}\text{He} + {}^{4}\text{He}.$$

Moreover, they measured the kinetic energy of the two outgoing ${}^{4}\text{He}$ nuclei. The masses of the proton, ${}^{7}\text{Li}$ and ${}^{4}\text{He}$ nuclei were already known from mass spectrometer measurements and so Cockcroft and Walton could calculate the change, Δm (0.018 622 u), in total mass that occurred in the reaction and hence the value of the equivalent energy Δmc^2. This value was in agreement with their measured value of the kinetic energy of the two ${}^{4}\text{He}$ nuclei, which was 17.2 ± 2.7 MeV. The experiment provided one of the earliest confirmations of Einstein's mass–energy relationship, and in 1951 Cockcroft and Walton were awarded the Nobel Prize in Physics for their pioneer work on the transmutation of atomic nuclei by artificially accelerated atomic particles.

Nuclear reactions continue to be extensively studied by researchers to investigate various aspects of *nuclear structure*. For example, there are experiments to determine the size and shape of a nucleus, to map out its excited states, and to synthesise exotic nuclei that lie off the line of stability. These experiments have been accompanied by a great deal of development in various types of particle accelerator, and accelerators can now produce beams of charged projectiles with energies of up to about 25 MeV for these studies of nuclear structure. There are also particle accelerators that produce beams with much greater energies, >1 TeV (10^{12} eV), such as the Large Hadron Collider at CERN. But these machines are used to probe the internal structure of individual nucleons rather than investigate the structure of the nuclides.

3.2.2 *Q*-value of a nuclear reaction

The *Q-value* of a nuclear reaction is the difference between the total mass of the initial reactants and the total mass of the final products of the reaction, in energy units in accordance with the mass–energy relationship $E = mc^2$. (The *Q*-value of a nuclear reaction is analogous to the Q-value we encountered in our discussion of α decay, in Section 2.3.3.) If particles a and X interact to form final particles b and Y, then

$$Q = [(M_a + M_X) - (M_b + M_Y)]\, c^2. \qquad (3.3)$$

As usual, atomic masses are used in order to balance the masses of the electrons. When the total mass of the products is less than the total mass of the incoming particles $Q > 0$, and the reaction is called *exothermic*. On the other hand, if the total mass of the products is greater than the total mass of the incoming particles, $Q < 0$ and the reaction is called *endothermic*. In this case, energy must be supplied to initiate the reaction.

Worked example

(a) Determine the Q-value of the reaction $p + {}^{7}\text{Li} \rightarrow {}^{4}\text{He} + {}^{4}\text{He}$ and show that the reaction is exothermic. The atomic masses for ${}^{1}\text{H}$, ${}^{7}\text{Li}$ and ${}^{4}\text{He}$ are 1.007 825, 7.016 003 and 4.002 603 u, respectively.
(b) Determine the Q-value of the reaction $\alpha + {}^{14}\text{N} \rightarrow {}^{17}\text{O} + p$. The atomic masses for ${}^{14}\text{N}$ and ${}^{17}\text{O}$ are 14.003 074 and 16.999 131 u, respectively.

Solution

(a) The total mass of the reacting particles is $(1.007\,825 + 7.016\,003)$ u and the total mass of the products is $(2 \times 4.002\,603)$ u. We see that

$$(M_a + M_X) - (M_b + M_Y) = +0.018\,622 \text{ u},$$

and hence

$$Q = 0.018\,622 \times 931.5 = +17.35 \text{ MeV}.$$

Q is a positive quantity, confirming that the reaction is exothermic.

(b) In this case, $(M_a + M_X) - (M_b + M_Y) = -0.001\,279$ u, i.e. the mass *increases* by 0.001 279 u and $Q = -1.191$ MeV. This reaction is endothermic and energy must be supplied for the reaction to take place. Momentum must be conserved in any collision. The calculated value of 1.191 MeV is for a head-on collision in which the total momentum of the two particles is zero. If, on the other hand, the α particle is incident upon a stationary ${}^{14}\text{N}$ target, the energy of the incident α particle must be greater than 1.191 MeV to conserve momentum; if all the kinetic energy of the α particle went solely into increasing rest mass energy, the final kinetic energy would be zero and so momentum could not be conserved. The amount of energy that the α particle must have in this case to initiate the reaction is called the *threshold energy* and is always larger than the Q-value. For this reaction, the threshold energy is 1.533 MeV.

Measuring the Q-value of a nuclear reaction provides an alternative means of determining the atomic mass of an isotope; in particular, for radioactive isotopes that do not live long enough to pass through a mass spectrometer. For example, the ^{9}Be isotope of beryllium is stable and its mass can be accurately measured in a mass spectrometer, but the ^{8}Be isotope is unstable with a half-life of 7×10^{-17} s. However, the Q-value of the reaction

$$\text{p} + {}^9\text{Be} \rightarrow {}^8\text{Be} + {}^2\text{D}$$

is measured to be 560 keV. The atomic masses of ^{1}H, ^{2}H and the ^{9}Be isotope are 1.007 825, 2.014 102 and 9.012 182 u, respectively. Hence, using Equation (3.3), the atomic mass of ^{8}Be is found to be

$$(1.007\,825 + 9.012\,182)\ \text{u} - 2.014\,102\ \text{u}$$
$$- (0.560/931.494)\text{u} = 8.005\,304\ \text{u}.$$

3.2.3 Reaction cross-sections and reaction rates

When we visualise projectiles being aimed at targets we naturally think in terms of the cross-sectional area of the target. So it is when a particle is fired at a nucleus or, indeed, where a nucleus collides with another nucleus. We imagine that the target nucleus presents a cross-sectional area or *cross-section* to the incident particle. A cross-section is usually denoted by the symbol σ. Although it is tempting to think of the cross-section as the geometrical cross-section of the nucleus, i.e. πR^2, where R is the nuclear radius, this is *not* the case. Some reactions have a cross-section much larger than this and some have a cross-section much smaller. For example, the cross-section for the absorption of a low-energy neutron by a cadmium nucleus is $\sim 10^3$ times larger than the geometric cross-section of the nucleus. On the other hand, more improbable reactions may have values of σ that are much smaller than the geometric cross-section of the target nucleus. Thus, although σ has units of area, it is more appropriate to think of the cross-section for a particular reaction *as a measure of the probability that an incident particle will cause that reaction to occur*.

Imagine that we have a beam of particles incident upon a thin slab of material, as shown in Figure 3.2. We define the *flux Φ* of the beam as the number of incident particles crossing a unit area perpendicular to the direction of motion per unit time. The reaction rate R_b, i.e. the number of nuclear reactions per second, depends on the Φ flux and also on the number N of nuclei that are exposed to the incident beam, assuming that the nuclei act independently of each other. Hence we can write

$$R_\text{b} = \sigma N \Phi, \tag{3.4}$$

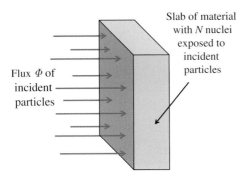

Slab of material with N nuclei exposed to incident particles

Flux Φ of incident particles

Figure 3.2 The figure shows a beam of particles that is incident upon a slab of material. The reaction rate depends on the flux, Φ, of the incident beam and also on the number, N, of nuclei that are exposed to the beam.

where the constant of proportionality is just the reaction cross-section σ. Thus, we can define σ as

$$\sigma = \frac{\text{reaction rate per nucleus}}{\text{incident flux}}. \tag{3.5}$$

This definition is consistent with σ representing the probability of a particular reaction occurring. Inspection of Equation (3.4) shows that σ has the dimensions of area. If an incident particle can cause more than one kind of reaction, then the total cross-section, σ_{tot}, is the sum of the cross-sections for the individual reactions. For example, if the incident particle may be scattered or absorbed by the target nucleus, σ_{tot} is the sum of the individual cross-sections, σ_s and σ_a, respectively:

$$\sigma_{\text{tot}} = \sigma_s + \sigma_a. \tag{3.6}$$

We have seen that nuclei diameters are $\sim 10^{-14}$ m and so a convenient unit for reaction cross-sections is the *barn*, which is defined as

$$1\,\text{barn} = 1\,\text{b} = 10^{-28}\,\text{m}^2 = 100\,\text{fm}^2.$$

(There are various stories concerning the origin of the unit 'barn'. It arose in the Second World War during the development of the atomic bomb. The scientists involved sought to disguise their work by using a very different unit for cross-sectional area, and apparently some scientists who were scattering neutrons off uranium nuclei described the uranium nucleus as being as 'big as a barn'.)

Worked example

The total mass of uranium in a certain nuclear reactor is 50 metric tonnes and the neutron flux is 5.0×10^{17} neutrons/m^2/s. Calculate the rate of fission reactions assuming that the cross-section σ_f for fission in ^{235}U is 584 b. The atomic mass of uranium is 238.0 u and the natural abundance of ^{235}U is 0.72%.

Solution

We have the reaction rate $R_f = N\sigma_f \Phi$ (Equation 3.4), where N is number of nuclei presented to the neutrons, σ_f is the fission cross-section and Φ is the neutron flux. The natural abundance of ^{235}U is 0.72% and 238.0 g of natural uranium contains Avagadro's number of atoms. Therefore the number of ^{235}U nuclei is given by

$$N = \frac{50 \times 10^3 \times 10^3 \times 6.02 \times 10^{23} \times 0.0072}{238.0} = 9.10 \times 10^{26} \text{ nuclei.}$$

Therefore,

$$R_f = 9.10 \times 10^{26} \times 584 \times 10^{-28} \times 5.0 \times 10^{17} = 2.7 \times 10^{19} \text{ reactions/s.}$$

This example also illustrates that a fission reactor produces intense neutron fluxes. These can be used to produce *radioisotopes*, which are unstable isotopes that are not normally found in nature. They are produced by exposing the appropriate stable nuclide to the intense neutron flux in the nuclear reactor. The nuclide absorbs neutrons and transmutes to the radioisotope. Radioisotopes have many practical applications, e.g. as tracers in medicine, biology, chemistry and engineering and also in radiation therapy.

Beam attenuation and mean free path

As a beam of particles travels through a slab of material, the beam becomes attenuated as the particles are scattered or absorbed by the nuclei of the material. Suppose that a particle beam with flux Φ_0 passes through a slab of material and is attenuated by a factor of 2 so that the emerging beam has flux $\Phi_0/2$ (see Figure 3.3). If that beam then passes through a second, identical slab of material, it will again be attenuated by a factor of 2 so that the emerging beam has flux $\Phi_0/4$. Similarly, after passing through three such slabs, the flux will be reduced to $\Phi_0/8$. This picture is consistent with the flux reducing *exponentially* with distance travelled in the slabs. More formally, Figure 3.4 shows a beam of particles with flux Φ_0 that is incident upon a slab of absorbing material. We consider the attenuation of

Figure 3.3 The attenuation of a beam of particles as it passes through three slabs of absorbing material. This picture is consistent with the beam flux Φ reducing exponentially with distance travelled in the slabs.

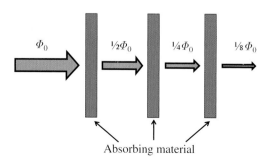

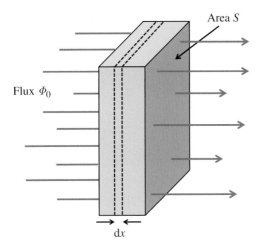

Flux Φ_0

Area S

dx

Figure 3.4 A slab of absorbing material of cross-sectional area S that is uniformly illuminated by a beam of particles. The flux, Φ, of the beam reduces steadily as it passes through the slab. We consider the attenuation of the beam by a thin slice of the material that has width dx.

the beam by a thin slice of width dx and cross-sectional area S. The view that the slice would present to the incident beam of particles is shown in Figure 3.5. The width dx is sufficiently small that it is unlikely that there will be any overlapping nuclei. If the slice contains n atoms per unit volume, the total number of nuclei in the slice is $nSdx$. Suppose that each nucleus has an absorption cross-section σ_a. Then the ratio of the total absorbing area of the nuclei to the cross-sectional area S of the slice is

$$\frac{nSdx\sigma_a}{S} = n\sigma_a dx. \qquad (3.7)$$

Hence, the flux of the beam will be reduced by dΦ, where

$$\frac{d\Phi}{\Phi} = -n\sigma_a dx,$$

which gives

$$\Phi(x) = \Phi_0 e^{-n\sigma_a x}; \qquad (3.8)$$

the beam flux reduces exponentially with distance x. The quantity $n\sigma_a$ is called the *linear attenuation coefficient* and is usually denoted by the symbol μ.

We can write Equation (3.8) as

$$\Phi(x) = \Phi_0 e^{-x/\lambda_a},$$

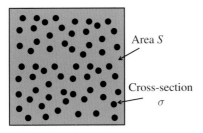

Area S

Cross-section
σ

Figure 3.5 The view that the thin slice of the slab in Figure 3.4 would present to the beam of particles. The slice has a cross-sectional area S and each nucleus has a cross-section σ.

where

$$\lambda_a = 1/n\sigma_a. \tag{3.9}$$

The decay of radioactive nuclei is described by the equation

$$N(t) = N_0 e^{-t/\tau}$$

(see Equations 2.53 and 2.61). In this case, t is the variable and we found that τ is the mean lifetime. Equation (3.8) has the same form as this equation, but it has distance x as the variable. By analogy we can immediately say that the parameter λ_a represents the mean distance that a particle travels before striking a nucleus. λ_a is called the *mean free path*. It is inversely proportional to n, the number of nuclei per unit volume and, in this case, to the absorption cross-section σ_a. We can also define the *mean time* τ_a as the time it takes the particle to travel distance λ_a before it is absorbed:

$$\tau_a = \frac{\lambda}{\bar{v}}, \tag{3.10}$$

where $\bar{v}$ is the mean speed of the particle.

More generally, we describe the mean free path for a specific reaction as the average distance that the particle travels before that reaction occurs. For example, it may be the mean free path λ_f that a neutron travels before it reacts with a nucleus causing it to fission. If σ_f is the cross-section for the fission reaction and n_B is the number density of the nuclei, $\lambda_f = 1/n_B\sigma_f$, and correspondingly $\tau_f = \frac{\lambda_f}{\bar{v}}$.

Worked example

The cross-section for absorbing thermal neutrons in cadmium is 2.45×10^3 b. (a) What thickness should a slab of cadmium have so that it absorbs 90% of an incident beam of thermal neutrons? Cadmium has atomic mass 112.4 u and a density of 8.64×10^3 kg/m^3. (b) Determine the mean free path of thermal neutrons in cadmium.

Solution

(a) The number density of nuclei in cadmium is

$$n = \frac{8.64 \times 10^3}{112.4 \times 1.66 \times 10^{-27}} = 4.63 \times 10^{28} \text{ atoms/m}^3.$$

Hence, $n\sigma_a = 4.63 \times 10^{28} \times 2.45 \times 10^{-25} = 1.13 \times 10^4$ atoms/m. Since $\Phi(x) = \Phi_0 \exp(-n\sigma_a x)$ from Equation (3.8), we have

$$x = \frac{1}{n\sigma} \ln\left[\Phi_0/\Phi(x)\right] = \frac{1}{1.18 \times 10^4} \ln\left(\frac{1}{0.1}\right)$$

$$= 2.0 \times 10^{-4} \text{ m} = 0.20 \text{ mm}.$$

Clearly, cadmium absorbs thermal neutrons very effectively and, indeed, cadmium rods are used to control the neutron flux and hence the rate of fission reactions in a nuclear reactor (see Section 3.4.2).

(b) As $n\sigma_a = 1.13 \times 10^4$ atoms/m, the mean free path

$$\lambda_a = \frac{1}{1.18 \times 10^4} = 8.8 \times 10^{-5} \text{ m}.$$

Energy dependence of a reaction cross-section

The cross-section of a particular reaction usually depends on the energy of the incident particle. This is exemplified by the cross-section for the absorption of neutrons by ^{6}Li to form the compound nucleus ^{7}Li:

$$n + {}^6\text{Li} \rightarrow {}^7\text{Li}.$$

The cross-section for this reaction is plotted as a function of neutron incident energy in Figure 3.6. Note that this is a log–log plot. Two particular features of this plot are immediately apparent. The absorption cross-section rises steadily as the neutron energy decreases; in fact, it varies inversely as the neutron velocity v. And there is a sharp peak in the cross-section at high neutron energy. The first feature is due to the so-called '$1/v$' law which was discovered by Italian physicist Enrico Fermi. He found that reaction rates for neutron-induced reactions increased dramatically when he slowed down the incident neutrons. We will see that this effect is of great importance in the operation of a fission nuclear reactor where low-energy *thermal* neutrons are used to induce fission in ^{235}U. (When we talk about thermal neutrons we usually mean neutrons with energy $kT = 1/40$ eV $\equiv 0.025$ eV, where k is the Boltzmann constant and temperature $T = 300$ K.) A physical interpretation of the $1/v$ law is as follows. The probability that the neutron will be absorbed by a nucleus depends on the time the neutron spends in the vicinity of

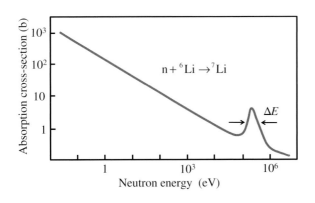

Figure 3.6 The cross-section for the absorption of a neutron by lithium, plotted as a function of incident neutron energy. The cross-section rises steadily as the neutron energy decreases, varying inversely as the neutron velocity v. There is a sharp peak in the cross-section due to a resonance at the neutron energy of about 250 keV. The width of the peak is ΔE.

the nucleus, which increases as the velocity of the neutron decreases. This time is $\sim 2R/v$, where R is the radius of the nucleus, giving the $1/v$ dependence of the neutron absorption cross-section.

The peak in the cross-section at about 250 keV is due to a *resonance* in the process of neutron absorption by the ^{6}Li target. This energy corresponds exactly to one of the excited states of the ^{7}Li compound nucleus. When an incident neutron has this energy, the probability that it will be absorbed by a ^{6}Li nucleus is considerably enhanced. We recall that a mechanical oscillator absorbs energy from the source that drives the oscillations. Moreover, the amount of energy the oscillator absorbs increases, usually sharply, when it is driven at its resonance frequency. By analogy, the sharp increase in the probability for particle absorption by a nucleus at a particular incident energy is called a resonance.

An excited compound state may exist for only a very short time before it decays. This is reflected in the energy width ΔE of the resonance peak in the cross-section (see Figure 3.6). The mean lifetime τ of an excited state is related to the energy width ΔE by the Heisenberg uncertainty principle in the form

$$\Delta E \times \tau \sim \frac{\hbar}{2\pi}, \tag{3.11}$$

where $\hbar = h/2\pi$ and h is Planck's constant (cf. Equation 2.29). This then provides us with a way of determining the mean lifetime of the excited compound state. A particularly striking example of a resonance occurs in the absorption of low-energy neutrons by cadmium. This reaction is dominated by a strong resonance with an energy width of 0.115 eV in the cross-section at a neutron energy of about 0.17 eV. Here the cross-section reaches a huge value of nearly 10,000 b. It follows that the mean lifetime of the excited compound state associated with the resonance is

$$\sim \frac{1.054 \times 10^{-34}}{2\pi \times 0.115 \times 1.60 \times 10^{-19}} \sim 10^{-15} \text{ s.}$$

3.3   Nuclear fission

Nuclear fission was discovered in 1939 by Otto Hahn and Fritz Strassman, and it could be argued that this scientific discovery changed the modern world more than any other. Hahn and Strassman bombarded a piece of uranium ($Z = 92$) with neutrons. Meticulous chemical analysis of the reaction products showed that an isotope of barium ($Z = 56$) had been produced. Following this discovery, Lise Meitner and Otto Frisch proposed that the incident neutron had split the uranium nuclei into two less massive nuclei. This was a truly astonishing proposal. How could a relatively light, low-energy neutron split a very much heavier nucleus? But their conclusion was correct and soon afterwards Niels Bohr and John A. Wheeler explained fission in terms of the *liquid drop model* of the nucleus.

The characteristics of a fission reaction are as follows:

- A heavy nucleus, with $A \sim 240$, splits into two less massive nuclei that lie towards the middle of the periodic table (with A within the range 90 to 145).

- A large amount of energy, $\sim$200 MeV, is released.

- Typically, two or three neutrons are emitted in the reaction.

The fact that several neutrons are emitted in a fission reaction suggests the possibility of using these neutrons to induce fission in other nuclei, and in turn using the neutrons that would be emitted in these fissions to produce further fissions, and so on, in a *chain reaction*. This may happen very rapidly and without control in a fission explosion, as happens in an atomic bomb. Alternatively, it may happen in a nuclear reactor at a carefully controlled rate that produces a steady output of power.

We can see why a fission reaction releases energy by considering the binding energy curve of the nuclides (see Figure 2.12). A heavy nucleus like uranium with $A \sim 240$ has binding energy per nucleon $B/A \sim 7.5$ MeV. Suppose that this nucleus splits into two nuclei with $A \sim 120$, which have $B/A \sim 8.5$ MeV. Going to a more tightly bound nucleus means that energy must be released. The total binding energy of two nuclei with $A \sim 120$ is greater than the binding energy of a nucleus with $A \sim 240$ by the amount $(2 \times 120 \times 8.5) - (240 \times 7.5)$ MeV $= 240$ MeV. This is the amount of energy released in the fission process and it appears primarily as kinetic energy of the fragment nuclei as Coulomb repulsion drives them apart.

3.3.1 Liquid-drop model of nuclear fission

The surface of a liquid behaves like an elastic membrane containing the liquid; the surface is in a state of tension, rather like the surface of an inflated balloon. This surface tension can be understood in terms of the forces acting between the molecules of the liquid. A molecule in the body of a liquid is surrounded by other molecules, which pull on it more or less uniformly in all directions. By contrast, a molecule at the surface of the liquid does not have any molecules above it. It thus experiences a net force that is directed inward and normal to the surface. Consequently, the surface sheet of molecules assumes the least surface area it can. The sphere contains the most volume for the least surface area and thus a liquid drop assumes a spherical shape. Surface tension gives rise to *surface energy*, as work must be done to increase the surface area of the drop. Thus, as we

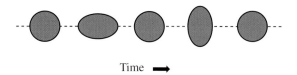

Time ➡

stretch a drop, its potential energy increases just as the potential
energy of a spring increases when it is stretched. Indeed, we can
cause a liquid drop to undergo harmonic oscillations as we can for
a mass on a spring. These oscillations are shown schematically in
Figure 3.7. As it oscillates, the liquid drop successively becomes a
sphere, a prolate spheroid, a sphere, an oblate spheroid, a sphere
and so on. The surface tension acts as a restoring force that returns
the drop towards equilibrium. Of course, if the drop is stretched too
much, it will break up into smaller droplets.

In Section 2.1.8 we developed the semi-empirical mass formula.
We saw that a nucleus can also be regarded as having surface tension.
In addition, there is the Coulomb repulsion between the protons. In
the liquid drop model, a nucleus is considered to be like a *charged and
incompressible spherical drop of liquid.* And nuclear fission involves a
delicate balance between the attractive force due to surface tension
and the repulsive Coulomb force.

If a nucleus receives some energy, say from the absorption of
a neutron, it may become distorted from its spherical shape. The
restoring force of the surface tension acts to return the nucleus back
to its equilibrium shape, but it must also counteract the Coulomb
repulsive force. If the distortion is not too large, the surface tension is
sufficient to hold the nucleus together and the nucleus performs oscillations. However, if the distortion becomes sufficiently large, a 'neck'
may appear, separating the nucleus into two regions, as shown pictorially in Figure 3.8. Coulomb repulsion causes the neck to stretch
further and finally two fragment nuclei are formed and these rapidly
fly apart, driven by their Coulomb repulsion. The nucleus splits into
two fragments, i.e. it fissions. The parameter r that is indicated in
Figure 3.8 is a rough measure of the distortion of the nucleus and
the subsequent separation of the two fission fragments.

The potential energy, $V(r)$, of a nucleus as a function of the deformation parameter r is illustrated schematically in Figure 3.9. As the
nucleus is initially stretched, its surface area increases and so does
its surface energy. On the other hand, the Coulomb energy due to
the electrostatic repulsion of the protons does not change as much.
The fact that the net energy (surface plus Coulomb) increases as the
nucleus stretches means that the potential energy of the nucleus initially increases as shown in Figure 3.9. When the deformation reaches
a certain critical point, the short-range nuclear attraction is no longer
able to compensate the Coulomb repulsion. At this point the nucleus

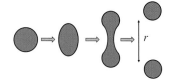

Figure 3.8 Pictorial
representation of the fission
of a heavy nucleus. The
parameter r is a rough
measure of the distortion of
the nucleus and the
subsequent separation of the
two fission fragments.

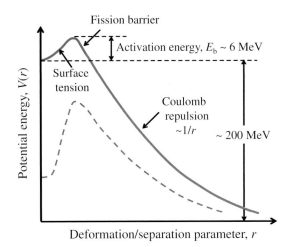

Figure 3.9 The solid blue curve represents the potential energy, $V(r)$, of a nucleus as a function of the parameter r, where r represents the deformation of the nucleus or its separation into two fragments. The figure shows the presence of a fission barrier, whose height E_b is ~6 MeV. This barrier inhibits the splitting of the nucleus into two fragments. The dashed blue curve represents $V(r)$ for a nucleus with $A \sim 100$. For such a nucleus the height E_b of the fission barrier is ~50 MeV.

splits into two fragments, which fly apart because of their Coulomb repulsion. The form of the potential energy curve $V(r)$ then follows the familiar $1/r$ Coulomb dependence. The fact that $V(r)$ first rises and then falls means that it must pass through a maximum. This results in a potential barrier, as shown in Figure 3.9. This is the *fission barrier* whose height E_b is called the *activation energy*. We previously estimated that the energy released in a fission process is ~200 MeV. This corresponds to the total kinetic energy of the two fission fragments when they are completely separated, i.e. when r approaches infinity. We can use this value to set the vertical scale of the potential energy diagram, as indicated on Figure 3.9.

For nuclei with $A \sim 240$ such as uranium the activation energy is ~6 MeV. The nuclei sit at the bottom of the local minimum in the potential energy function $V(r)$ at $r = 0$ and so are stable to spontaneous decay by fission; the fission barrier inhibits the nucleus splitting in two. The fission process *can* take place by barrier penetration but the fragment nuclei are so massive that the probability of barrier penetration is very small; see also Section 2.3.3. If ^{238}U decayed only via barrier penetration, its lifetime would be ~10^{16} years. Instead, it is far more likely, by a factor of ~10^6, for a ^{238}U nucleus to decay by α particle emission. When, however, we supply sufficient additional energy to a nucleus like ^{238}U, it can surmount the fission barrier and split into two fragments. This is what happens in *induced fission*, as we will describe shortly.

For very heavy nuclei ($A > 240$), Coulomb repulsion dominates over short-range nuclear attraction. In that case there is no fission barrier and if we were to form such a nucleus in the laboratory, it would immediately fly apart into fragments. According to the binding energy formula (Equation 2.27), surface energy varies as $A^{2/3}$, while Coulomb repulsion varies as $Z(Z-1)/A^{1/3} \sim Z^2/A^{1/3}$ for large Z. Hence the ratio Z^2/A is the important parameter for determining whether a particular nucleus will instantaneously decay by fission. From detailed considerations of the liquid-drop model and from experimental observations, Bohr and Wheeler estimated that a nucleus would instantly fission if $Z^2/A \geq 47.8$. For lighter nuclei, below about $A = 230$, the height of the fission barrier increases rapidly. This is illustrated in Figure 3.9 by the dotted curve, which represents the shape of $V(r)$ for a nucleus with $A \sim 100$. For such a nucleus, the activation energy is ~ 50 MeV. This means that it would take a huge amount of externally applied energy to induce fission in these nuclides.

In fact, there are only a few nuclides that have a fission barrier with a height of about 6 MeV. This value is significant because this is about the amount of energy that a heavy nucleus gains when it absorbs a neutron. This means that there is only a small number of nuclides for which neutron-induced fission is a practical source of energy. These lie in the atomic mass range between 230 and 240 u.

3.3.2   Induced nuclear fission

We see that it is possible to induce fission in some heavy nuclides, like uranium, by supplying the nucleus with enough energy to surmount the fission barrier. In the case of neutron-induced fission of ^{235}U, the reaction is

$$^{235}\text{U} + \text{n} \rightarrow {}^{236}\text{U}^* \rightarrow \text{A} + \text{B} + \text{few n's}, \qquad (3.12)$$

where A and B are fragment nuclei. The fission process proceeds in two steps. In the first step, the ^{235}U nucleus absorbs a neutron to form a compound nucleus ^{236}U* and furthermore a ^{236}U nucleus that is in an excited state, as indicated by the symbol *. The amount of energy gained by the ^{236}U nucleus in absorbing a neutron is sufficient for it to surmount its fission barrier, and in the second step of the process, the ^{236}U nucleus fissions into the two fragment nuclei, A and B (although we invariably speak of the fission of ^{235}U, we are in fact dealing with the fission of ^{236}U). We can readily determine the value of the excitation energy of the ^{236}U nucleus from the mass difference in the first reaction step: $^{235}\text{U} + \text{n} \rightarrow {}^{236}\text{U}^*$.

Worked example

Determine the energy that is made available when a thermal neutron is absorbed by a ^{235}U nucleus.

Solution

Thermal neutrons have an energy of ~ 0.025 eV, which can be neglected in this calculation. From the mass difference and using

$E = mc^2$, we see that the energy that becomes available from the absorption of the neutron is

$$E = \left[\left({}^{235}_{92}M + m_{\mathrm{n}} \right) - \left({}^{236}_{92}M \right) \right] c^2.$$

The atomic masses of ^{235}U, ^{236}U and the neutron are 235.043 924, 236.045 563 and 1.008 665 u, respectively. Using the conversion factor $c^2 = 931.5$ MeV/u, we obtain:

$$E = 0.007\,026 \times 931.5 \text{ MeV} = 6.54 \text{ MeV}.$$

The activation energy for ^{236}U is 6.2 MeV. Hence, the energy the ^{236}U nucleus gains in absorbing a neutron is more than enough for it to surmount the fission barrier. And note that the incident neutron was not required to supply any extra energy in the form of kinetic energy for fission to occur.

We can repeat the above exercise for the absorption of a neutron by a ^{238}U nucleus, where the excited compound nuclear state is ^{239}U* (the atomic masses of ^{238}U and ^{239}U are 238.050 785 and 239.054 290 u, respectively). In this case the ^{239}U nucleus is formed with an excitation energy of 4.8 MeV. This is less than the activation energy in ^{239}U, which is 6.6. MeV. Consequently, if we wish to induce fission in ^{238}U by the absorption of a neutron, the neutron must have sufficient kinetic energy to supply the extra 1.8 MeV.

Natural uranium contains 0.72% of ^{235}U and 99.3% of ^{238}U. Importantly, as we have seen, thermal neutrons can induce fission in the relatively rare isotope ^{235}U but not in the much more abundant isotope ^{238}U. The main reason for this is the substantial difference in the energy that is released when they absorb a thermal neutron. The difference in their respective activation energies is not so large: 6.2 MeV for ^{236}U and 6.6 MeV for ^{239}U. The origin of the difference in the released energies can be understood from consideration of the semi-empirical mass formula (SEMF). The compound nucleus ^{236}U has even Z and even N and so the pairing term in the SEMF is positive, which increases the binding energy of the nucleus; we recall that nucleons like to pair together. On the other hand, the compound nucleus ^{239}U has an odd value of A and hence the pairing term is zero and does not contribute to the binding energy of the nucleus. The net result is that it is more difficult to separate a neutron from ^{236}U than from ^{239}U. This means, by energy balance, that when we add a neutron to ^{235}U we release more energy than when we add a neutron to ^{238}U.

3.3.3   Fission cross-sections

Figure 3.10 shows the cross-sections for neutron-induced fission of ^{235}U and ^{238}U as a function of the energy of the incident neutron.

Figure 3.10 The cross-sections for neutron-induced fission of ^{235}U and ^{238}U. The cross-section for ^{235}U steadily increases as the neutron energy reduces while, in contrast, the cross-section for ^{238}U remains zero until the neutron energy is ~1 MeV. The sharp structures in the ^{235}U cross-section are due to resonances that occur at energies of excited states of the compound nucleus ^{236}U.

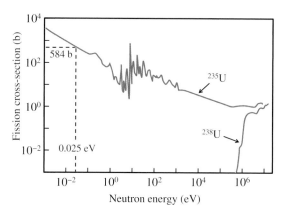

These cross-sections show some striking features. The fission cross-section of ^{235}U varies dramatically with neutron energy; note the logarithmic scales of the axes. At the thermal energy of 0.025 eV the fission cross-section for ^{235}U is 584 b, which is three orders of magnitude larger than it is at a few MeV. Indeed, the cross-section for ^{235}U increases as the neutron energy reduces in accordance with the $1/v$ law (see Section 3.2.3). The fission cross-section of ^{235}U also shows numerous sharp peaks between 1 and 100 eV. These peaks are due to resonances that occur at energies of excited states of the compound nucleus ^{236}U. In contrast, the cross-section for ^{238}U remains zero until the neutron energy is ~1 MeV. This is consistent with our previous discussion of induced fission in ^{235}U and ^{238}U. There we saw that absorption of a thermal neutron provided sufficient energy for the fission of ^{235}U, while ^{238}U required the neutron to provide an additional 1.8 MeV of kinetic energy to induce fission. These features of the two fission cross-sections explain why the fission of ^{235}U by thermal neutrons is the reaction that is usually used to produce energy in commercial nuclear reactors.

3.3.4   Fission reactions and products

When a neutron induces fission in a heavy nucleus, we might expect the nucleus to break up into two nuclei that have about the same atomic mass as each other. This would minimise Coulomb repulsion. However, it is found that fission into two nuclei of nearly equal mass is very unlikely. Moreover, the fission fragments are not uniquely defined and there is a distribution of masses for the two fission nuclei. The form of this mass distribution is shown schematically in Figure 3.11. It is symmetric about the centre of the distribution as, for every heavy fragment, there must be a corresponding light fragment. We see that there are maxima in the distribution at about $A = 95$ and $A = 135$ and again that fission into fragments of equal or nearly equal mass is much less probable.

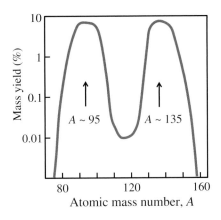

Figure 3.11 The form of the mass distribution of fragment nuclei following the fission of a heavy nucleus. The vertical scale corresponding to percentage mass yield is logarithmic. There are maxima in the distribution at $A \sim 95$ and $A \sim 135$, and fission into fragment nuclei of equal or nearly equal mass is very unlikely.

Typical fission reactions in ^{235}U are:

$$^{235}_{92}U + n \rightarrow {}^{144}_{56}Ba + {}^{89}_{36}Kr + 3n, \tag{3.13}$$

$$^{235}_{92}U + n \rightarrow {}^{141}_{55}Cs + {}^{93}_{37}Rb + 2n. \tag{3.14}$$

The fission fragments always have too many neutrons to be stable. For example, the $^{144}_{56}$Ba product nucleus in Reaction (3.13) has 88 neutrons, which is six more than for the heaviest stable isotope of barium, $^{138}_{56}$Ba. More generally, the neutron–proton ratio, N/Z, for stable nuclides around $A = 240$ is almost 1.6. On the other hand, it is about 1.3 for stable nuclides around $A = 135$, because of the reduced influence of the Coulomb repulsion of the protons. When a heavy nucleus fissions, the fission fragments initially have about the same N/Z ratio, i.e. $\sim$1.6. These fragments must reduce their N/Z ratios to achieve greater nuclear stability and they do so in two ways. Firstly, some neutrons are shed at the instant of fission, (within about 10^{-16} s) as indicated by Reactions (3.13) and (3.14). These neutrons are called *prompt* neutrons and the number of these prompt neutrons depends on the particular reaction path. The *average number* of prompt neutrons, denoted by the symbol ν, typically lies between 2 and 3. For example, $\nu = 2.42$ for the fission of ^{235}U induced by thermal neutrons.

Even after the emission of the prompt neutrons, however, the nuclear fragments are still *neutron rich*. So they undergo a series of β^- decays until a stable value of N/Z is reached. Each of these β^- decays converts a neutron into a proton, increasing Z by 1 and decreasing N by 1. An example of such a decay chain is

$$^{93}_{37}Rb_{56} \xrightarrow{\beta^-} {}^{93}_{38}Sr_{55} \xrightarrow{\beta^-} {}^{93}_{39}Y_{54} \xrightarrow{\beta^-} {}^{93}_{40}Zr_{53} \xrightarrow{\beta^-} {}^{93}_{41}Nb_{52}.$$

In some decay chains, the β^- decay may leave the product nuclei so highly excited that neutron emission becomes an alternative decay route. An example of this occurs when the fragment nucleus

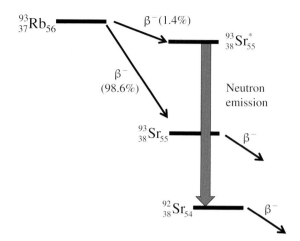

Figure 3.12 When a fragment nucleus $^{93}_{37}\text{Rb}_{56}$ β^- decays to $^{93}_{38}\text{Sr}_{55}$, there is a 1.4% probability that the $^{93}_{38}\text{Sr}_{55}$ nucleus will be left in a highly excited state that has enough energy to decay by the emission of a neutron to $^{92}_{38}\text{Sr}_{54}$.

$^{93}_{37}\text{Rb}_{56}\beta^-$ decays to $^{93}_{38}\text{Sr}_{55}$. There is a 1.4% probability that the ^{93}Sr nucleus will be left in a highly excited state that has enough energy to decay by the emission of a neutron. The decay scheme is illustrated in Figure 3.12. The excited state is denoted as $^{93}_{38}\text{Sr}^*_{55}$ and it decays to $^{92}_{38}\text{Sr}_{54}$. Such emitted neutrons are called *delayed neutrons*. The lifetime of the excited state in $^{93}_{38}\text{Sr}^*_{55}$ is 6 s, which explains why these neutrons are delayed with respect to the prompt neutrons. The number of delayed neutrons is small compared with the number of prompt neutrons, ~1 per 100 fissions. However, these delayed neutrons play a crucial role in the control of a nuclear reactor, as we shall see in Section 3.4.2.

3.3.5 Energy in fission

We can use Reaction (3.14) to determine the energy released in a typical fission reaction. The atomic masses of ^{235}U, ^{93}Rb, ^{141}Cs and the neutron are 235.043 925, 92.922 039, 140.920 045 and 1.008 665 u, respectively. Thus the mass defect of the reaction is +0.193 176 u, giving a Q-value of +180 MeV. There are other possible final products, which may have a different number of emitted neutrons and the value of 200 MeV is often take as the average energy released in a fission reaction. The fission energy is distributed between the various reaction products: the two fragment nuclei, prompt and delayed neutrons, prompt γ radiation that is emitted at the instant of fission, β particles and anti-neutrinos and radiation from the radioactive decay of the reaction products. As an example, the average energies of the reaction products from the fission of ^{235}U are given in Table 3.1.

Most of the released energy (~80%) goes into the kinetic energy of the two fragment nuclei. The neutrons take just a few per cent of the total energy and, as there are typically two or three of them, each neutron has ~2 MeV of energy. Most of the reaction products can be intercepted in a nuclear reactor and their energy collected.

Table 3.1 The distribution of energy between the various products in the fission of ^{235}U (average values)

	Energy (MeV)
• Kinetic energy of fragment nuclei	165
• Prompt and delayed neutrons	5
• Prompt γ radiation	8
• γ radiation from radioactive decays	7
• β particles	7
• Anti-neutrinos	12
Total	**204**

The exception to this are the anti-neutrinos, which have extremely little interaction with matter and escape the reactor. Their energy of ~12 MeV is lost and contributes nothing to energy production.

Worked example

Compare the amount of energy that is released from the fission of 1.0 kg of ^{235}U with the amount of energy that is released when 1.0 kg of coal is burnt. Assume that the coal is composed entirely of carbon and that 4.0 eV is released in the reaction: $C + O_2 \rightarrow CO_2$.

Solution

1.0 kg of ^{235}U contains

$$\frac{6.02 \times 10^{23} \times 1000}{235.0} = 2.56 \times 10^{24} \text{ atoms.}$$

Taking the energy released per fission reaction to be 200 MeV,

total energy release $= 2.56 \times 10^{24} \times 200 \times 10^6 \text{ eV} = 5.1 \times 10^{32} \text{ eV}$

$$= 8.2 \times 10^{13} \text{ J.}$$

Similarly, the amount of energy released by burning 1.0 kg of carbon is

$$\frac{6.02 \times 10^{23} \times 1000 \times 4.0}{12} = 2.0 \times 10^{27} \text{ eV} = 3.2 \times 10^8 \text{ J.}$$

This gives a ratio of

$$\frac{5.1 \times 10^{32}}{2.0 \times 10^{27}} \approx 2.5 \times 10^5.$$

The factor 10^5 that appears in this result is typical of the factors that occur when nuclear energies are compared with electronic energies.

3.3.6 Moderation of fast neutrons

The prompt neutrons emitted in a fission reaction have an average energy of ~2 MeV and are called *fast* neutrons. As we have seen, however, the fission cross-section for ^{235}U is three orders of magnitude larger at thermal energy 1/40 eV. It is therefore necessary to slow down or *moderate* the prompt neutrons in a nuclear reactor. This effect was first observed by Enrico Fermi. He found that inserting a sheet of paraffin wax into a neutron beam increased its ability to induce fission by a factor of ~100.

Moderation of fast neutrons is achieved by passing the neutrons through a material of low atomic mass; the molecules in the paraffin wax used by Fermi consisted mainly of hydrogen nuclei. The neutrons make numerous elastic collisions with the nuclei in the material and steadily lose their energy until they become in thermal equilibrium with the material at an energy of $\sim kT \sim 1/40$ eV. The neutrons are said to *thermalise*. We consider the simplest case where a neutron of mass m_n and velocity u_n collides head-on with a stationary nucleus of mass M and is scattered through $180°$. As in most nuclear collision processes, $u_n \ll c$ and so a non-relativistic treatment is valid. After the collision, the velocities of the neutron and nucleus are v_n and v_M respectively. Conservation of energy and momentum gives

$$\frac{1}{2}m_n u_n^2 = \frac{1}{2}m_n v_n^2 + \frac{1}{2}M v_M^2; \qquad (3.15)$$

$$m_n u_n = -m_n v_n + M_M v_M. \qquad (3.16)$$

Solving these two equations we readily obtain

$$v_n(M + m_n) = u_n(M - m_n). \qquad (3.17)$$

If the kinetic energy of the neutron is E before the collision and E' after the collision:

$$\frac{E'}{E} = \frac{v_n^2}{u_n^2} = \left(\frac{M - m_n}{M + m_n}\right)^2. \qquad (3.18)$$

Using atomic mass numbers for M and m_n, we obtain

$$\frac{E'}{E} = \left(\frac{A - 1}{A + 1}\right)^2. \qquad (3.19)$$

If the atomic mass $A \gg 1$, $E' \approx E$, and the neutron simply 'bounces' off the nucleus with very little loss of kinetic energy. To make the ratio E'/E as small as possible, the atomic mass A of the moderator material should be as small as possible, and if $A = 1$, the neutron gives all its energy to the nucleus.

The first nuclear reactor was built by Fermi and his team in 1942 in Chicago, USA. The reactor used carbon in the form of graphite as the moderator. For ^{12}C, Equation (3.19) gives $E'/E = (11/13)^2 = 0.72$, and after N collisions the ratio of final to initial kinetic energies of the neutron is $(11/13)^{2N}$. To reduce the energy of a neutron from 1 MeV to thermal energy:

$$\left(\frac{11}{13}\right)^{2N} = \frac{0.025}{10^6},$$

which gives $N \approx 50$. We chose the most favourable case for maximum energy transfer, i.e. a head-on collision where the neutron is scattered through $180°$. If we average over all scattering angles, we find that $\sim 2N$ (~ 100) collisions are required.

Graphite continues to be used as the moderator in some types of commercial fission reactor, while ordinary, *light* water (H_2O) is used as a moderator in some other types. In the latter case, the neutrons lose their energy to the hydrogen nuclei, and as hydrogen nuclei are light, only ~ 20 collisions are required to thermalise the neutrons. However, light water has the disadvantage that hydrogen nuclei can absorb a neutron to produce deuterium with the emission of a γ-ray:

$$n + {}^1H \rightarrow {}^2D + \gamma,$$

and the neutron is lost to the chain reaction. To compensate for this loss of neutrons, the percentage of ^{235}U in the uranium fuel has to be increased, i.e. enriched in ^{235}U. On the other hand, the deuterons in *heavy* water (D_2O) have a small absorption cross-section and so neutrons are not lost in this way. Consequently, natural, i.e. non-enriched, uranium fuel can be used with a heavy water moderator. Heavy water also has the advantage of a relatively small mass, but it is considerably more expensive than ordinary, light water. Nevertheless some types of commercial reactors do use heavy water as the moderator.

3.3.7   Uranium enrichment

We have seen that nuclear reactors using ordinary, light water as moderator need to use enriched uranium fuel. Typically, the fraction of the isotope ^{235}U is increased to ~ 2–5%. The different isotopes of an element all have the same number of electrons and nearly identical chemical properties. Consequently, isotope separation techniques use various physical processes that depend on the *masses* of the isotopes. The main methods for uranium enrichment are the gas diffusion technique and the gas centrifuge technique, but there are also some more recent methods that use lasers for isotopic separation.

Gas diffusion technique

In this technique, uranium hexafluoride, UF_6 gas is passed through a series of porous membranes. As $^{235}UF_6$ molecules are lighter than $^{238}UF_6$ molecules their mean velocity is greater and therefore they have a greater rate of diffusion through the pores in the membrane.

Figure 3.13 The process of
effusion through a tiny hole.
The left-hand chamber
contains gas at a relatively
high pressure, while the
pressure in the right-hand
chamber is much lower. The
effusion rate of the gas is
inversely proportional to the
square root of its molecular
mass.

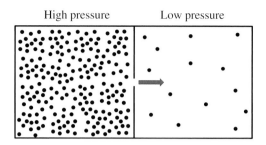

The UF$_6$ gas that diffuses through the membrane is thus slightly
enriched in ^{235}U.

The process of diffusion through a small hole is called *effusion*,
which is illustrated in Figure 3.13. This shows two chambers that
are separated by a tiny hole. The left-hand chamber contains gas
at a relatively high pressure, while the right-hand chamber is at
a much lower pressure. The diameter of the hole is much smaller
than the *mean free path* of the molecules in the left-hand chamber,
i.e. the average distance between molecular collisions. Under these
conditions, according to the kinetic theory of gases, the number of
molecules in the left-hand chamber striking the hole per second is
$n\bar{v}S/4$, where n is the number of molecules per unit volume, $\bar{v}$ is
their mean speed and S is the area of the hole. The mean speed $\bar{v}$
of the molecules is derived from the *Maxwell–Boltzmann distribution
of speeds* and is given by $\bar{v} = \sqrt{8\,kT/\pi M}$, where k is the Boltzmann
constant, T is temperature and M is the molecular mass. Thus the
rate R at which molecules enter the right-hand-chamber is given by

$$R = \frac{nS}{4}\left(\frac{8kT}{\pi M}\right)^{1/2} \text{ molecules/s.} \tag{3.20}$$

We see that the effusion rate is inversely proportional to the square
root of the molecular mass.

The molecular masses of ^{235}UF$_6$ and ^{238}UF$_6$ are 349 and 352 u,
respectively, which is a mass difference of about 1%. As all the
molecules are at the same temperature, the ratio of their effusion
rates is

$$\frac{R_{349}}{R_{352}} = \frac{n_{349}}{n_{352}}\left(\frac{352}{349}\right)^{1/2}.$$

Hence, after one stage of effusion, the ratio of the number densities
of the two gases will also be equal to

$$\frac{n_{349}}{n_{352}}\left(\frac{352}{349}\right)^{1/2}.$$

The ^{235}UF$_6$:^{238}UF$_6$ density ratio is increased by the factor
$\sqrt{352/349} = 1.0043$. This increase is clearly very modest and so
many hundreds of successive stages are required to achieve significant enrichment. After p such stages, the ratio is increased by the

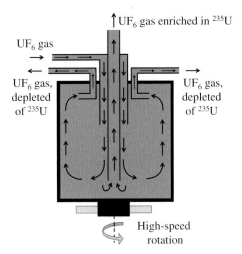

Figure 3.14 The gas centrifuge technique to enrich uranium. Uranium hexafluoride gas is passed into a cylinder that rotates at a very high rate, ~500 000 rpm. The rotation creates a strong centrifugal force that causes the heavier $^{238}UF_6$ molecules to move preferentially towards the outer wall of the cylinder, while the lighter $^{235}UF_6$ molecules collect closer to the centre. The $^{235}UF_6$ molecules are then drawn off the rotating cylinder close to the axis of rotation. In practice, the cylinder is ~3–5 m tall with a diameter of ~0.2 m.

factor $(1.0043)^{p/2}$. The large number of stages required makes this separation technique very expensive due to the high cost of vacuum pumping. However, it was the only large-scale method available until the 1970s.

Gas centrifuge technique

In this technique, UF_6 gas is passed into a cylinder that rotates at a very high rate, ~500 000 rpm, as illustrated in Figure 3.14. The rotation of the cylinder creates a strong centrifugal force which causes the heavier $^{238}UF_6$ molecules to move toward the outer wall of the cylinder while the lighter $^{235}UF_6$ molecules collect closer to the centre. The $^{235}UF_6$ molecules are then drawn off the rotating cylinder close to the axis of rotation. The rotating cylinder is designed so that gas flows continuously in and out of the centrifuge and this allows multiple centrifuge stages to be cascaded in series. Again many stages are required to increase the fraction of the $^{235}UF_6$ molecules substantially. However, each centrifuge stage increases the $^{235}UF_6$:$^{238}UF_6$ ratio by the factor 1.3, which is much greater than for the gas diffusion technique. This means that far fewer centrifuge stages are required for the same degree of enrichment: ~15 stages for a 5% enhancement of ^{235}U. Currently, the gas centrifuge technique it is the main uranium enrichment process.

Laser ionisation

In one enrichment technique using laser ionisation, uranium is heated in an oven to produce a beam of uranium atoms. This atomic beam is crossed with a monochromatic laser beam with photon energy $h\nu_{exc}$. This matches the energy E of one of the excited levels of ^{235}U, as illustrated in Figure 3.15, and excited $^{235}U^*$ atoms are produced.

Figure 3.15 The principle of a laser ionisation technique for uranium enrichment. Monochromatic laser light with energy $h\nu_{exc}$ excites only atoms of the ^{235}U isotope because the photon energy only matches energy E. A second laser with energy $h\nu_{ion}$ then ionises the excited ^{235}U* atoms so that a beam of ^{235}U$^+$ ions is obtained that can be separated from the neutral ^{238}U atoms in the beam by electrostatic deflection.

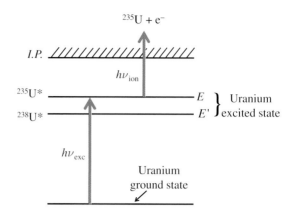

However, because the two different isotopes of uranium have slightly different nuclear masses, the energy E' of the corresponding excited level in ^{238}U is *very* slightly different from the energy E. The photon energy of the monochromatic laser does not match the energy E' and so excited, ^{238}U* atoms are not formed. A second laser with photon energy $h\nu_{ion}$ ionises the excited, ^{235}U* atoms so that a beam of ^{235}U$^+$ ions is obtained. This second step is represented by the reaction

$$^{235}\text{U}^* + h\nu_{ion} \rightarrow {}^{235}\text{U}^+ + \text{e}^-.$$

It is then easy to separate the positively charged ^{235}U$^+$ ions from the neutral ^{238}U atoms in the beam by electrostatic deflection.

Worked example

The $n = 3$ to 2 emission line in the *Balmer* series of atomic hydrogen has the wavelength 656.29 nm. Use the Bohr model to determine the wavelength of the corresponding emission line in the deuterium isotope.

Solution

According to the Bohr model of the hydrogen atom, the wavelength of an emission line connecting levels n_i and n_f is given by

$$\frac{hc}{\lambda} = E_i - E_f = \frac{m_e Z^2 e^4}{\varepsilon_0^2 8 h^2}\left(\frac{1}{n_f^2} - \frac{1}{n_i^2}\right), \qquad (3.21)$$

where we have used Equations (2.14) and (2.16), and the symbols have their usual meanings. This expression assumes that the nucleus is infinitely massive compared with the mass m_e of the electron, which is a very good assumption for the hydrogen atom, as the proton is 1836 times heavier than the electron mass. For the required accuracy of the present example, however, we have to take into account the finite mass of the nucleus, which we do by replacing m_e with the *reduced mass* μ of the nucleus–electron system:

$$\mu = \frac{m_e M}{m_e + M}, \tag{3.22}$$

where M is the mass of the nucleus. Replacing m_e by μ in Equation (3.21), we obtain

$$\frac{hc}{\lambda} = \frac{\mu Z^2 e^4}{\varepsilon_0^2 8h^2}\left(\frac{1}{n_f^2} - \frac{1}{n_i^2}\right). \tag{3.23}$$

As n_i and n_f are the same for both cases, we have

$$\frac{\lambda_D}{\lambda_H} = \frac{\mu_H}{\mu_D}.$$

Therefore,

$$\lambda_D = \lambda_H \times \frac{m_e M_H}{(m_e + M_H)} \times \frac{(m_e + M_D)}{m_e M_D}$$

$$= 656.29 \times \frac{(1 \times 1836)}{1837} \times \frac{(1 + 2 \times 1836)}{(2 \times 1836)} = 656.11\,\text{nm},$$

where we have taken the nuclear masses of hydrogen and deuterium to be 1 and 2 u, respectively, and the mass of the electron to be $(1/1836)$ u. The difference in wavelength is small (0.18 nm) but can be observed with a high-resolution optical spectrometer. Indeed, deuterium was discovered in 1932 by Harold C. Urey, who observed this shift between hydrogen and deuterium spectral lines. In 1934 he won the Nobel Prize in Chemistry for his discovery of deuterium.

3.4 Controlled fission reactions

3.4.1 Chain reactions

We have seen that the average number, ν, of prompt neutrons emitted in a fission reaction is ~2.5 and that these neutrons may induce further fission reactions in other nuclei in a chain reaction. This suggests that the number of neutrons can, in principle, increase exponentially. This would lead to an exponentially increasing reaction rate, resulting in an explosion of energy. Indeed, this is what happens in an atomic bomb. On the other hand, it is possible to control the number of neutrons so that their number remains constant from one *generation* of fission reactions to the next. This produces a steady output of energy and this is the situation in the core of a nuclear reactor. Whichever happens, atomic bomb or energy source, depends on how many of the two or three emitted neutrons go on to produce fission in another nucleus.

We centre our discussion on the fission of ^{235}U by thermal neutrons because this is the process that is used in most commercial reactors. We might expect that each thermal neutron will, on average, lead to the production of ν thermal neutrons to induce the following generation of fission reactions in the chain. However, this is not the case. Some of the neutrons are lost to the chain reaction in various ways as described below. We define the *neutron reproduction factor* k as the ratio of the number of thermal neutrons in one generation of fission reactions to the number of thermal neutrons in the *preceding* generation. The important point is that k is less than ν, i.e. only a fraction of the thermal neutrons go on to induce further fissions. Clearly for a chain reaction to occur, $k \geq 1$. If a nuclear reactor has $k > 1$ it is said to be *supercritical*. If $k = 1$, it is said to be *critical*, which is the condition for the steady production of energy. For $k < 1$, it is said to be *subcritical*. To determine the value of k, we must follow the fate of a collection of thermal neutrons from one generation of fission reactions to the next. Calculating what happens inside a real nuclear reactor is a very complicated problem and sophisticated computer programs have been developed for this purpose. However, we will consider the main factors that combine to determine the value of k.

Suppose that we start with N *thermal* neutrons at a particular point in a chain reaction. We consider the following four factors that reduce the probability that these thermal neutrons will cause further fission.

Fission versus radiative capture

When a thermal neutron is absorbed by a ^{235}U nucleus, a compound nucleus ^{236}U* is formed. This compound nucleus may fission as in Reactions (3.13) and (3.14). However, the compound nucleus may instead decay by an alternative mode called *radiative capture*. In this decay mode the compound nucleus ^{236}U* decays by emitting a γ-ray:

$$^{235}\text{U} + \text{n} \rightarrow \,^{236}\text{U}^* \rightarrow \,^{236}\text{U} + \gamma. \tag{3.24}$$

In this case, the neutron is lost to the chain reaction. The fraction of thermal neutrons that do cause fission of the ^{235}U nuclei is given by

$$\text{fraction} = \frac{\text{probability for fission}}{\text{probability for fission} + \text{probability for radiative capture}}. \tag{3.25}$$

In terms of the cross-sections for fission σ_f and radiative capture σ_rc, the fraction is given by

$$\text{fraction} = \frac{\sigma_\text{f}}{\sigma_\text{f} + \sigma_\text{rc}}. \tag{3.26}$$

On average, each fission reaction produces ν prompt neutrons. Hence, we define the parameter η as the number of prompt neutrons produced per original thermal neutron:

$$\eta = \nu \frac{\sigma_f}{\sigma_f + \sigma_{rc}}. \tag{3.27}$$

Straight away, we can see that $\eta < \nu$. The values of the cross-sections for ^{235}U at thermal energy are $\sigma_f = 584$ b and $\sigma_{rc} = 96$ b, respectively, and ν has the value 2.42. So if the reactor core consisted solely of the ^{235}U isotope,

$$\eta = 2.42 \frac{584}{584 + 96} = 2.08,$$

which is substantially greater than 1. However, if the fuel is naturally occurring uranium, the value of η is smaller. We recall that the natural abundances of ^{238}U and ^{238}U are 99.28% and 0.72%, respectively. For a natural mixture of uranium, the effective fission cross-section $\sigma_{f,eff}$ is given by

$$\sigma_{f,eff} = \frac{0.72}{100} \sigma_{f,235} + \frac{99.28}{100} \sigma_{f,238},$$

with a similar expression for the effective radiative capture cross-section. We saw in Figure 3.10 that ^{238}U has a zero cross-section for fission at thermal energy. Hence, the effective fission cross-section for natural uranium is

$$\sigma_{f,eff} = \frac{0.72}{100} \times 584 = 4.20 \text{ b.}$$

However, ^{238}U does have a radiative-capture cross-section of 2.72 b at this energy. For a natural mixture of uranium, the effective radiative-capture cross-sections is then

$$\sigma_{rc,eff} = \frac{0.72}{100} \times 96 + \frac{99.28}{100} \times 2.72 = 3.39 \text{ b.}$$

Substituting these cross-sections in Equation (3.27) with $\nu = 2.42$ gives $\eta = 1.34$, which is getting close to 1. If, however, we increase the fraction of ^{235}U in the fuel, the reduction in the value of η will not be so large. For example, if the fraction of ^{235}U is increased to 3%, the value of η is 1.84.

Fast fission

We see that the original N thermal neutrons produce ηN prompt neutrons in fission reactions, and these prompt neutrons have a mean energy of ~ 2 MeV. ^{238}U has a small but nevertheless finite fission cross-section at this energy and so these fast, prompt neutrons may cause fission in ^{238}U. This *increases* the number of neutrons by a factor ε, called the *fast fission factor*. However, because this fission cross-section in ^{238}U is relatively small, ~ 1 b, ε is close to unity and is typically ~ 1.03. Following fission by these fast neutrons, we therefore have $\varepsilon \eta N$ neutrons, and these have a mean energy of ~ 2 MeV.

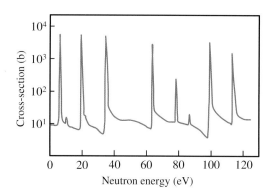

Figure 3.16 The sharp peaks in the spectrum correspond to resonances in the radiative-capture cross-section of ^{238}U, in the energy range between 10 and 120 eV. At the resonance energies, the cross-section reaches peak values of 10^3–10^4 b. As fast, prompt neutrons are moderated, they steadily lose energy and have to pass through the energy region containing the resonances. If a neutron encounters a ^{238}U nucleus when its kinetic energy matches a resonance energy, it is highly likely to initiate a radiative-capture reaction and be lost to the chain reaction.

Moderation of the fast neutrons

We saw above that the cross-section $\sigma_{\rm rc}$ for *thermal* neutrons causing radiative-capture reactions in ^{238}U is 2.72 b, and this reduces the value of η, the number of prompt neutrons produced per original thermal neutron. However, the cross-section for radiative capture in ^{238}U has numerous resonances in the energy range between 10 and 120 eV, which appear as the sharp peaks in the cross-section as illustrated in Figure 3.16. The radiative-capture cross-section $\sigma_{\rm rc}$ rises sharply at the resonance energies to $\sim$$10^3$–$10^4$ b. The fast, prompt neutrons steadily lose energy as they are moderated and have to pass through the energy region containing the resonances. If a neutron encounters a ^{238}U nucleus when its kinetic energy matches a resonance energy, it is highly likely to initiate a radiative-capture reaction and be lost to the chain reaction. To account for this loss, there is a factor p called the *resonance escape probability*. This loss of neutrons can be minimised by the design of the reactor and a typical value of p is $\sim$0.9. After moderation, we thus end up with $p\varepsilon\eta N$ thermal neutrons.

Thermal utilisation factor

Thermal neutrons may also be absorbed by the moderator or by structural components in the reactor and lost to the chain reaction. This is taken into account by the *thermal utilisation factor f*, which typically has a value of $\sim$0.9.

Thus the number of thermal neutrons that finally survive to produce the succeeding generation of fission reactions is equal to $fp\varepsilon\eta N$. Whether this is greater or smaller than the original number of thermal neutrons, N, determines the criticality of the reactor. It readily follows that the neutron reproduction factor k is

$$k = fp\varepsilon\eta, \qquad (3.28)$$

which is called the *four-factor formula*.

In addition to these factors, neutrons may leak out of the sides of the reactor core. This loss of neutrons is proportional to the surface area of the core, $\sim L^2$, where L characterises its linear size, while the number of generated neutrons depends on the volume of the core, $\sim L^3$. This simple dimensional argument shows that the larger the reactor core, the smaller the surface area to volume ratio ($\propto 1/L$) and the smaller the fraction of neutrons that leak (A similar dimensional argument, with respect to loss of thermal energy, explains why mice have fur and elephants do not). The dimensional argument also indicates that there is a critical minimum size for a nuclear reactor. A spherical reactor consisting of natural uranium and a graphite moderator has a critical radius of about 5 m. The size of a fission reactor is reduced somewhat by surrounding the reactor core with a neutron *reflector* made from suitable material like beryllium. Beryllium has a high cross-section for scattering neutrons and reflects escaping neutrons back into the core. In this way, neutron leakage can be kept below about 3%.

3.4.2 Control of fission reactions

The ability to control the neutron reproduction factor k is essential for the safe operation of a nuclear reactor. This is achieved by inserting *control rods* into the reactor core. These rods are made from a material such as cadmium that has a very large cross-section for the absorption of thermal neutrons and can therefore remove neutrons from the chain reaction very effectively; the absorption cross-section of cadmium for thermal neutrons is $\sim 2 \times 10^3$ b. When the reactor is started up, the control rods are inserted so that $k < 1$. The rods are then steadily withdrawn so that fewer neutrons are absorbed and k rises to 1, which is the condition for the steady output of energy. If k becomes > 1, the rods are inserted further. Furthermore, the rods can be used to stop the chain reaction if necessary. Over time there will inevitably be variations in k and it is necessary to manipulate the control rods to take account of these variations. For example, the amount of fission fuel steadily reduces and so the rods are slowly removed to maintain a constant output power.

In addition to the reproduction factor k, we must also consider the time interval τ between one generation of fission reactions and the next. If we have N thermal neutrons in one generation then the number of thermal neutrons in the following generation will be kN, and this occurs over a characteristic time interval τ. Thus the number of thermal neutrons increases at the rate

$$\frac{\mathrm{d}N}{\mathrm{d}t} \approx \frac{(Nk - N)}{\tau} = N\frac{(k - 1)}{\tau}, \qquad (3.29)$$

from which we obtain

$$N(t) = N_0 \exp\left[\frac{t(k-1)}{\tau}\right]. \tag{3.30}$$

If $k = 1$, then $N(t)$ is constant, which is the desired operating mode of a reactor. If $k < 1$, the fission rate decreases exponentially. For $k > 1$, we have the familiar result that the number of neutrons rises exponentially, with a time constant characterised by $\tau/(k-1)$.

Consider for a moment the role of the fast, prompt neutrons. The time it takes these fast neutrons to be moderated plus the time it takes the resultant thermal neutrons to diffuse through the reactor core before the next fission reaction is $\sim 10^{-3}$ s. Then if k is slightly above 1, say $k = 1.001$, $\tau/(k-1) = 1.0$ s and the time it would take for $N(t)$ to double is given by

$$\frac{N(t)}{N_0} = 2 = \exp\left[\frac{t(0.001)}{0.001}\right],$$

from which $t = 0.693$ s. This *doubling time* is not long enough for the necessary mechanical manipulation of the control rods to control the fission reactions and it would not be safe to operate the reactor. This is where the *delayed* neutrons from the fission reactions play a crucially important role in the control process. As we saw in Section 3.3.4, the delayed neutrons are emitted from fission fragments long after the prompt neutrons are emitted. They are relatively low in number; in the case of ^{235}U, about 0.7% of the emitted neutrons are delayed. However, they serve the essential purpose of increasing the effective time interval between one generation of fission reactions and the next. To get the essential idea of how this works, we use for the characteristic time interval τ a weighted mean that takes into account the fast, prompt neutrons and the delayed neutrons as shown in the following example.

Worked example

If 0.7% of the fission neutrons from ^{235}U are delayed by 15 s, determine the time it takes the reaction rate to double. Assume that the time it takes the fast neutrons to be thermalised and diffuse through the reactor core before the next fission reaction is 10^{-3} s, and $k = 1.001$.

Solution

The weighted mean time between successive fission generations is

$$(0.993 \times 10^{-3}) + (0.007 \times 15) = 0.106 \text{ s}.$$

Taking this to be the characteristic time interval between successive generations of fission reactions, we obtain

$$\frac{N(t)}{N_0} = 2 = \exp\left(\frac{t(1.001 - 1)}{0.106}\right),$$

which gives $t = 73.5$ s. Even though the fraction of delayed neutrons is less than 1%, their presence has a large effect on the doubling time and allows plenty of time for mechanical operation of the control rods. This is necessarily a much simplified description of the control process in a real reactor but it demonstrates the key role played by the delayed neutrons.

3.4.3 Fission reactors

A schematic diagram of a nuclear reactor based on thermal fission of uranium is presented in Figure 3.17, and shows its main components. The core of the reactor contains the nuclear fuel, the moderator and the control rods. It is surrounded by a neutron reflector that reflects escaping neutrons back into the core. The whole assembly is housed in a steel pressure vessel. The kinetic energy of the fission fragments is absorbed in the reactor core and transformed into thermal energy. The thermal energy is extracted by a coolant, which circulates through the core. The thermal energy is used to produce steam that drives a turbine that, in turn, drives an electrical generator. A reactor is a powerful source of nuclear radiation and so it is contained within a biological shield. In particular, γ-rays and neutrons are highly penetrating. The biological shield is typically a thick layer of concrete, which contains low hydrogen-bearing materials that are effective in slowing and stopping neutrons. The shield also contains heavy metals such as lead that absorb the γ-rays.

Figure 3.17 A schematic diagram of a nuclear fission reactor based on thermal fission of uranium, showing its main components. The core of the reactor contains the nuclear fuel, the moderator and the control rods. Thermal energy is extracted by a coolant that circulates through the core and the thermal energy is used to produce steam that drives a turbine, which in turn drives an electrical generator. The core is surrounded by the neutron reflector and housed in a steel pressure vessel, and the reactor is contained within a biological shield.

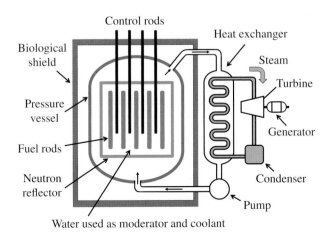

Water used as moderator and coolant

Nuclear fuel

The most commonly used fuel is enriched uranium with typically 2–5% of ^{235}U. The fuel is usually in the form of uranium oxide pellets, which are inserted into long hollow metal tubes. Uranium oxide has a much higher melting point than uranium metal so that the core can be maintained at a higher temperature. This serves to increase the efficiency of power generation. It is possible to use naturally occurring uranium, with 0.72% of ^{235}U, but this requires the use of heavy water (D_2O) or carbon (graphite) as the moderator, instead of light water (H_2O).

Moderator

The moderator should have low atomic mass, have a large cross-section for scattering neutrons and a small cross-section for absorbing thermal neutrons. It should also be a liquid or a solid so that the number of nuclei per unit volume is high, it should be chemically stable and ideally it should be cheap and abundant. Ordinary water (H_2O) is cheap and hydrogen nuclei are light, with $A = 1$. However, because hydrogen has a large absorption cross-section for thermal neutrons, enriched uranium must be used for the fuel to compensate for the loss of neutrons. On the other hand, heavy water does not have a large absorption cross-section and so natural uranium can be used. Carbon has a reasonably large cross-section for neutron scattering but a reasonably small cross-section for neutron absorption so it, too, can be used with natural uranium fuel. Moreover, it is relatively light and chemically stable.

The moderator must be incorporated with the uranium fuel in the core of the reactor in some way. The fuel consists mostly of the ^{238}U isotope and, as we saw in Section 3.4.1, this isotope has numerous resonances in its radiative-capture cross-section in the energy region between 10 and 120 eV. Also, at these resonance energies there is a large probability that a neutron will be captured by the uranium nucleus and lost to the chain reaction. This loss is minimised by the *clumping* of the uranium fuel in the form of rods that sit in the intervening moderator, as depicted in Figure 3.17. This has the advantage that the neutrons travel mainly in the moderator and less in a fuel rod.

Coolant

The coolant must transfer the thermal energy away from the core efficiently. The coolant maybe a gas, e.g. carbon dioxide, or a liquid, e.g. water or liquid sodium. In the reactor shown in Figure 3.17, water in a *primary loop* is circulated through the core at high temperature, say ~600 K, and also under high pressure (~150 atm) to maintain it as a liquid. The thermal energy is transferred to a *secondary loop* in a heat exchanger.

Thermal efficiency of reactor

Steam is generated in the heat exchanger and this steam drives a turbine, which in turn drives an electrical generator. The steam is then condensed to liquid water and pumped back to the heat exchanger to start the cycle again. A device that converts thermal energy into mechanical energy in this cyclical manner is called a *heat engine*. In Section 4.6.5 we discuss heat engines in more detail and here we just give some essential results. We define the *thermal efficiency ε* of a heat engine as the work output W of the engine divided by the thermal energy Q_H absorbed from the source of heat:

$$\varepsilon = \frac{W}{Q_H}. \tag{4.80}$$

The *maximum* thermal efficiency of a heat engine is

$$\varepsilon = 1 - \frac{T_C}{T_H}, \tag{4.91}$$

where, in the present case, T_C is the temperature of the condenser and T_H is the temperature of the core. For example, if we consider the reactor to be a source of thermal energy at 600 K and take the temperature of the condenser to be 300 K, then $\varepsilon = 0.5$. Of course, practical heat engines are never ideal and their efficiencies will be lower than that given by Equation (4.91). The important point is that the higher the temperature of the reactor core, the higher is the thermal efficiency. This explains the use of uranium oxide as fuel rather than uranium metal, which has a lower melting point. We also see that a substantial fraction of the thermal energy that is produced is not converted into useful output power, which is the case for any heat engine.

Poisoning of a reactor by fission products

In the operation of a reactor, certain fission fragments are produced that have a large cross-section for the absorption of thermal neutrons. A particular example is ^{135}Xe, which has an absorption cross-section of 2.7×10^6 b. These *reactor poisons* absorb neutrons, causing the neutron reproduction factor, k, to decrease. This loss of neutrons is compensated by withdrawal of the control rods to restore equilibrium at $k = 1$.

3.4.4 Commercial nuclear reactors

Nuclear reactors became economically viable in the 1960s. As of 2017, 30 countries worldwide operate about 440 commercial nuclear reactors, with a total capacity of almost 400 GW, and about 65 more reactors are under construction in 15 countries. These reactors provide about 11% of the world's electricity and there are some 13 countries that rely on nuclear energy for at least a quarter of their electricity needs. In addition there are about 240 research reactors around the world, while nuclear reactors power some 140 ships and submarines. Currently, there are a number of different types of reactor that are based upon thermal fission of uranium and here we outline the main characteristics of some of the more common types.[1]

Magnox and the advanced gas-cooled reactor (AGR)

The Magnox reactor was an early, British design. It had a graphite moderator and used CO_2 gas as coolant. The reactor used uranium in metal form as fuel, which was encased in tubes of a magnesium alloy called Magnox. The CO_2 gas circulated through the core with an outlet temperature of 350°C. The reactor delivered a thermal output of about 900 MW and an electric output power of 250 MW, i.e. at an efficiency of about 28%. Because of improvements in reactor design and the availability of enriched uranium fuel, this type of reactor is no longer being built. The last Magnox reactor, at Wylfa power station in Anglesey, Wales, ceased operation in 2015 after 45 years of service.

The advanced gas-cooled reactor (AGR) is the second generation of British gas-cooled reactors. It improves the efficiency of the gas-cooled reactor by having a higher coolant temperature. This involved changing the fuel from uranium metal to uranium oxide, which has a much high melting point, and increasing the percentage of ^{235}U to ~3%. It also involved using materials for the structural components and fuel cladding that could withstand operation at the higher temperature. The moderator and coolant are graphite and CO_2 gas, respectively, as in the Magnox reactor. The CO_2 gas circulates through the core, reaching a temperature of 650°C. Heat exchangers are located in the space between the core and the pressure vessel containing the reactor. The thermal output of a typical AGR is about 1500 MW and it operates with a thermal efficiency of about 42%, which is considerably higher than that of a Magnox reactor.

Pressurised water reactor (PWR)

This is the most common type of commercial reactor with several hundred in use. The design was developed in the USA where enriched uranium was more readily available after the Second World War. A PWR uses light water as the moderator. Light water is a better

---

[1] For details of a wider range of nuclear reactor types, see John Lilley, *Nuclear Physics; Principles and Applications*, John Wiley & Sons, 2001.

moderator than graphite and this enables the size of the core to be considerably reduced, to one that is typically 3 m in height and diameter. However, enriched fuel must be used to offset the tendency for light water to absorb neutrons. The fuel consists of 4.5% enriched uranium in the form of uranium oxide pellets encased in tubes. Water is also used as the coolant, in a dual loop system, as shown in Figure 3.17. The water circulates through the core of the reactor in a primary cooling circuit at high temperature ($\sim$325°C) and under very high pressure ($\sim$150 atm) to prevent it from boiling. Electrical power is produced at an efficiency of about 32%. This efficiency is less than that of the AGR, but the PWR is cheaper to build because of the smaller size of the core. The ability to build a reactor core of small size and moderated by water led to the use of the PWR in naval vessels. They enabled submarines to remain at sea for long periods without the need for refuelling.

Boiling water reactor (BWR)

This design has many similarities to the PWR, with light water serving as both moderator and coolant. However, in the BWR, there is only a single cooling circuit in which the water is at lower pressure ($\sim$75 atm) so that it boils in the core at $\sim$325°C. The steam passes directly to the turbines, after first being dried. This eliminates the need for heat exchangers with their inevitable reduction in thermal efficiency. Water in the core of a reactor can become contaminated with traces of radioactive nuclides, which can reach the turbines. This effect is minimised by using water with a very high level of purity and by shielding the turbines. A BWR provides electric power with an efficiency of about 35%.

The CANDU reactor

This reactor is distinguished by using heavy water as the moderator. As we have seen, this is a more efficient moderator than ordinary water and allows natural uranium to be used as fuel. After the Second World War, Canada had the capability of producing heavy water in sufficient quantities to moderate a reactor by D_2O and this led to the development of the CANDU reactor. As in the PWR, the primary coolant (D_2O) generates steam (H_2O) in a secondary circuit to drive the turbines. This type of reactor produces more energy per kilogram of mined uranium than other designs, but also produces a larger amount of used fuel per unit output.

3.4.5 Nuclear waste

The fuel in a nuclear reactor steadily gets consumed and, in addition, it becomes poisoned by the build-up of fission products such as ^{135}Xe that absorb the neutrons. Hence the spent fuel rods must be replaced after several years in a reactor. This produces a great deal of *nuclear waste* that must be safely stored until its activity has decayed to a safe level. Most of the radioactive nuclides in the spent rods are fragment nuclei such as ^{90}Sr and ^{137}Cs. These fragments predominantly emit β particles and γ radiation and have half-lives of typically tens of years; ^{90}Sr and ^{137}Cs have half-lives of 28.8 and 30.2 years, respectively. However, there are also other reaction products called *transuranic* nuclides. These nuclides, such as ^{239}U, ^{242}Pu and ^{243}Am, are heavier than uranium and are formed by successive absorption of neutrons by the uranium fuel. They decay predominantly by α-particle emission and have very long half-lives; ^{239}U, ^{242}Pu and ^{243}Am have half-lives of 2.4×10^4, 3.7×10^5 and 7.4×10^3 years, respectively. Because of their long lifetimes, these fission products need to be safely stored for many thousands of years.

The spent fuel rods are initially put into deep storage ponds, which allow ∼8 m of water to cover the fuel rods. The water provides effective biological shielding from the nuclear radiation and also cools the rods, which continue to produce thermal energy. Pumps circulate the water from a pond to heat exchangers and then back to the pond. The fuel rods remain in the pond for at least 5 years to allow them to cool down and the radiation levels to decrease. They may then be removed from the storage ponds for more permanent storage in steel reinforced concrete casks. These casks are usually stored close to the nuclear reactor.

An alternative way to deal with spent fuel rods is to *reprocess* them, as is done, for example, at the Sellafield nuclear facility in the UK. The spent fuel rods still contain ∼95% of the original uranium, although it is significantly depleted of ^{235}U. The remaining ∼5% of the fuel rods are the fragment nuclei such as ^{90}Sr and ^{137}Cs, and the transuranic nuclides such as ^{239}U, ^{242}Pu and ^{243}Am. The uranium and the fission products are separated from each other by a series of chemical processes. In particular, the depleted uranium and the plutonium nuclide ^{239}Pu are recovered from the spent fuel rods. ^{239}Pu, which amounts to about 1% of the spent fuel rods, is a fissionable nuclide; its fission cross-section for thermal neutrons is 742 b and the height of its fission barrier is 6.0 MeV. Thus it can be mixed with the extracted uranium to produce a nuclear fuel that can be recycled in another nuclear reactor. This so-called *mixed-oxide (MOX)* fuel reacts similarly although not identically to enriched uranium fuel. The remaining material, amounting to about 3% of the original fuel rods, is highly toxic radioactive waste, although the volume of high-level waste that has to be stored is considerably reduced. This toxic waste is treated by a process called *vitrification*. Here the nuclear waste in powder form is mixed with glass powder and the mix is heated to about 1100°C. This produces a highly stable material that should not degrade over extended periods of time. This material is stored in shielded stainless steel tanks.

Secure storage of nuclear waste on a timescale of many thousands of years is a challenge and so far no method of doing this has been established. One possibility is to bury the storage casks and vitrified waste in deep geological repositories.

3.5   Nuclear fusion

In nuclear fusion we bring together two light nuclei to form a single, more massive nucleus. We again climb the binding energy curve towards more stable nuclei, but from the opposite direction to fission (see Figure 3.1). The binding energy per nucleon after the fusion reaction is greater than before and so energy is released. The energy released *per unit mass of material* (MeV/u) is comparable to that obtained from fission and indeed, in some cases, exceeds it. Fusion, moreover, has some significant advantages as an energy source. The light nuclei are plentiful and relatively easy to obtain and nuclear fusion produces much less radioactive waste than nuclear fission. Moreover, any radioactivity that is produced dies away relatively rapidly. Fusion, however, presents a major technical challenge. All nuclei are positively charged. If we want to bring them sufficiently close that the nuclear force binds them together, we first have to overcome their mutual Coulomb repulsion. And this repulsion is considerable. One way to achieve this is to heat the nuclei to an extremely high temperature. Then the nuclei have a large amount of thermal energy and they may approach each other to a close enough proximity that the nuclear force takes over and they fuse together. Because thermal energy is used to overcome the Coulomb repulsion, this process is called *thermonuclear* fusion. As we will see this entails heating the nuclei to a temperature of $\sim 10^8$ K. At such a high temperature, light atoms are fully ionised and the nuclear fuel is in the form of a *plasma* consisting of electrons and positively charged ions.

Another major challenge is to confine the plasma long enough and at sufficiently high density that the fusion energy produced exceeds any loss of energy to the surroundings. Ordinary materials, like stainless steel, cannot be used to confine the fusion plasma as they cannot stand the extremely high temperatures involved. Instead, two main alternative approaches have been taken. One approach uses magnetic fields to confine the plasma and is called *magnetic confinement*. In the other approach, a tiny pellet of nuclear fuel is bombarded on all sides by laser pulses of high energy, $\sim 10^4$ J, and short duration, $\sim 10^{-8}$ s. The fuel particles are confined to the pellet during the exceedingly short time of the fusion reaction because of their inertia. Hence this approach is called *inertial confinement*.

In this section we discuss the physical process of fusion and how it may be possible to harness fusion energy in a reactor using magnetic or inertial confinement. The technical challenges involved are so great that the achievement of obtaining usable power from fusion has not yet been attained. Indeed, it is expected that it will be several decades before a commercial fusion reactor becomes available. Despite the technical challenges, however, the huge potential and advantages of nuclear fusion as an energy source ensure that it is a field of vigorous research and development and the goal of fusion power is being actively pursued in a number of countries around the world.

3.5.1 Fusion reactions

As already emphasised, the Coulomb repulsion between nuclei inhibits them from fusing together. More generally, this is a very desirable feature, as otherwise nuclides with $A < \sim 56$, corresponding to the maximum in the binding energy curve, would not exist! However, for the purpose of producing energy from nuclear fusion we need to minimise this Coulomb repulsion. As the strength of this repulsion depends on the product of nuclear charges, this strongly suggests the choice of isotopes of hydrogen, with $Z = 1$. In addition, the greater the binding energy of the product nucleus, the greater will be the energy release. We recall that ^{4}He nuclei are particularly strongly bound, giving rise to a peak at $A = 4$ in the binding energy curve of the nuclides (see Figure 2.12). Hence, any fusion reaction that produces ^{4}He will release a particularly large amount of energy.

The available isotopes, with $Z = 1$, are hydrogen (^{1}H), deuterium (^{2}D) and tritium (^{3}T). The most elementary fusion reaction ^{1}H $+ ^1$H $\rightarrow ^2$He does not occur because the nuclear force between two protons is not strong enough to produce a ^{2}He bound state. (There is an alternative process involving ^{1}H that leads to ^{2}D and which occurs in thermonuclear fusion in the Sun, as we will describe in Section 4.1.2. However, this reaction proceeds at too slow a rate under conditions achievable here on Earth to provide a useful source of energy.). Of the possible combinations of the hydrogen isotopes, the reaction

$$^2\text{D} + {}^3\text{T} \rightarrow {}^4\text{He} + \text{n} + 17.6 \text{ MeV} \tag{3.31}$$

appears to be the most favourable. This is because it has a relatively large fusion cross-section at the working temperature that is envisaged for practical reactors. It fuses deuterium and tritium and is called the D-T reaction; it is used in most experimental fusion reactors.

Deuterium is a naturally occurring isotope. Its isotopic abundance of 0.015% is not high, but nevertheless every litre of seawater contains about 30 mg of deuterium. So it is a widely available, harmless and virtually inexhaustible resource. Tritium by contrast is radioactive with a half-life of 12.3 years and so cannot be obtained from natural sources. However, tritium can be produced by neutron bombardment of the two isotopes of natural lithium:

$$^6\mathrm{Li} + \mathrm{n} \rightarrow {}^3\mathrm{T} + {}^4\mathrm{He}, \tag{3.32}$$

$$^7\mathrm{Li} + \mathrm{n} \rightarrow {}^3\mathrm{T} + {}^4\mathrm{He} + \mathrm{n}. \tag{3.33}$$

As neutrons are produced in the D-T reaction, Reaction (3.31), the tritium fuel can be produced, or *bred* by the fusion reactor itself. This is done by surrounding the reactor with a blanket of lithium that absorbs neutrons produced in the fusion reactions. Lithium is plentiful in the Earth's crust and known reserves of lithium would be sufficient for use in fusion reactors for many hundreds of years.

3.5.2   Energy in fusion

Calculation of the energy release in a fusion reaction is quite straightforward. Under normal conditions, the kinetic energies, $\sim$keV, of the reacting nuclei are negligible compared with the Q-values of the fusion reactions, which are $\sim$MeV. Hence, the energy release, which is shared between the reaction products, is essentially equal to the Q-value of the reaction. If the fusion reaction produces particles a and b, the sum of their kinetic energies is given by

$$\frac{1}{2} m_\mathrm{a} v_\mathrm{a}^2 + \frac{1}{2} m_\mathrm{b} v_\mathrm{b}^2 = Q. \tag{3.34}$$

Again, neglecting the initial kinetic energies, the final momenta of the reaction products are equal and opposite:

$$m_\mathrm{a} v_\mathrm{a} = m_\mathrm{b} v_\mathrm{b}. \tag{3.35}$$

The kinetic energies of the two products are then given by:

$$\frac{1}{2} m_\mathrm{a} v_\mathrm{a}^2 = \frac{Q}{1 + m_\mathrm{a}/m_b}; \quad \frac{1}{2} m_\mathrm{b} v_\mathrm{b}^2 = \frac{Q}{1 + m_\mathrm{b}/m_\mathrm{a}}. \tag{3.36}$$

One consequence of this energy sharing is that the lighter reaction product takes away the larger share of the available energy. The ratio of kinetic energies is, from Equation (3.36),

$$\frac{\frac{1}{2} m_\mathrm{a} v_\mathrm{a}^2}{\frac{1}{2} m_\mathrm{b} v_\mathrm{b}^2} = \frac{m_\mathrm{b}}{m_\mathrm{a}}. \tag{3.37}$$

Worked example

Determine the Q-value for the D-T reaction and the sharing of energy between the reaction products. The atomic masses of deuterium, tritium and helium are 2.014 102, 3.016 049 and 4.002 603 u, respectively, and the neutron mass is 1.008 665 u.

Solution

As usual, we use atomic masses to take account of the masses of the atomic electrons. Then the mass defect is

$$(4.002\,603 + 1.008\,665)\,\text{u} - (2.014\,102 + 3.016\,049)\,\text{u} = 0.018\,883\,\text{u},$$

which is equivalent to an energy of

$$0.018\,883 \times 931.5\,\text{MeV} = 17.59\,\text{MeV}.$$

Taking the masses of the α particle (^{4}He) and the neutron to be 4 and 1 u, respectively, and using Equation (3.36) we find:

$$\text{kinetic energy of } \alpha \text{ particle} = \frac{17.59}{1 + 4/1}\,\text{MeV} = 3.5\,\text{MeV},$$

$$\text{kinetic energy of neutron} = \frac{17.59}{1 + 1/4}\,\text{MeV} = 14.1\,\text{MeV}.$$

The products of the D-T reaction share 17.6 MeV of energy. Although this is less than the energy released in a fission reaction, ~200 MeV, it is a greater amount of energy per nucleon (17.6/5 ~ 3.5 MeV per nucleon) than it is for fission, (~1 MeV per nucleon). The emitted neutron receives 80% of the released energy, while the α particle receives just 20%. The α particles have a positive charge $+2e$ and can be confined to the fusion plasma by a magnetic field. They transfer their energy to plasma particles in collisions with them, and this helps to maintain the operating temperature of the plasma. On the other hand, the uncharged neutrons are unaffected by the magnetic field and escape from the plasma. (This situation is quite different from that in a fission reactor, where relatively little of the energy release is given to the emitted neutrons and where the kinetic energy of the fission fragments is readily deposited in the reactor core.) In a fusion reactor, the neutrons can be captured by a blanket of lithium that surrounds the reaction zone – the same blanket that is used to breed the tritium fuel. The capture reactions are ^{6}Li + n → ^{4}He + ^{3}T and ^{7}Li + n → ^{4}He + ^{3}T + n. The energetic ^{4}He and ^{3}T nuclei then deposit their energy in the lithium blanket as thermal energy, which can be used to drive a steam turbine. Thermal energy can then be used to produce steam to drive a turbine.

3.5.3   Coulomb barrier for nuclear fusion

The nuclear force is attractive, while the Coulomb force is repulsive. The combination of these two forces results in a Coulomb barrier that must be surmounted if two nuclei are to fuse together. We encountered such a barrier in our discussion of α particle decay in Section 2.3.3; in that case, the barrier confined the α particle to the nucleus. We can deduce the height of the Coulomb barrier for fusion from the potential energy V_{Coul} due to the electrostatic repulsion of the two nuclei when they just touch at their surfaces. If the nuclei have radii R_a and R_b, respectively, then

$$V_{Coul} = \frac{1}{4\pi\varepsilon_0} \frac{Z_a Z_b}{(R_a + R_b)}, \qquad (3.38)$$

where R_a and R_b are given by $R = 1.2A^{1/3}$ (Equation 2.6). A favourable characteristic of the isotopes ^{2}D and ^{3}T, with respect to minimising the Coulomb barrier, is that their extra neutrons increase the value of $(R_a + R_b)$. For the case of the D-T reaction, Equation (3.38) gives a value for V_{Coul} of 0.4 MeV. In order for the two interacting nuclei to surmount the Coulomb barrier they must have a total kinetic energy of 0.4 MeV. Hence, for the case of a head-on collision, each must have a kinetic energy of least 0.2 MeV. Such energies can be obtained in a particle accelerator. However, the energy required to run the accelerator would be much larger than the energy released by the fusion reaction! Instead, the fusion fuel is heated to an extremely high temperature so that fusion reactions occur as a result of random thermal collisions.

According to the *equipartition of energy theorem*, the mean kinetic energy of a particle in thermal equilibrium in a gas at temperature T is equal to $\frac{3}{2}kT$, where k is the Boltzmann constant (see Section 4.6.2). If the mean kinetic energy of the nuclei needs to be $\sim$0.2 MeV, this implies a plasma temperature of $\sim 2 \times 10^8$ K. This is more than an order of magnitude *greater* than the temperature at the centre of the Sun. Fortunately, the temperature need not to be as high as this in practice for two reasons: (i) even though the nuclei may not have energies greater than the Coulomb barrier, they may still *tunnel* through this barrier; and (ii) the particles in the plasma have a distribution of kinetic energies and some will have an energy much greater than $\frac{3}{2}kT$. Consequently, fusion can occur at plasma temperatures below 2×10^8 K.

3.5.4   Fusion reaction rates

We saw in Section 3.2.3 that a nuclear reaction rate R_b is given by

$$R_b = \sigma N \Phi, \qquad (3.4)$$

where N is the number of target nuclei, σ is the reaction cross-section and Φ is the flux of incident particles. We consider the case of a plasma of tritium and deuterium nuclei with number densities, n_T and n_D, respectively and a fusion cross-section σ_f. For the sake of simplicity, we assume that the tritium nuclei are fixed in space and that we have an incident flux of deuterium nuclei with speed v. Then the flux of deuterium nuclei is $n_D \times v$ and so the fusion rate R_f is

$$R_f = N_T \sigma_f n_D v, \qquad (3.39)$$

and the fusion rate per unit volume is

$$R_f = n_T n_D \sigma_f v. \qquad (3.40)$$

We see that the fusion rate depends on the product of the tritium and deuterium densities, the fusion cross-section σ_f and the relative speed v of the nuclei. In reality, tritium and deuterium nuclei have a distribution of speeds, as we have previously noted. Consequently, we must use a value for $\sigma_f v$ that is averaged over all relative speeds, which we denote as $\langle \sigma_f v \rangle$. The reaction rate is then

$$R_f = n_T n_D \langle \sigma_f v \rangle. \qquad (3.41)$$

Figure 3.18 shows a plot, as the blue curve, of $\langle \sigma_f v \rangle$ against plasma temperature for the D-T reaction. (It is usual practice to give the temperature of a plasma in terms of kT, where T is the plasma temperature and kT is the characteristic energy of the particles in the plasma. Thus the *plasma temperature* of 10 keV corresponds to an absolute temperature of 1.2×10^8 K.) We see that for the D-T reaction, $\langle \sigma_f v \rangle$ has the value of $\sim 10^{-22}$ m^3/s at the plasma temperature of 10 keV, which is envisaged for practical fusion reactors. Note also that the value of $\langle \sigma_f v \rangle$ for the D-T reaction at this temperature is at least two orders of magnitude larger than for alternative fusion reactions: the D-D reaction in which two ^{2}D nuclei are fused together and the D-^{3}He reaction in which a ^{2}D nucleus is fused with a ^{3}He nucleus; plots of $\langle \sigma_f v \rangle$ for these reactions are also shown in Figure 3.18.

Figure 3.18 The blue curve is a plot of $\langle \sigma_f v \rangle$ against plasma temperature for the D-T reaction, where $\langle \sigma_f v \rangle$ denotes the value for $\sigma_f v$ averaged over all relative speeds; σ_f is the fusion cross-section and v is the relative speed of the participating nuclei. Plots of $\langle \sigma_f v \rangle$ for alternative fusion reactions: D-D and D-^{3}He are also shown.

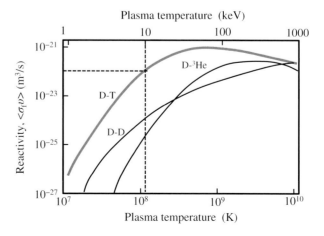

3.5.5   Performance criteria

There are a number of intermediate stages in the realisation of usable energy from a nuclear fusion reactor. They all relate to the energy that is released by the fusion reactions compared with the energy that must be supplied to heat the plasma and replace any energy losses that it may suffer. Clearly, we want to get more energy out of the fusion reactions than we put in.

The hot fusion plasma consists of charged particles: electrons and ions. These particles collide with each other and, in doing so, undergo accelerations. When charged particles accelerate (or decelerate) they emit electromagnetic radiation. Hence, the plasma particles emit radiation and this carries energy away from the plasma. This radiation is called *Bremsstrahlung*, which means 'breaking radiation' in German. It also occurs, for example, when high-energy electrons strike a copper plate and decelerate sharply, producing radiation in the form of X-rays. The largest accelerations are suffered by the lightest particles in the plasma, which are the electrons. However, the electrons are in thermal equilibrium with the ions in the plasma and consequently any energy loss suffered by the electrons is also suffered by the ions. The ions, i.e. bare nuclei, are then less able to surmount the Coulomb barrier and fuse together. The bremsstrahlung power loss per unit volume in a fusion plasma at temperature T is proportional to $\sqrt{T}$, and to Z^2, where Z is the charge of the ions. With their low value of Z, this Z^2 dependence further favours the isotopes of hydrogen as fusion fuel. Bremsstrahling power loss exceeds the power produced by fusion below a certain plasma temperature and so the plasma must be maintained at a higher temperature than this. For the D-T reaction, this plasma temperature is ~4 keV. In practice, however, a fusion reactor is operated at a plasma temperature of ~10 keV, at which temperature the energy produced by the fusion reactions far exceeds the bremsstrahling losses.

In the case of the D-T reaction, the fusion products are energetic α particles and neutrons. The positively charged α particles are confined to the fusion plasma by a magnetic field and their energy contributes to the heating of the plasma. However, the uncharged neutrons escape from the plasma and do not contribute to the heating; their energy is lost to the plasma. To maintain the plasma temperature, the lost energy must be replaced by external auxiliary heating. The so-called *break-even* point occurs when the *total* fusion power produced by the plasma is equal to the external heating power that is needed to maintain the plasma temperature. The break-even point does not take into account the energy that is lost from the plasma by the escaping neutrons. So even when the break-even point is reached, external energy must still be applied to replace this lost energy and maintain the plasma temperature. A more advanced stage, called the *ignition point*, occurs at higher plasma temperature and ion densities than break-even. At the ignition point, the fusion reactions occur so frequently that the energy deposited by the α particles in the plasma is sufficient to compensate for all energy losses. External heating of the plasma is then no longer necessary and the reactor becomes *self-sustaining*. A preliminary stage on the way to break-even and ignition is to be able to confine the hot fusion plasma long enough and at sufficiently high density that the fusion energy produced exceeds the energy required to create the plasma. This requirement is known as the *Lawson criterion*.

The Lawson criterion

The British engineer and physicist John D. Lawson considered the conditions necessary for the production of fusion energy, and in a paper published in 1957 he presented the Lawson criterion. It gives the value of the product of *confinement time* τ and plasma density n that is necessary for the released fusion energy to be greater than the thermal energy that was used to create the plasma.

The Lawson criterion for the case of the D-T reaction can be derived as follows. At temperature T, a plasma of tritium ions, deuterium ions and electrons has a total mean energy per unit volume of

$$\frac{3}{2}kT(n_T + n_D + n_e),\qquad(3.42)$$

where n_T, n_D and n_e are the number densities of the tritium ions, deuterium ions and electrons, respectively. Assuming $n_T = n_D = n/2$, where n is the ion density, $n_e = n$. Then the total energy per unit volume is $3kTn$. This is the amount of energy we must put into the fuel to heat it to temperature T. If the plasma lasts for time τ and each fusion reaction releases energy Q, then the total energy released is

$$\text{reaction rate} \times \tau \times Q,$$

which is, using Equation (3.41),

$$n_T n_D \langle \sigma_f v \rangle Q\tau = \frac{1}{4}n^2 \langle \sigma_f v \rangle Q\tau.\qquad(3.43)$$

For the fusion energy to be greater than the thermal energy supplied,

$$\frac{1}{4}n^2 \langle \sigma_f v \rangle Q\tau > 3kTn,$$

or

$$n\tau > \frac{12kT}{\langle \sigma_f v \rangle Q}.\qquad(3.44)$$

This is the Lawson criterion for the product $n\tau$ (plasma density $\times$ plasma confinement time). We see that there is the choice of confining a large number of participating nuclei for a short time or fewer nuclei for a longer time. Beyond this criterion it is still necessary for the temperature of the plasma to be high enough for the nuclei to come together and fuse. At an operating temperature of 10 keV for the D-T reaction, with $\langle \sigma_f v \rangle \sim 10^{-22}\,\mathrm{m}^3/\mathrm{s}$,

$$n\tau > 10^{20}\,\mathrm{s}/\mathrm{m}^3.$$

The ratio of the power produced by a nuclear fusion reactor to the input power required to maintain the plasma in its steady state is called the *fusion energy gain factor*. It is usually denoted by the symbol Q (not to be confused with the Q-value of a nuclear reaction). Q = 1 is the break-even condition. For ignition, the Q-factor must be greater than about 5, to account for energy losses from the plasma. To obtain usable energy from a commercial reactor, the value of the Q-factor has to be at least ~ 20.

3.5.6   Controlled thermonuclear fusion

A fusion plasma is at an extremely high temperature, $\sim 10^8$ K, and in a practical fusion reactor it must be confined in some way. As already mentioned, there have been two main approaches to this. One approach is to confine the plasma within a magnetic field so that it is kept away from the walls of any containing vessel and so is thermally insulated from the walls. The alternative approach is to bombard a pellet of fuel with a pulse of high energy and exploit the inertia of the fuel particles to confine them for a sufficiently long time.

Magnetic confinement fusion

A charged particle moving at velocity $\boldsymbol{v}$ in a uniform magnetic field $\boldsymbol{B}$ experiences a Lorentz force $\boldsymbol{F} = q\boldsymbol{v} \times \boldsymbol{B}$, where q is the charge of the particle. If $\boldsymbol{v}$ is perpendicular to $\boldsymbol{B}$, the path of the charged particle is a closed circle. When $\boldsymbol{v}$ is not perpendicular to $\boldsymbol{B}$, the charged particle follows a helical path along the direction of the field lines, as shown in Figure 3.19. The frequency at which a particle spirals

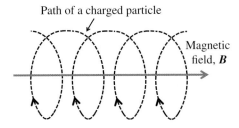

Figure 3.19 The helical path of a charged particle in a magnetic field $\boldsymbol{B}$.

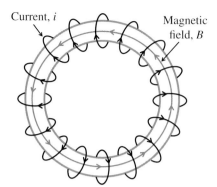

Figure 3.20 A toroidal magnetic field, in which the field lines are joined. This field is obtained with a coil that is wound in the form of a doughnut.

around the magnetic field lines is called its *cyclotron frequency*. In a fusion reactor we want to trap the charged particles in the magnetic field so that they remain in the plasma. We can do this by joining the magnetic field lines together. Such a closed magnetic field is obtained using a *toroidal solenoid*, in which the current-carrying coil is wound in the form of a doughnut, as illustrated in Figure 3.20. In principle the charged particles would spiral endlessly around the field lines. However, a simple toroidal field is non-uniform and becomes weaker with increasing radial distance from the median axis of the toroid. The consequence of this is that charged particles are steadily lost from the plasma. To prevent this loss, a second magnetic field, called a *poloidal* field, is added. Such a poloidal field is shown in Figure 3.21 and is generated by passing a current around the median axis of the toroid. In practice this can be achieved by passing a current through the plasma itself. This is the method used in a reactor design called a TOKAMAK, which is a Russian acronym for 'torroidal magnetic chamber'. This design was conceived in Russia and has been adopted in most devices using magnetic field confinement. The current is induced in the plasma using a transformer arrangement, where the

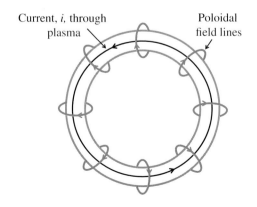

Figure 3.21 A poloidal field, which is generated by passing a current through the fusion plasma itself.

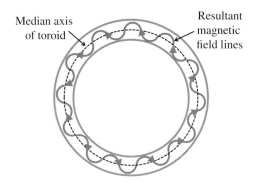

Median axis
of toroid

Resultant
magnetic
field lines

Figure 3.22 The addition
of the poloidal field gives a
twist to the toroidal field and
the resultant magnetic field
lines trace out helical paths.

plasma acts as the secondary coil. The addition of the poloidal field
gives a twist to the toroidal field and the resultant magnetic field
lines trace out helical paths as shown in Figure 3.22. The charged
particles spiral around these magnetic field lines. And the action of
these twisted toroidal field lines is to eliminate the tendency of the
charged particles to drift away from the axis of the toroid.

Plasma heating   The plasma in the reaction zone of a TOKAMAK
can be heated in several ways. First, the magnetic fields that control the plasma produce an electrical current i that travels through
the plasma. As the electrons and ions move through the plasma,
they collide with each other. These collisions create a 'resistance' R,
which results in an 'ohmic' heating effect, which transfers energy to
the plasma. As the temperature of the plasma rises, this resistance
and therefore the power input $(i^2 R)$ decreases and this sets a limit
to this heating effect. In order to obtain still higher temperatures,
thermal energy from outside the TOKAMAK must be supplied. This
can be accomplished by *neutral beam injection* or by the application
of *high-frequency electromagnetic waves*. Neutral beam injection consists of shooting high-energy particles into the plasma. To do this,
positively charged H^+ or D^+ ions are accelerated to high energy,
$\sim$100 keV, outside the TOKAMAK. These accelerated ions then pass
through an *ion beam neutraliser* where they capture electrons and
become neutral atoms, although still with a large amount of kinetic
energy. These energetic neutral atoms pass into the plasma, unaffected by the magnetic fields of the TOKAMAK. They collide with
the plasma particles, transferring their energy to them. In the other
heating method, a powerful radiofrequency source bathes the plasma
in electromagnetic waves. The frequency of the radiation is tuned to
the cyclotron frequency of either the electrons or the positive ions.
Under this resonant condition, the absorption of the radiation is
maximised and the electromagnetic waves heat the plasma particles,
just as microwaves heat food in a microwave oven.

Production of tritium fuel   Tritium is produced by neutron bombardment of the two isotopes of natural lithium (see Reactions 3.33 and 3.34). Neutrons are produced in the D-T reaction, so that the tritium fuel can be bred by the fusion reactor itself. This is done by surrounding the reactor with a lithium blanket to absorb the emitted neutrons, as we have already described. The tritium gas can then be removed from the blanket and recycled into the plasma as fuel. Moreover, the lithium can be in liquid form and used to transfer the thermal energy from the reactor to a steam generator.

A practical source of fusion energy from magnetic confinement has not been realised yet, but there have been experimental reactors that pave the way to a commercial reactor. A particular example, which uses a TOKAMAK, is the Joint European Torus (JET) at Culham in the UK. JET is essentially a research facility that is used to study various aspects of fusion reactors and, in particular, the behaviour of the fusion plasma. It is anticipated that the fusion conditions obtained in JET can be *scaled up* to those required in a commercial fusion reactor. In 1997 JET produced 16 MW of fusion power from a total input power of 24 MW, a world record and an energy gain factor Q of ~0.7. Figure 3.23 is a split image showing an interior view of the JET vacuum vessel, with a superimposed image of an actual plasma taken with a visible light camera. Only the cold edges of the plasma can be seen, as the centre is so hot that it radiates only in the ultraviolet part of the spectrum. Following in the footsteps of JET, a much larger machine, called ITER (Latin for 'the way'), is being built. ITER is an international project designed to prove the scientific and technological feasibility of a full-scale fusion power reactor. The heart of ITER is a superconducting TOKAMAK with

Figure 3.23 A split image showing an interior view of the Joint European Torus vacuum vessel, with a superimposed image of an actual plasma taken with a visible light camera. Only the cold edges of the plasma can be seen, as the centre is so hot that it radiates only in the ultraviolet part of the spectrum. Courtesy of EUROfusion. https://www.euro-fusion.org/2011/08/the-virtual-vessel-5/?view=gallery-428

design similarities to JET but with twice the linear dimensions. It is designed to deliver 10 times more power than the amount required to heat the plasma: an output power of approximately 500 MW, sustained for more than 500 s. ITER is based in Cadarache, France, and it is planned to have a plasma in the reactor by 2025.

Inertial confinement fusion

In this approach to nuclear fusion, a tiny pellet containing a mix of solid tritium and deuterium is struck from all sides by a short pulse of intense laser radiation, as illustrated in Figure 3.24. This both heats the pellet and compresses it to high density. The aim is to achieve densities and temperatures that are high enough for fusion to occur before the pellet blows apart. Computer simulations indicate that the pellet could be compressed to $\sim 10^3$ times its normal density and heated to a temperature of $\sim 10^8$ K. Laser radiation appears to be the best form of excitation, but pulses of high-energy charged particles can also been used.

The sequence of events in inertial confinement is as follows. The fuel pellet is struck by a large number of intense laser beams from many directions. This ensures that it is illuminated uniformly on all sides, because otherwise the compression of the target is uneven and fusion does not occur. The energy is delivered at such a rate that material from the heated pellet is violently ejected from its surface. This drives a compression shock wave back into the remaining core of the pellet. The pellet is accurately spherical to ensure that the inward shock wave is also highly symmetrical in shape. The shock wave compresses the core of the pellet and raises its temperature to the point where fusion occurs. The α particles from the fusion reactions rapidly lose their energy in collisions with particles in the dense plasma and this contributes additional heating. Finally the pellet blows apart and the fusion reactions cease. The pellet would be replaced with each excitation pulse so that there would be essentially a series of carefully controlled micro-thermonuclear explosions. It is envisaged that ~ 10 pellets/s would be used in cycles of fuel injection, compression, ignition and power output. The major technical challenges of inertial confinement are generating and focusing sufficient laser power to trigger fusion and understanding how the pellets behave when the excitation pulse strikes them. In addition to experimental studies, extensive use is made of computer programs to simulate the behaviour of the pellet during the various stages of the fusion process.

We can obtain an order of magnitude estimate for the time in inertial confinement by considering the size of the pellet and the speed of the particles in the plasma. The speed of mechanical waves

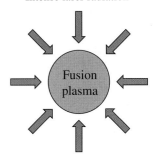

Intense laser radiation

Figure 3.24 The principle of inertial confinement fusion is to bombard a pellet of fusion fuel from all sides by a short pulse of intense laser radiation. The aim is to compress the pellet to $\sim 10^3$ times its normal density and heat it to a temperature of $\sim 10^8$ K so that thermonuclear fusion occurs.

in a gas is of the same order as the mean speed of the molecules in the gas. For example, at room temperature the mean speed of nitrogen molecules is ~ 300 m/s, while the speed of sound is 330 m/s. Treating the plasma as a gas at a temperature of ~ 10 keV, the particle mean speed is $\sim 10^6$ m/s. Using this value as the speed of the shock wave in a pellet of radius ~ 0.1 mm, the time it takes for the shock wave to travel through the pellet, i.e. the confinement time $\tau \sim$ radius/speed, $\sim 10^{-10}$ s. Applying the Lawson criterion for the D-T reaction with a value of 10^{20} s/m^3 for $n\tau$, we obtain a value of $n \sim 10^{30}$/m^3. This is about two orders of magnitude greater than the density of normal liquids or solids and illustrates the extreme conditions of matter that will need to be achieved for inertial confinement. Moreover, this is a simplified picture of the process and there are substantial losses of laser energy that we have not taken into account. So to achieve the condition of ignition, the density of the fuel pellet will need to increase by perhaps three orders of magnitude.

Worked example

Estimate the amount of energy that is required to heat 10 D-T pellets each of mass 1.0 mg to a temperature corresponding to $kT \sim 10$ keV. Assume that the heating process is 10% efficient. How much fusion power would be produced if 10 D-T pellets were consumed per second? Assume that 30% of the nuclei undergo fusion reactions.

Solution

Assume a mean atomic mass of 2.5 u. Then the number of nuclei in each pellet is

$$\approx \frac{1.0 \times 10^{-6}}{2.5 \times 1.66 \times 10^{-27}} \approx 2.4 \times 10^{20}.$$

The total thermal energy of these nuclei $\approx 2.4 \times 10^{20} \times 10^4 \times 1.6 \times 10^{-19}$ J $\approx 4 \times 10^5$ J. Assuming that the heating process is 10% efficient, the required input energy for 10 pellets $\approx 10 \times 4 \times 10^6$ J = 40 MJ.

The number of reactions per second $\approx 10 \times \frac{2.4 \times 10^{20}}{2} \times 0.3 = 3.6 \times 10^{20}$, which each release 17.6 MeV of fusion energy, giving a total energy output per second of

$$3.6 \times 10^{20} \times 17.6 \times 10^6 \times 1.6 \times 10^{-19} \text{J} \approx 1 \text{ GJ}.$$

This very simplified picture of inertial confinement gives an energy gain of ~ 25.

Figure 3.25 Part of the optics that amplify the energy of laser beams as they make their way towards the target chamber at the National Ignition Facility, USA. Damien Jemison, courtesy of Lawrence Livermore National Laboratory. https://lasers .llnl.gov/media/photo -gallery?id=2012-037864

An example of a research facility for laser-driven inertial confinement is the National Ignition Facility (NIF) at the Lawrence Livermore National Laboratory in the USA. At NIF there are 192 lasers that are focused onto a D-T pellet. The number of laser beams is so large because, as noted above, it is essential for the pellet to be struck uniformly on all sides. The total energy in the laser beams is ~5 MJ and is delivered to the pellet in a light pulse ~10^{-8} s long. It is envisaged that the pellet will be compressed to 100 times the density of lead and the resulting fusion reactions will release 10–100 times more energy than the amount deposited by the laser beams. Figure 3.25 shows part of the optics at NIF that amplify the energy of laser beams as they make their way towards the target chamber. In 2012, NIF achieved a 500 TW (5×10^{14} W) laser shot. This is 1000 times more power than the United States uses at any instant in time, but of course it last for only a tiny fraction of a second.

The feasibility of laser-driven fusion has yet to be demonstrated, but research is continuing at a vigorous pace.

Problems 3

3.1   (a) Calculate the Q-values for the following reactions:
   (i) $^{233}\text{U} + \text{n} \rightarrow {}^{234}\text{U}$,
   (ii) $^{12}\text{C} + \alpha \rightarrow {}^{16}\text{O} + \gamma$, where the γ ray has an energy of 7.115 MeV,
   (iii) $^{16}\text{O} + {}^{16}\text{O} \rightarrow 2\alpha + {}^{24}\text{Mg}$.
  (b) In the reaction

$$^{6}\text{Li} + \text{p} \rightarrow \alpha + {}^{3}\text{He} + Q,$$

it is found that Q has the value 4.018 50 MeV in excess of the proton's kinetic energy. Determine the mass of the helium-3 atom. The atomic masses are: ^{233}U, 233.039 628 u; ^{234}U, 234.040 947 u;

^{16}O, 15.994 915 u; ^{4}He, 4.002 603 u; ^{24}Mg, 23.985 042 u; ^{6}Li, 6.015 121 u; ^{1}H, 1.007 825 u; and the neutron, 1.008 665 u.

3.2 Estimate the mass of ^{235}U that is required per day to produce 1 GW of thermal power. Assume that each fission reaction releases 200 MeV of recoverable energy.

3.3 Use the semi-empirical mass formula to deduce the energy released by the thermal fission reaction:

$$^{235}_{92}\text{U}_{143} + \text{n} \rightarrow {}^{87}_{35}\text{Br}_{52} + {}^{147}_{57}\text{La}_{90} + 2\text{n}.$$

Note that some masses will cancel in this calculation.

3.4 A thin foil of cobalt of mass 40 mg is irradiated by a beam of neutrons with a flux of 2.5×10^{13} neutrons/s. Given that the capture cross-section for the reaction $^{58}\text{Co} + \text{n} \rightarrow {}^{59}\text{Co}$ is 1700 b, determine how many ^{59}Co nuclei are formed after 1 hour of irradiation.

3.5 Calculate the number of fissions per second for a 2 kg quantity of natural uranium and a neutron flux of 1.0×10^{13} cm/s, given that the thermal fission cross-section for ^{235}U is 548 b. Estimate the power produced. Natural uranium has an atomic mass of 238.029 u and contains 0.72% of ^{235}U.

3.6 An expression for the moderation of neutrons that averages over all scattering angles is

$$\frac{E'}{E} = \left[\frac{M^2 + m_{\text{n}}^2}{(M + m_{\text{n}})^2} \right]^N,$$

where the symbols have their usual meanings. Calculate the number of scattering events required to reduce the energy of fast neutrons with energy 1 MeV to the thermal energy of 1/40 eV, using as moderator: (i) protons, (ii) deuterons, (iii) ^{12}C; and (iv) ^{238}U.

3.7 (a) Calculate the mean free path λ_{a} and the mean time τ_{a} for the absorption of fast neutrons of energy 1 MeV in a sample of pure ^{235}U if the cross-section for absorption is 1.4 b at that energy. (b) Calculate the mean free path λ_{s} and the mean time τ_{s} for the scattering of fast neutrons of energy 1 MeV in a sample of pure ^{235}U if the scattering cross-section is 6.9 b at that energy. (c) Calculate λ_{a} and λ_{s} and the mean times τ_{a} and τ_{s} for thermal neutrons in pure ^{235}U. The absorption and scattering cross-sections for thermal neutrons in pure ^{235}U are 680 and 10 b, respectively. Compare your results from parts (a) and (b). The density of uranium is 18.95 g/cm^3.

3.8 In the fission of a ^{235}U nucleus, the number of prompt neutrons per original thermal neutron is given by

$$\eta = \nu \frac{\sigma_{\text{f}}}{\sigma_{\text{f}} + \sigma_{\text{rc}}},$$

where the symbols have their usual meanings and ν has the value of 2.42. (a) Show that if the fraction of ^{235}U in naturally occurring uranium is increased to 3%, the resulting value of η is 1.84. The relevant cross-sections at thermal energy are: $\sigma_{\text{f},235} = 584$ b, $\sigma_{\text{rc},235} = 96$ b, $\sigma_{\text{f},238} = 0$ b, $\sigma_{\text{rc},238} = 2.72$ b. (b) Why can enriched uranium fuel be used with a water moderator, but natural uranium fuel cannot?

3.9   The table below shows the atomic masses and fission activation energies for four different nuclides. Deduce which nuclides are fissionable by thermal neutrons. The atomic masses of $^{241}_{94}\mathrm{Pu}_{147}$, $^{239}_{92}\mathrm{U}_{147}$ and $^{243}_{95}\mathrm{Am}_{148}$ are 241.056 846, 239.054 290 and 243.061 375 u, respectively.

Nuclide	Atomic mass (u)	Activation energy (MeV)
$^{240}_{94}\mathrm{Pu}_{146}$	240.053 808	6.3
$^{239}_{94}\mathrm{Pu}_{145}$	239.052 158	6.0
$^{238}_{92}\mathrm{U}_{146}$	238.050 785	6.6
$^{242}_{95}\mathrm{Am}_{147}$	242.059 542	6.2

3.10   Consider the fission reaction

$$^{235}_{92}\mathrm{U}_{143} + \mathrm{n} \rightarrow {}^{93}_{37}\mathrm{Rb}_{56} + {}^{141}_{55}\mathrm{Cs}_{86} + 2\mathrm{n}.$$

Assume that immediately after the fission of the uranium nucleus the $^{93}\mathrm{Rb}$ and $^{141}\mathrm{Cs}$ nuclei are just touching at their surfaces. (a) Assuming the nuclei to be spherical, calculate the electrostatic potential energy associated with the repulsion between the two nuclei. (b) Compare this energy with the energy released in the fission reaction. The atomic masses of $^{235}\mathrm{U}$, $^{93}\mathrm{Rb}$, $^{141}\mathrm{Cs}$ and the neutron are 235.043 924, 92.922 039, 140.920 045 and 1.008 665 u, respectively.

3.11   Calculate the Q-value for the fission of $^{98}\mathrm{Mo}$ into two nuclei of equal mass. Why is the spontaneous fission of $^{98}\mathrm{Mo}$ extremely unlikely? The atomic masses of $^{98}\mathrm{Mo}$ and $^{49}\mathrm{Sc}$ are 97.905 407 and 48.950 02 u, respectively.

3.12   Determine the number of stages of gaseous diffusion that are required to increase the $^{235}\mathrm{UF}_6{:}^{238}\mathrm{UF}_6$ ratio from 0.72% to 5%.

3.13   The Maxwell–Boltzmann distribution for the speed, v, of particles of mass m in a gas in thermal equilibrium at temperature T is

$$P(v) = \left(\frac{m}{2\pi kT}\right)^{3/2} 4\pi v^2 \exp\left(-\frac{mv^2}{kT}\right),$$

where k is the Boltzmann constant. Make a sketch of the distribution as a function of v. Obtain an expression for the most probable speed and hence show that the kinetic energy of a particle that has the most probable speed is kT. Verify that the value of kT at room temperature is $1/40$ eV. What is the de Broglie wavelength of a thermal neutron that has a kinetic energy of $1/40$ eV? Where might such thermal neutrons find application?

3.14   (a) Determine the energy released in the following fusion reactions and the percentage of the energy that is taken up by the product neutron or proton:

   (i) $^2_1\mathrm{D}_2 + {}^2_1\mathrm{D}_2 \rightarrow {}^3_2\mathrm{He}_1 + \mathrm{n}$,

   (ii) $^2_1\mathrm{D}_2 + {}^2_1\mathrm{D}_2 \rightarrow {}^3_1\mathrm{T}_2 + {}^1_1\mathrm{H}_1$.

The atomic masses of ^{2}D, ^{3}He, ^{3}T, ^{1}H and the neutron are 2.014 102, 3.016 029, 3.016 049, 1.007 825 and 1.008 665 u, respectively.

(b) Determine the required confinement time for the fusion reaction $^2_1$D$_2$ + $^2_1$D$_2$ → $^3_1$T$_2$ + $^1_1$H$_1$, assuming values of $kT = 10$ keV, $\langle \sigma_f v \rangle = 5 \times 10^{-25}$ m^3/s and a typical value of ion density.

(c) The fusion reaction ^{2}D + ^{2}D → ^{2}He + γ releases the relatively large amount of energy of 23.9 MeV. What limits this reaction as a practical energy source?

3.15 (a) Two identical particle accelerators are used to fire two proton beams at each other. Through what voltage must the protons be accelerated so that two protons are brought to rest just 'touching' each other? (b) At what temperature must a gas of protons be maintained for a proton to have a mean energy equal to the energy of the accelerated protons in part (a)?

3.16 The so-called *p–p cycle* is a set of fusion reactions that produces the Sun's energy. The overall cycle can be represented by

$$4p \rightarrow \alpha + 2e^+ + 2v + 2\gamma.$$

Determine the amount of energy liberated in the cycle. The atomic masses of ^{1}H and ^{4}He are 1.007 825 and 4.002 603 u, respectively, and the mass of the electron is 0.000 549 u. The energy of each γ ray is 1.02 MeV.

3.17 A fuel pellet in a laser inertial confinement device contains equal numbers of deuterium and tritium atoms and has a density of 200 kg/m^3. The action of the laser pulses increases the density of the pellet by a factor of 1000. (a) What is the number density of the atoms in the compressed pellet? (b) For how long must this density be maintained to achieve breakdown according to the Lawson criterion.

3.18 Compare the energy produced by the fusion of 1 kg of deuterium with the energy produced by the fission of 1 kg of uranium-235. For how long would these amounts of energy supply a 100 W light bulb?

3.19 A deuteron ion moves in a circular orbit in a magnetic field of 3 T. Determine the radius of the orbit and the cyclotron frequency of the deuteron ion. In what region of the electromagnetic frequency does this cyclotron frequency lie?

4

Solar power

The Sun is our main source of energy by far. Even oil and coal are the result of *photosynthesis* in ancient trees and vegetation. On a clear, cloudless day the incident solar power at the Earth's surface may be up to ~1 kW/m^2, delivering a substantial amount of energy. Indeed, the amount of solar energy falling on the Earth in 1 hour is as much as the total energy consumption of the Earth's entire population in 1 year. Moreover, the Sun provides us with renewable energy and will continue to deliver this energy for the next 5 billion years or so. The challenge is to harvest solar energy as effectively and efficiently as possible.

Solar energy comes to us in the form of electromagnetic radiation, which we readily experience as heat on our skin on a sunny day. This radiation is the result of thermonuclear fusion in the core of the Sun where hydrogen nuclei fuse together to form helium nuclei. The energy from these fusion reactions migrates to the surface of the Sun where it appears as sunlight. The Sun's surface behaves like a *blackbody* at a temperature of about 6000 K, providing an almost continuous distribution of wavelengths from the ultraviolet through the visible to the infrared. In the first part of this chapter we describe the creation of solar radiation and its main characteristics. We describe the journey of the photons from the core of the Sun to their arrival at the surface of Earth and how sunlight interacts with the Earth and with its atmosphere. Some of the solar energy penetrates into the ground and this so-called shallow geothermal energy provides a useful energy source, as we will describe. We also take the opportunity to describe deep geothermal energy.

Solar energy can be harvested in a number of ways. It can be used directly to heat a fluid and the second part of this chapter describes solar water heaters and solar thermal power systems. Solar water heaters are a common sight on roofs in hot climates and their use

Physics of Energy Sources, First Edition. George C. King.
© 2018 John Wiley & Sons, Ltd. Published 2018 by John Wiley & Sons, Ltd.

is becoming more widespread. These heat water to moderately high temperatures, $\sim$60–80°C, for domestic purposes such as washing. By contrast, solar thermal power systems concentrate solar radiation by mirrors. In this way they heat a fluid such as molten salt to much higher temperatures, >400°C. Thermal energy from a source at high temperature can be converted into mechanical energy efficiently, e.g. to drive an electrical generator. The conversion of thermal energy into mechanical energy involves the use of a *heat engine* and this is described in the third part of this chapter.

Solar energy can also be harvested in photo-chemical or photo-physical processes. The harvesting of solar energy utilising a photo-chemical process is well exemplified by the process of photosynthesis in Nature, which converts sunlight into chemical energy in plants. Photo-physical processes are exploited in photovoltaic solar cells, which are also a common sight on rooftops these days. Solar cells are the subject of Chapter 5. The Sun also provides us with energy in more indirect ways. For example, solar heating leads to wind currents in the atmosphere, whose energy can be captured by wind turbines. Furthermore, the winds generate ocean waves, whose energy can also be harvested. Chapters 6 and 7 deal with wind power and wave power respectively.

4.1 Stellar fusion

4.1.1 Star formation and evolution

It is the force of gravity that drives the formation and evolution of stars. Gravity leads to the compression of gas clouds that exist in the space between the stars in a galaxy and this compression leads to conditions where thermonuclear fusion can occur. The number densities of the particles in interstellar clouds are extremely low compared with, say, the number densities of molecules in the Earth's atmosphere. However, it can happen that the density of an interstellar gas cloud becomes sufficiently high that gravity causes the cloud to contract. This might arise from a shock wave from a nearby supernova. Roughly speaking, if the gravitational energy of the cloud dominates over its *internal energy*, i.e. the kinetic energies of the particles in the cloud, then gravitational contraction can occur. There are various steps in the subsequent formation of a star, and the ultimate fate of the star that is formed depends upon its mass. Here we present a brief outline of stellar evolution.[1] As the cloud of interstellar gas contracts, its temperature increases as gravitational potential energy is converted into kinetic energy. Eventually the temperature of the gas

---

[1] For more details, see A.C. Phillips, *The Physics of Stars*, John Wiley, 1999.

may reach $\sim 10^7$ K, and at this temperature the first thermonuclear reactions take place. In Section 3.5.3 we saw that nuclear fusion is hindered by the Coulomb repulsion of the participating nuclei. Consequently, the first to fuse together are hydrogen nuclei, with $Z = 1$. This converts hydrogen into helium and is called *hydrogen burning*. This occurs in a sequence of reactions known as the *proton–proton chain*, as described in Section 4.1.2. The net result is the transformation of four protons, i.e. hydrogen nuclei, into a helium nucleus with an energy release of about 27 MeV.

The energy released by the nuclear fusion reactions increases the pressure in the gas cloud and at some point the gravitational attractive force is balanced by the gas pressure and equilibrium is reached. This brings a temporary halt to the contraction of the star. Our Sun is presently at this stage. Eventually the hydrogen fuel is used up and gravitational collapse can resume. If the star is sufficiently massive, the temperature of the interior of the star gas rises to $\sim 10^8$ K, which is sufficiently high that the Coulomb barrier between helium nuclei (with $Z = 2$) can be more easily overcome and *helium burning* occurs. However, helium burning is hindered by the fact that the $^8_4$Be nucleus is not stable. The process by which $^4_2$He nuclei do combine is via a set of reactions called the *triple-alpha process*. The net effect of this is to combine three $^4_2$He nuclei to form stable $^{12}_6$C, via a sufficient population of unstable $^8_4$Be nuclei. The process is particularly fascinating because it depends on the existence of an otherwise unremarkable excited state of the carbon nucleus.

Triple-alpha process

The $^8_4$Be nucleus is unstable because its mass is greater than twice the mass of a $^4_2$He nucleus. The mass difference is 9.9×10^{-5} u $\equiv 92$ keV. So if a $^8_4$Be nucleus is formed, it rapidly decays into two $^4_2$He nuclei with a mean lifetime of 2.6×10^{-16} s. It *can* be formed if two $^4_2$He nuclei approach each other with a relative energy of 92 keV. In fact, the probability of interaction of the two $^4_2$He nuclei is greatly enhanced if they come together with exactly this amount of energy, i.e. there is a resonance at 92 keV in the cross-section for the reaction

$$^4_2\text{He} + {}^4_2\text{He} \Leftrightarrow {}^8_4\text{Be} \qquad (4.1)$$

(see Section 3.2.3). The symbol $\Leftrightarrow$ indicates that this reaction can occur in both directions. In the interior of a star at $\sim 10^8$ K, the $^4_2$He nuclei have a distribution of thermal energies and some of them have the required energy for Reaction (4.1) to occur. So we have the situation where $^8_4$Be nuclei are being continually created and destroyed, and at equilibrium there will be a finite population of unstable $^8_4$Be nuclei.

Usually, when a (short-lived) $^{8}_{4}$Be nucleus formed by Reaction (4.1) collides with another $^{4}_{2}$He nucleus, they do not combine, i.e. the reaction

$$^{4}_{2}\text{He} + ^{8}_{4}\text{Be} \rightarrow ^{12}_{6}\text{C} \qquad (4.2)$$

has a very low probability of occurring and certainly would not account for the amount of carbon that exists in the universe. Of course, carbon is very abundant and, indeed, 18% of our bodies comprise carbon. So there must be a way in which $^{4}_{2}$He and $^{8}_{4}$Be nuclei combine to produce $^{12}_{6}$C. The way they do so was first set out by Edwin Salpeter in 1952, and in 1954 Fred Hoyle pointed out how this sequence depended on the existence of a hitherto unknown excited state of carbon $^{12}_{6}$C at the excitation energy of 7.65 MeV. Shortly afterwards, William Fowler observed experimentally the state at exactly the predicted energy and received the Nobel Prize in Physics for this work. The sequence of steps is as follows. If they come together with exactly the right relative energy (288 keV), a $^{8}_{4}$Be nucleus can combine with a $^{4}_{2}$He nucleus to form a compound nuclear state (see Section 3.2.1). In fact, this compound nuclear state is just the excited state $^{12}_{6}$C* predicted by Hoyle:

$$^{4}_{2}\text{He} + ^{8}_{4}\text{Be} \Leftrightarrow ^{12}_{6}\text{C}^{*} \qquad (4.3)$$

The excited $^{12}_{6}$C* nucleus is very short-lived and almost all the excited $^{12}_{6}$C* nuclei decay back to a $^{8}_{4}$Be nucleus and a $^{4}_{2}$He nucleus in $\sim 10^{-16}$ s, with an energy release of 7.65 MeV. The reversibility of the reaction is again indicated by the symbol $\Leftrightarrow$. However, about 1 in 2500 of the excited $^{12}_{6}$C* nuclei take a different decay path and decay to the ground state of $^{12}_{6}$C in the following way:

$$^{12}_{6}\text{C}^{*} \rightarrow ^{12}_{6}\text{C} + \{2\gamma \text{ or } (\text{e}^{+} + \text{e}^{-})\}. \qquad (4.4)$$

Although the *branching ratio* for this decay path is only 0.04%, it is enough to account for the observed abundance of carbon in the universe.

The net result of the triple-alpha process is that three $^{4}_{2}$He nuclei combine to form a $^{12}_{6}$C nucleus. Once a $^{12}_{6}$C nucleus is produced, oxygen can be formed via the α-capture reaction

$$\alpha + ^{12}_{6}\text{C} \rightarrow ^{16}_{8}\text{O} + \gamma. \qquad (4.5)$$

Further α-capture reactions leading to $^{20}_{10}$Ne and $^{24}_{12}$Mg also occur, but with decreasing probability due to the increasing Coulomb repulsion.

Many stars come to the end of their lives when helium burning is complete. However, more massive stars, say several times the solar mass, may proceed through a further sequence of nuclear burning stages involving nuclei of increasing Z, which interrupt and delay gravitational contraction. This produces more massive elements. The heaviest nuclei that can be produced by fusion in this manner have $A \sim 56$, which corresponds to the maximum in the binding energy curve of the nuclides (see Figure 2.12). Elements with $A > 56$ are produced in neutron absorption reactions when there are a high number of neutrons. This can occur during the final stages of star evolution or in the explosion of a star in a supernova. Such an explosion throws out lots of matter, which consists of heavy elements as well as the light elements hydrogen and helium. This matter can then go on to produce the next generation of stars.

4.1.2   Thermonuclear fusion in the Sun: the proton–proton cycle

Thermonuclear fusion in the Sun occurs through a chain of nuclear fusion reactions called the proton–proton cycle. The dominant reactions, which occur 85% of the time, are

$$p + p \rightarrow {}^2_1D + e^+ + \nu \qquad (4.6)$$
$$p + {}^2_1D \rightarrow {}^3_2He + \gamma, \qquad (4.7)$$

and then

$$ {}^3_2He + {}^3_2He \rightarrow {}^4_2He + p + p. \qquad (4.8)$$

Each of these reactions is exothermic and the total amount of energy released is about 27 MeV per ${}^4_2He$ nucleus formed. The net effect is to transform four protons into a ${}^4_2He$ nucleus:

$$4p \rightarrow {}^4_2He + 2e^+ + 2\nu + 2\gamma. \qquad (4.9)$$

The direct process of the fusion of two protons in the reaction

$$p + p \rightarrow {}^2_2He \qquad (4.10)$$

does *not* occur because the nuclear force between two protons is not strong enough to produce a ${}^2_2He$ bound state.

The first step in the proton–proton cycle, Reaction (4.6), occurs via a *weak* nuclear interaction. The underlying mechanism is that one of the interacting protons undergoes inverse β decay, $p \rightarrow n + e^+ + \nu$, and the neutron that is produced becomes bound to another proton to form a deuteron, ${}^2_1D$. This requires the inverse β decay to occur during the time that the protons are interacting with each other. The probability of inverse β decay during this short period is extremely small and so Reaction (4.6) proceeds extremely slowly. It can be calculated that a proton in the core of the Sun has to wait $\sim 10^{10}$ years before it fuses with another proton. However, once a deuteron is produced, it reacts almost immediately with another proton, as there are lots of protons around; a deuteron in the centre of the Sun survives for about a second before it reacts with a proton. The ${}^3_2He$ nuclei formed by this proton–deuteron reaction can then react in several ways, but the dominant one is Reaction (4.8). A $_2^3$He nucleus cannot combine with a proton because the isotope $_3^4$Li is not a bound system. Since the first step, Reaction (4.6), is by far the slowest link in the chain, it governs the rate at which energy is released by the chain as a whole. Its astronomically long timescale sets the timescale for the hydrogen-burning stage of the Sun. Fortunately for us, the Sun is expected to burn hydrogen for a total of about 10 billion years, and we are about halfway through this period. After the hydrogen burning stops in about 5 billion years, the Sun will turn into a *red dwarf*, as it has insufficient mass to proceed to helium burning.

4.1.3 Solar radiation

The p–p fusion reactions occur in the Sun's core, which consists of a highly dense plasma of electrons and positive ions at a temperature of $\sim 10^7$ K. A plasma at this temperature radiates energy in the X-ray region of the electromagnetic spectrum. On the other hand, the Sun appears to us as a blackbody at a temperature of about 6000 K, emitting visible light. This transformation in wavelength occurs because the photons from the Sun's interior steadily lose energy as they migrate to the surface. Consequently, and fortunately for us, the X-rays are trapped to a large extent within the interior of the Sun.

The internal structure of the Sun is illustrated in Figure 4.1, which shows the *core*, the *radiative zone*, the *convection zone* and the *photosphere*. It is from the photosphere that the solar radiation is emitted. The temperature difference between the core and surface of the Sun sets up a temperature gradient and the radiant energy migrates towards the surface. The underlying mechanism for the transport of radiant energy in the core and radiative zone is *radiative diffusion*. This is a *random walk* process in which the photons

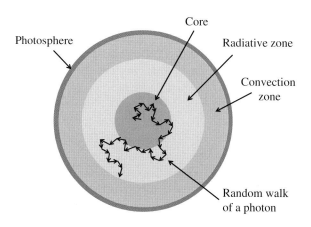

Figure 4.1 Schematic diagram of the internal structure of the Sun, showing the core, the radiative zone, the convection zone and the photosphere from where the solar radiation is emitted. Also illustrated is the random walk of a photon through the core and radiation zone. In a random walk, each successive step occurs in a random direction with respect to the previous step.

are continuously scattered, predominantly from electrons, as they make their way to the surface. In a random walk, each successive step occurs in a random direction with respect to the previous step as illustrated schematically in Figure 4.1. (Random walk processes turn up in many physical situations. A familiar example is the diffusion of molecules in a gas.) If a photon takes N steps, each of length l, then the vector displacement $\boldsymbol{D}$ from its starting point is given by

$$\boldsymbol{D} = \boldsymbol{l}_1 + \boldsymbol{l}_2 + \ldots + \boldsymbol{l}_N. \tag{4.11}$$

The square of the net distance travelled after N steps is

$$D^2 = l_1^2 + l_2^2 + \cdots\cdots + l_N^2 + 2(\boldsymbol{l}_1 \cdot \boldsymbol{l}_2 + \boldsymbol{l}_1 \cdot \boldsymbol{l}_3 + \cdots). \tag{4.12}$$

If we average over many steps, the terms involving scalar products cancel because the direction of each step is random. Hence,

$$D^2 = l_1^2 + l_2^2 + \cdots\cdots + l_N^2 = Nl^2, \tag{4.13}$$

and

$$D = \sqrt{N}l. \tag{4.14}$$

We see that in a random walk, the net distance travelled is equal to the step length times the square root of the number of steps.

To reach beyond the radiative zone, a photon must diffuse a distance that is comparable to the solar radius R_S and so the number of steps is $\sim R_S^2/l^2$. As the time for each step is l/c, where c is the speed of light, the random walk escape time is approximately R_S^2/cl. In practice there is a distribution of mean free paths for the photons, i.e. a distribution of l values. For the sake of simplicity we take l to have the fixed value of 1 mm, which is a reasonable estimate. Taking R_S to be 7.0×10^8 m, we find the time for a photon to diffuse out of the radiative zone is about 50 000 years. It is the long timescale of radiative diffusion that restricts the flow of radiation energy and prevents the Sun from losing heat catastrophically.

When the photons reach the convection zone, the radiation energy is transported via mass movement of plasma within the zone, where heated plasma ascends and cooler plasma descends. This process is much faster than radiative diffusion and so the time the photons take to reach the Sun's surface is dominated by the time they spend in the core and radiative zone. The above order of magnitude calculation shows that a photon born in the core of the Sun takes many thousands of years to reach the surface and escape as sunlight. By comparison, the neutrinos from the fusion reactions in the core interact extremely weakly with matter and escape from the Sun within a time $\sim R_S/c$. This amounts to just a few seconds.

4.2   Blackbody radiation

Any object that is above the absolute zero of temperature emits *thermal radiation*. It does so because the charged particles in the surface of the object have thermal energy and this motion causes the charged particles to emit electromagnetic radiation. Even the reader of this book is emitting radiation; it lies in the infrared region of the electromagnetic spectrum. We see a vivid example of thermal radiation when we heat an iron bar in a fire. When the heated bar is at a relatively low temperature, it certainly radiates heat but is not *visibly* hot. However, as its temperature increases, the bar first becomes dull red in colour and then later becomes bright red. Eventually at very high temperatures, it becomes a blue-white colour. This change in colour with temperature has practical applications. We can deduce the temperature of a star from its observed colour and we can take a *thermograph* of an object, which maps out the variation in temperature of its surface. Such thermographs can be used, for example, to detect cancerous tumours that have a higher temperature than surrounding tissue, or to see where heat is being lost from areas of a house that has insufficient thermal insulation.

If an object is in thermal equilibrium with its surroundings, it must emit as much radiant energy as it absorbs; otherwise its temperature would change. It follows that any object that is a good absorber of radiation is also a good emitter of radiation and vice versa (see also Section 4.2.2). A perfect absorber, which absorbs all radiation that falls upon it regardless of wavelength, is called a blackbody. In turn, a blackbody is a perfect emitter of thermal radiation; it emits the maximum possible amount of thermal radiation for a source at a given temperature. Blackbody radiation is relevant to our present discussion because the Sun is a blackbody radiator to a good approximation. However, blackbody radiation has much wider significance. It was the study of blackbody radiation that first led Max Planck to the concept of energy quantisation, as we will see in Section 4.2.3. More recently, the observation of blackbody radiation left over from the *big bang* is providing valuable information about the origin of the universe.

A good approximation to a blackbody can be constructed by making a tiny hole in the side of a cavity, as illustrated in Figure 4.2. Any radiation entering the hole is reflected many times and at each reflection its intensity reduces. If the area of the hole is very small compared with the area of the cavity walls, the radiation has little chance of being reflected out of the cavity. The *hole* acts as a nearly perfect absorber. If we were to look at the hole it would indeed appear to be black, just as the pupils of our eyes appear to be black. If we now heat the cavity, the walls will emit thermal radiation that fills

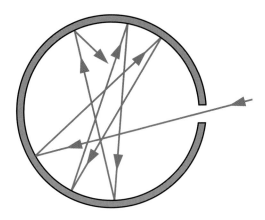

the cavity. A small fraction of this radiation will pass through the tiny hole. As the hole is acting as a blackbody, it must be emitting blackbody radiation. It follows that the radiation inside the cavity must also be blackbody radiation.

4.2.1 Laws of blackbody radiation

Towards the end of the 18th century the Austrian physicist Josef Stefan measured the rate at which energy is emitted by a blackbody. He found that it was proportional to the fourth power of the absolute temperature:

$$W = \sigma T^4, \qquad (4.15)$$

where W is the total radiated energy per second per unit area of the blackbody and σ is Stefan's constant, which has the value $5.6703 \times 10^{-8} \, \text{W/m}^2 \, \text{K}^4$. W is called the *radiant exitance* and has units of watts per unit area (W/m^2). A few years later Ludwig Boltzmann, also an Austrian physicist, used a combination of classical electromagnetic theory and thermodynamic arguments to derive this expression theoretically and Equation (4.15) is known as the *Stefan–Boltzmann law*.

As well as the total power emitted by a blackbody radiator, we are also interested in how that power is distributed between the emitted wavelengths. This distribution is called the *spectral power distribution* and is specified by the function $W_e(\lambda)$, which is called the *spectral radiant exitance*. $W_e(\lambda)$ is defined as the radiated power per unit area of the radiator per unit wavelength and has units $(\text{W/m}^2 \, \text{m})$. The radiated power of a blackbody is distributed over a *continuous* range of wavelengths. This is in contrast to the sharp lines in the emission spectrum of an excited atom, e.g. the familiar sight of the yellow, monochromatic light emitted by sodium street lamps. Figure 4.3 shows the general shape of the spectral power distribution of a blackbody radiator, where the spectral radiant exitance $W_e(\lambda)$ is

Figure 4.3 Distribution of the radiated power of a blackbody radiator with respect to wavelength λ. The distribution is specified by the spectral radiant exitance $W_e(\lambda)$, which has units of watts per unit area per unit wavelength (W/m^2 m). The radiated power due to the wavelength band of width $\delta\lambda$ centred at wavelength λ_0 is given by the elemental area $W_e(\lambda_0) \times \delta\lambda$. The total area under the curve is the total power radiated by the blackbody. The figure also defines λ_{max}, which is the wavelength at which the maximum in the spectral distribution occurs.

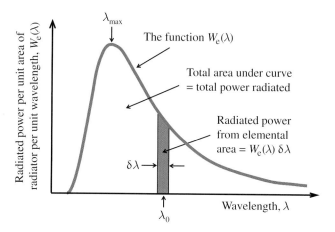

plotted against wavelength λ. As this distribution contains a continuum of wavelengths, we cannot speak in terms of a single wavelength, say 500.00 nm. Furthermore, any practical spectrometer that is used to measure the spectral distribution has a finite resolution $\delta\lambda$, so that it measures the power radiated over a finite range of wavelengths. We must therefore speak in terms of a *band* of wavelengths. Such a band is illustrated by the shaded slice of width $\delta\lambda$ centred at wavelength λ_0 in the spectral distribution in Figure 4.3. The radiated power due to this wavelength band is given by the elemental area $W_e(\lambda_0) \times \delta\lambda$, while the total power radiated by the blackbody is given by the total area under the curve. Figure 4.3 also defines λ_{max}, which is the wavelength at which the maximum in the spectral distribution occurs.

The spectral power distribution of blackbody radiation was first measured by Otto Lummer and Ernst Prinsheim in 1899 using an optical spectrometer. As the emitted radiation from their blackbody source extended into the infrared, they used optical components that transmitted this radiation as well as visible radiation, and detected the infrared radiation using an instrument called a *bolometer*. Spectral power distributions for blackbody radiators at a range of temperatures are shown in Figure 4.4. As can be seen, the wavelength λ_{max} at which the maximum of a distribution occurs decreases with increasing temperature. This is consistent with our observations of a heated iron bar. It was found empirically by Wilhelm Wien that λ_{max} and T are related by

$$\lambda_{max} T = \text{constant}. \tag{4.16}$$

This is called *Wien's displacement law* because λ_{max} is displaced as the temperature increases. The constant has the value $2.898 \times 10^{-3} \, \text{m K}$. So we can deduce the temperature of the Sun's surface simply by measuring the wavelength at which the maximum in the solar power distribution occurs.

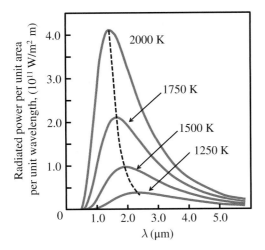

Figure 4.4 The spectral power distribution of blackbody radiation for various temperatures. The maxima of the distributions are displaced towards shorter wavelengths as the temperature increases.

Planck initially sought to find an empirical expression that fitted the experimental data for the spectral power distribution of a blackbody radiator and obtained the expression

$$W_e(\lambda)\mathrm{d}\lambda = \frac{2\pi hc^2}{\lambda^5(\mathrm{e}^{hc/\lambda kT} - 1)}\mathrm{d}\lambda, \qquad (4.17)$$

which did fit the data; h is Planck's constant and c is the speed of light. Subsequently, he used thermodynamic arguments to find a physical justification for his empirical expression. In doing so, Planck introduced the idea of energy quantisation, which has been described as the birth of quantum physics. It is fascinating that such an apparently mundane study as the investigation of the radiation from a cavity could lead to such a major revolution in physics.

The Stefan–Boltzmann law (Equation 4.15) and Wien's displacement law, Equation (4.16), are both contained in Planck's formula. To obtain Wien's law we take the derivative of Equation (4.17) and set it equal to zero to find the value of λ at which $W_e(\lambda)$ is a maximum. To obtain the Stefan–Boltzmann law we integrate Equation (4.17) over all wavelengths to find the total power radiated.

4.2.2 Emissivity

Electromagnetic radiation is characterised by the energy per unit time (power) per unit area received by a surface that is exposed to the radiation. Traditionally this has been called the *intensity* of the radiation but the term intensity has been replaced by *irradiance*; both have units of W/m^2. For example, the irradiance of solar radiation falling on a surface might be 500 W/m^2.

Suppose we have a blackbody A that is in thermal equilibrium with an enclosure at a temperature T. We envisage the enclosure to be filled with blackbody radiation with irradiance I. As the black-body is in equilibrium with the enclosure, it must emit a power I per unit area. Suppose we have in the same enclosure another body B, which is not a blackbody. The irradiance falling on B is again I. Suppose that B absorbs a fraction a of this radiation, where a is called the *absorption factor*. Again for equilibrium, the total power absorbed by B, which is aI, must be equal to the total power it emits, and so the total power it emits per unit area is also aI. The ratio of the emitted power of a given body to that of a blackbody at the same temperature is called the *emissivity e* of the body. Hence,

$$e = \frac{aI}{I} = a. \tag{4.18}$$

We see that the emissivity e of a body is equal to its absorption factor a. When we combine this result with the Stefan–Boltzmann law (Equation 4.15), we find that the total power W radiated per unit area by a body of emissivity e at temperature T is

$$W = e\sigma T^4. \tag{4.19}$$

A body that has a value of e that is less than unity but is independent of wavelength is called a *greybody*. Its emission spectrum will have the same *shape* as that of a blackbody at the same temperature, but its spectral power distribution $W_e(\lambda)$ will be reduced by the constant factor e. The hot tungsten filament in an incandescent light bulb is a good example of a greybody where the value of e is $\sim$0.3.

Most objects, however, are coloured, which means that they absorb some wavelengths better than others. Leaves are green because they absorb blue and red light much more strongly than they absorb green light; the green light is mostly reflected. We can repeat the preceding discussion but restrict it to a narrow band of wavelengths between λ and $\lambda + \delta\lambda$ to show that the emissivity at a particular wavelength is equal to the absorption factor at that wavelength, and define the *spectral emissivity e_λ* by

$$e_\lambda = \frac{\text{energy radiated by a body in range } \lambda \text{ to } \lambda + \delta\lambda}{\begin{array}{c}\text{energy radiated in same range by a}\\\text{blackbody at same temperature}\end{array}},$$

with

$$e_\lambda = a_\lambda, \tag{4.20}$$

where a_λ is the *spectral absorption factor*. Equation (4.20) expresses a law due to Kirchhoff: the spectral emissivity of a body, for a given wavelength, is equal to its spectral absorption factor at the same wavelength. The important point for us is that e_λ, and hence a_λ, may vary with wavelength and this can be exploited in, for example, solar water heaters (see Section 4.5.1).

Worked example

The tungsten filament of an incandescent light bulb operates at a temperature of 2500 K and has a surface area of 40 mm^2. It has an emissivity that is independent of wavelength and equal to 0.30. (a) Determine the total power radiated by the light bulb. (b) Determine the power radiated between the wavelengths of 500 and 504 nm. (c) Estimate the fraction of the radiated power that falls within the visible region.

Solution

(a) Using Equation (4.19), the total power radiated is

$$0.30 \times 5.67 \times 10^{-8} \times 40 \times 10^{-6} \times (2500)^4 = 27\,\text{W}.$$

(b) As the wavelength range 500–504 nm is small compared with the total wavelength range of the emitted radiation, we do not need to integrate Planck's formula. Instead we can simply multiply the value of $W_e(\lambda)$ at the wavelength midway between 500 and 504 nm by $\delta\lambda$, which is 4 nm. Using Equation (4.17) and taking into account the emissivity e of the filament, we have

$$W_e(\lambda = 502\,\text{nm}) = e \times \frac{2\pi hc^2}{\lambda^5 \left(e^{hc/\lambda kT} - 1\right)}$$

$$= \frac{0.3 \times 6.63 \times 10^{-34} \times 2\pi \left(3 \times 10^8\right)^2}{\left(502 \times 10^{-9}\right)^5 \times \left[\begin{array}{c} \exp\left(6.63 \times 10^{-34} \times 3 \times 10^8 /\right. \\ \left. 502 \times 10^{-9} \times 1.38 \times 10^{-23} \times 2500\right) - 1\end{array}\right]}\,\text{W/m}^2\,\text{m}$$

$$= 3.63 \times 10^{10}\,\text{W/m}^2\,\text{m}.$$

As the area of the filament is 40×10^{-6} m, the power radiated between 500 and 504 nm is, to a very good approximation,

$$3.63 \times 10^{10} \times 40 \times 10^{-6} \times 4 \times 10^{-9}\,\text{W}$$
$$= 5.8\,\text{mW}.$$

(c) We can make a plot of the power per wavelength radiated by the filament using a spreadsheet program to evaluate $W_e(\lambda)$ over a range of wavelengths. As the emissivity e of the filament is wavelength-independent, the shape of the spectral distribution will be the same as for a blackbody, but $W_e(\lambda)$ will be reduced by the factor e:

$$W_e(\lambda) = e \times \frac{2\pi hc^2}{\lambda^5 \left(e^{hc/\lambda kT} - 1\right)}\,\text{W/m}^2\,\text{m}.$$

Substituting for the fundamental constants in this equation and measuring wavelengths in microns (μm) rather than nanometers (nm) for convenience, we obtain

$$W_{\mathrm{e}}(\lambda) = e \times \frac{3.75 \times 10^{14}}{\lambda^5 [\exp(1.44 \times 10^{-2}/\lambda T) - 1]} \times 10^{-6} \; \mathrm{W/m^2 \mu m}.$$

Finally, we multiply $W_{\mathrm{e}}(\lambda)$ by area A to obtain the power per wavelength radiated by the filament:

$$A \times W_{\mathrm{e}}(\lambda) = A \times e \times \frac{3.75 \times 10^{14}}{\lambda^5 [\exp(1.44 \times 10^{-2}/\lambda T) - 1]} \times 10^{-6} \; \mathrm{W/\mu m}.$$

Taking $A = 40 \times 10^{-6} \, \mathrm{m^2}$, $e = 0.30$ and $T = 2500$ K, this equation has been evaluated for values of λ from 0.3 to 5.0 μm. The results are shown in Figure 4.5. We may note that the power distribution peaks at a wavelength $\lambda_{\max} = 1.16$ μm, which lies in the infrared, well above the visible range.

We can estimate the fraction of radiated power in the visible region by counting the squares under the curve between 0.4 and 0.7 μm and comparing this number with the total number of squares under the curve. We count incomplete squares as half a square and ignore the region of the distribution above 5.0 μm. This gives an approximate ratio of $18/570 \sim 0.03$. We see that just about 3% of the total radiated power lies in the visible. The shape of the spectral distribution in Figure 4.5 is roughly triangular, with a height of ~ 15 W/μm and a base width ~ 4 μm. The area of this triangle is $^1/_2 (15 \times 4)$W $= 30$ W. This gives an estimate of the total radiated power that is consistent with the answer in part (a).

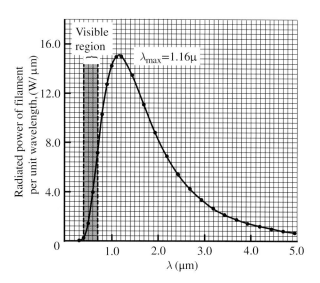

Figure 4.5 Spectral power distribution of an incandescent light bulb at a filament temperature of 2500 K. Only about 3% of the emitted radiation falls within the visible region.

The relatively small amount of radiation in the visible region arises because the power distribution peaks in the infrared and because of the relatively low emissivity of the tungsten surface, 0.3. Indeed, almost all of the electrical energy supplied to the bulb goes into thermal energy. This explains the drive to replace tungsten filament light bulbs with higher-efficiency devices such as LED lights.

4.2.3   Birth of the photon

The blackbody radiation that is emitted by a tiny hole in a cavity has the same properties as the radiation inside the cavity. Thus by measuring the properties of the emitted radiation, we can deduce the properties of the cavity radiation. This is important because the properties of cavity radiation can be analysed theoretically and the theoretical predictions can then be compared with the experimentally observed spectrum of blackbody radiation. Around 1900, Lord Rayleigh and James Jeans analysed cavity radiation using classical physics. They used electromagnetic theory to show that the radiation must exist as *standing waves* in the cavity. Using geometrical arguments, they counted the number of standing waves in the frequency range ν to $\nu + d\nu$. They then used results of classical kinetic theory to calculate the average total energy of these waves when the system is in thermal equilibrium. In this way they were able to determine the energy per unit volume of the cavity in the frequency interval ν to $\nu + d\nu$. To illustrate this procedure we consider the simpler case of standing waves in one dimension.

A familiar example of a standing wave occurs when a stretched string is plucked. The condition for such a standing wave is

$$n\frac{\lambda}{2} = L, \tag{4.21}$$

where λ is wavelength, L is the length of the string and $n = 1, 2, 3, \ldots$. This condition on n arises because the displacement of the string must be zero at the two fixed ends of the string, i.e. there must be nodes at those two points. In terms of frequency ν, we have

$$\nu = n\frac{v}{2L}, \tag{4.22}$$

since $\lambda\nu = v$, where v is the wave velocity. Electromagnetic waves in a laser cavity also provide an example of standing waves in one dimension. The laser cavity that is illustrated in Figure 4.6 consists of two plane mirrors separated by distance L. As illustrated schematically, nodes again exist at the two mirrors and we have the analogous conditions

$$n\frac{\lambda}{2} = L; \tag{4.23}$$

$$\nu = n\frac{c}{2L}, \tag{4.24}$$

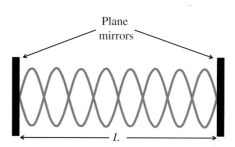

Figure 4.6 Electromagnetic standing waves, shown as the blue curves, between two plane mirrors in a laser cavity. The separation of the mirrors is L.

where c is the speed of light. In a laser, $L \sim 0.5$ m, and is very much larger than the wavelength λ of the light, which is ~ 500 nm, and so n is a very large number, $\sim 10^6$. Thus the difference between successive values of n, which of course is 1, is very small compared with the value of n, i.e. $[(n+1) - n] \ll n$. This allows us to treat n as a continuous variable in Equation (4.24). By differentiating this equation, we obtain the quantity $dn/d\nu$, which is the number of n values *per unit frequency interval*:

$$\frac{dn}{d\nu} = \frac{2L}{c}. \tag{4.25}$$

Hence, the number of allowed values of n in the frequency range ν to $\nu + d\nu$, which we call $N(\nu)d\nu$, is $(2L/c)d\nu$. For each of the allowed frequencies, there are two possible states of polarisation of the electromagnetic waves. Thus we have for this one-dimensional case:

$$N(\nu)d\nu = \left(\frac{4L}{c}\right)d\nu. \tag{4.26}$$

Now, for the case of a three-dimensional cavity with equal sides of length L, the corresponding result for $N(\nu)d\nu$ is

$$N(\nu)d\nu = \pi\left(\frac{2L}{c}\right)^3 \nu^2 d\nu. \tag{4.27}$$

It can be shown that $N(\nu)$ is independent of the shape of the cavity and depends only on its volume. We now apply this result to a cavity filled with blackbody radiation.

For a system that contains a large number of entities in thermal equilibrium, the *equipartition of energy theorem* predicts the *average* energy of the entities (see Section 4.6.2). In the present case the entities are the standing waves in the cavity and the equipartition theorem predicts that the mean energy $\langle E \rangle$ of a standing wave is kT, where k is the Boltzmann constant. The energy per unit volume $\rho(\nu)d\nu$ in the frequency range ν to $\nu + d\nu$ is the product of the average energy per standing wave times the number of standing waves in that frequency range, divided by the volume of the cavity:

$$\rho(\nu)d\nu = \frac{kT}{L^3} \times \pi\left(\frac{2L}{c}\right)^3 \nu^2 d\nu,$$

giving

$$\rho(\nu)d\nu = \frac{8\pi\nu^2 kT}{c^3}d\nu. \tag{4.28}$$

This is the *Rayleigh–Jeans formula for blackbody radiation*. This formula gives a shape for the blackbody spectrum that agrees with experimental data at low frequencies. However, as ν increases, the theoretical prediction goes to infinity! This is clearly unphysical and this absurd prediction of classical physics is known as the *ultraviolet catastrophe*.

It was Planck who resolved this discrepancy. He considered the walls of the cavity to consist of simple harmonic oscillators that are in equilibrium with the standing waves in the cavity. Again he used thermodynamic arguments, but his breakthrough was to *quantise* the energy of the oscillators. He proposed that the energy of an oscillator can only take certain discrete values: 0, $h\nu$, $2h\nu$, $3h\nu$, ..., where h is Planck's constant and ν is frequency; in contrast, classical theory allows a continuous distribution of energy. Planck's reasoning led to a different form for the mean energy of a standing wave. Instead of it being kT, Planck obtained the following expression for the mean energy $\langle E \rangle$:

$$\langle E \rangle = \frac{h\nu}{\left(e^{h\nu/kT} - 1\right)}. \tag{4.29}$$

(For the case where $h\nu \ll kT$, which corresponds to the classical regime, Equation 4.29 reduces to $\langle E \rangle = kT$.) Using this result for $\langle E \rangle$, the energy density in the cavity is

$$\rho(\nu)d\nu = \frac{8\pi\nu^2}{c^3}\frac{h\nu}{\left(e^{h\nu/kT} - 1\right)}d\nu. \tag{4.30}$$

This is the celebrated Planck's radiation law. Planck assumed quantisation of the energy of the oscillators in the walls of the cavity, rather than quantisation of the energy of the standing waves in the cavity. It was some years later, in 1905, that Einstein applied quantisation of energy to the electromagnetic radiation itself and so the *photon*, a discrete amount of electromagnetic energy, was born.

Equation (4.30) gives the *energy density per unit frequency*, $\rho(\nu)d\nu$, in a cavity. We can instead have an expression for the *energy density per unit wavelength*, $\rho(\lambda)d\lambda$. In fact, it is often more convenient to express Planck's radiation law in terms of wavelength λ for comparison with experimental data. Suppose that we have an elemental volume δV within the cavity. The energy contained in that volume, within frequency range $d\nu$, is $\delta V \rho(\nu) d\nu$. The energy contained within volume δV must be the same whether we express the energy density in terms of frequency or wavelength. Hence we can say that the energy contained within volume δV is $\delta V \rho(\lambda) d\lambda$, and so

$$\delta V \rho(\lambda) d\lambda = \delta V \rho(\nu) d\nu,$$

or

$$\rho(\lambda) d\lambda = \rho(\nu) d\nu,$$

so long as the interval in wavelength $d\lambda$ is equivalent to the interval in frequency $d\nu$. λ and ν are connected through the relation $\nu = c/\lambda$, and hence

$$d\nu = -\frac{c}{\lambda^2} d\lambda.$$

The minus sign arises because an increase in frequency ν gives rise to a decrease in wavelength λ, and since we are only interested in the absolute values of $d\lambda$ and $d\nu$, we may neglect it. Then substituting $d\nu = \left(c/\lambda^2\right) d\lambda$ and $\nu = c/\lambda$ in Equation (4.30), we obtain

$$\rho(\lambda) d\lambda = \frac{8\pi hc}{\lambda^5 \left(e^{hc/\lambda kT} - 1\right)} d\lambda. \tag{4.31}$$

The American physicist and astronomer William C. Coblentz made detailed measurements of blackbody radiation and verified Planck's law. Moreover, by fitting his experimental data to Equation (4.31) he obtained a value for Planck's constant h that is close to the presently accepted value. This is an example of the interplay between experimental and theoretical physics. Theory seeks to explain experimental observations. In turn, theory makes predictions, which are then tested by experiment.

We can relate the energy density $\rho(\lambda)$ within the cavity to the spectral radiant exitance, $W_e(\lambda)$, of the radiation emanating from the hole in the cavity. For this we use a photon representation where we consider the radiation in the cavity to be a *photon gas*. For the case of molecules in a vessel that is pierced by a tiny hole, we know from kinetic theory that the rate of effusion, R, of gas molecules through the hole per unit area is given by

$$R = \frac{1}{4} n \bar{v}, \tag{4.32}$$

where n is the number density of the molecules and $\bar{v}$ is their mean speed (see also Section 3.3.7). We picture photons escaping from the cavity like gas molecules effusing through a hole. If we identify the energy density $\rho(\lambda)$ in the cavity with number density n and spectral radiant exitance $W_e(\lambda)$ with effusion rate R, we have

$$W_e(\lambda) = \frac{1}{4}\rho(\lambda)c, \qquad (4.33)$$

where c is the speed of light. Note that this equation is dimensionally correct. Substituting for $\rho(\lambda)$ from Equation (4.31), we obtain

$$W_e(\lambda)d\lambda = \frac{2\pi hc^2}{\lambda^5}\frac{1}{\left(e^{hc/\lambda kT} - 1\right)}d\lambda,$$

in agreement with Equation (4.17).

4.3   Solar radiation and its interaction with the Earth

4.3.1   Characteristics of solar radiation

As noted previously, the Sun can be considered to be a blackbody radiator and observations on Earth show that the solar spectrum peaks at a wavelength of about 500 nm. Using this value, together with the results of Section 4.2.1, we can readily obtain a value for the surface temperature of the Sun, T_S, the total power that it radiates and we can estimate the power per unit area received at the Earth. From Wien's displacement law (Equation 4.16), we have

$$T_S = \frac{2.898 \times 10^{-3}}{500 \times 10^{-9}} = 5800\,\mathrm{K},$$

and from the Stefan–Boltzmann law (Equation 4.15), the power radiated per unit area is

$$5.6703 \times 10^8 \times (5800)^4 = 6.4 \times 10^7\,\mathrm{W/m^2}.$$

The total power radiated by the Sun is called its *luminosity* and is given the symbol $L_\odot$. Taking a value for the radius R_S of the Sun to be 7.0×10^8 m, we find that $L_\odot$ has the value

$$4\pi(7 \times 10^8)^2 \times 6.4 \times 10^7 = 3.9 \times 10^{26}\,\mathrm{W}.$$

The mean distance from the Sun to the Earth is 1.5×10^{11} m. Assuming the radiation from the Sun is emitted isotropically, the power per unit area delivered to Earth, which is called the *solar constant* is then

$$\frac{3.9 \times 10^{26}}{4\pi(1.5 \times 10^{11})^2} = 1.38\,\mathrm{kW/m^2}.$$

This is close to the accepted value of $1.367\,\mathrm{kW/m^2}$, measured when the Earth is at its mean distance from the Sun. As the Sun–Earth distance varies over the Earth's elliptical orbit, the power density varies by approximately ±4%. There is also a variation due to the changing activity of sunspots, but this variation is small, <1%. The figure of 1.367 kW/m² is the power density, i.e. the solar irradiance at the top of Earth's atmosphere. The atmosphere attenuates the Sun's radiation, as we shall see, and the maximum solar irradiance at the Earth's surface is about 1 kW/m². This value might be approached in the middle of a desert with the Sun directly overhead.

The solar spectrum before the radiation enters Earth's atmosphere is called the *extraterrestrial spectrum.* This is presented in Figure 4.7, where we plot *spectral irradiance* against wavelength, and where spectral irradiance is defined as irradiance per unit wavelength with units kW/m² μm in this case. The total area under the extraterrestrial spectrum is the solar constant 1.367 kW/m². The solar spectrum peaks over the approximate wavelength range 0.4–0.7 μm. This coincides with the visible part of the electromagnetic spectrum to which the human eye has adapted. The ultraviolet and infrared regions are also indicated in Figure 4.7. Also shown in the figure is the spectral irradiance curve produced by a blackbody at a temperature of 5800 K. Its close resemblance to the extraterrestrial spectrum is apparent. The dips in the extraterrestrial spectrum are due to Fraunhofer lines named after the German physicist Joseph von Fraunhofer who made a systematic study of them. These lines are due to absorption of solar radiation by elements in the solar atmosphere.

It is also useful to know the range of photon energies in the solar spectrum. For example, operation of a solar cell requires photons to have a minimum energy of ~1 eV. The upper horizontal scale in Figure 4.7 corresponds to photon energy E in units of eV; the conversion relation is $E(\text{eV}) = 1.240/\lambda(\mu m)$. We see that the solar spectrum covers the photon energy range from about 0.5 to 4 eV,

Figure 4.7 Plot of the spectral irradiance against wavelength λ for solar radiation before it enters the Earth's atmosphere, called the extraterrestrial spectrum. Also shown is the corresponding blackbody spectrum for a temperature of 5800 K. The lower horizontal scale is photon wavelength (μm), while the upper horizontal scale is photon energy (eV).

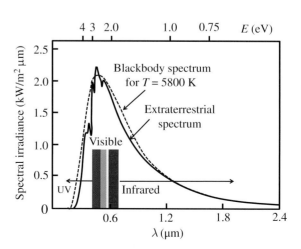

and it is useful to remember that the energy of a visible photon is $\sim$2 eV.

Worked example

Estimate the surface temperature T_E of the Earth, assuming that it is in radiative equilibrium with the Sun. Assume that the Earth absorbs 70% of the radiation falling on it and that it has an emissivity of 1.0. The radii of the Sun, R_S, and of Earth, R_E, are 7.0×10^8 and 6.3×10^6 m respectively, and the mean distance d from the Sun to Earth is 1.5×10^{11} m.

Solution

Total power radiated from the Sun is $\sigma \times 4\pi R_S^2 \times T_S^4$, as we can take it to be a blackbody. Assuming that the Earth presents a flat disc to the Sun, the total power absorbed by the Earth is

$$0.7 \times \left(\frac{\pi R_E^2}{4\pi d^2} \right) \times \left(\sigma \times 4\pi R_S^2 \times T_S^4 \right).$$

At equilibrium, the power absorbed by the Earth is equal to the power it radiates. Hence,

$$e\sigma 4\pi R_E^2 T_E^4 = 0.7 \times \left(\frac{\pi R_E^2}{4\pi d^2} \right) \times \left(\sigma \times 4\pi R_S^2 \times T_S^4 \right).$$

Simplifying this result, and taking the Earth to also be a blackbody, with $e = 1$, we obtain:

$$= 5800 \times 0.7^{1/4} \left(\frac{7.0 \times 10^8}{2 \times 1.5 \times 10^{11}} \right)^{1/2} = 256 \, \text{K}.$$

This is a very chilly $-17°$C! Fortunately, because of the *greenhouse effect* (see Section 4.3.2), the atmosphere acts like a blanket to trap solar heat and the actual mean temperature of the Earth is $\sim$14°C, providing an adequate environmental temperature for water to remain liquid and for life to thrive.

4.3.2   Interaction of solar radiation with Earth and its atmosphere

A number of things may happen to solar radiation as it passes through Earth's atmosphere, as illustrated schematically in Figure 4.8. The radiation may be *reflected* by clouds or by the surface of the Earth, especially from snow and ice. It may be *scattered* by atmospheric molecules or by particles such as dust and pollen in the atmosphere. These processes result in approximately 30% of solar radiation being reflected back into space. The fraction of the incident

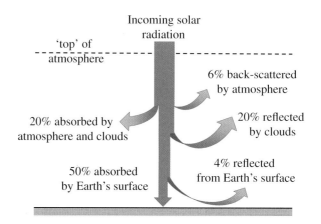

Figure 4.8 Illustration of the various ways in which solar radiation is reflected, scattered or absorbed by the Earth and its atmosphere. About 70% of incoming solar energy is absorbed and the rest is reflected back into space.

radiation that is reflected is called the *albedo* of Earth. The solar radiation may also be *absorbed* by atmospheric molecules such as water and carbon dioxide and this may account for 20% of the incident radiation. So about 50% of solar radiation is eventually absorbed at the Earth's surface. All these processes, which we will now describe in more detail, modify the extraterrestrial spectrum. The resulting spectrum of solar radiation at the surface of Earth is shown in Figure 4.9.

Absorption processes

Molecules have three different kinds of motion: *electronic, vibrational* and *rotational.* By electronic motion we mean the promotion of an electron from one electronic orbital to another. This is analogous to electronic transitions in atoms, as for example, where we promote

Figure 4.9 Comparison of the extraterrestrial spectrum, at the top of the atmosphere, with the solar spectrum at the surface of Earth. The attenuation and dips in the spectrum at Earth's surface are due to various absorption and scattering processes as indicated.

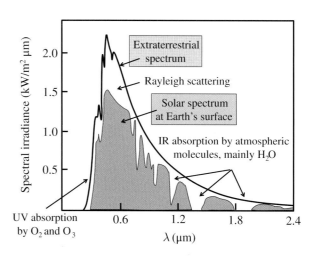

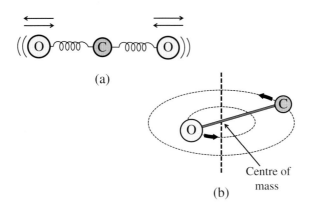

(a)

(b)

Centre of
mass

Figure 4.10 Schematic diagrams of (a) vibrational motion of a CO_2 molecule and (b) rotational motion of a CO molecule.

an electron from the $n = 1$ to the $n = 2$ orbital of atomic hydrogen. Vibrational motion is illustrated schematically in Figure 4.10(a) by the example of the carbon dioxide molecule CO_2. The vibrating molecule is modelled as three masses that are connected by springs and which perform harmonic motion when displaced from equilibrium. Rotational motion is illustrated in Figure 4.10(b) by the example of the carbon monoxide molecule CO. The molecule is modelled as a dumbbell, which rotates in a perpendicular plane about its centre of mass. The photon energies involved in the three different molecular motions are quite different. Excitation of electronic transitions needs photon energies of around several eV, vibrational excitation requires energies of ~0.1 eV and rotational excitation requires energies of ~0.001 eV. Converting these energies into wavelengths, electronic excitations typically lie in the visible or ultraviolet, vibrational excitations lie in the infrared, and rotational excitations lie in the microwave region of the electromagnetic spectrum. When a particular molecular motion is excited by solar radiation of the appropriate wavelength, that radiation is absorbed and the spectral irradiance at that wavelength is reduced. The solar spectrum at Earth's surface (Figure 4.9) exhibits a number of dips in the region between 0.7 and 2.4 μm. These are due to absorption of infrared radiation by atmospheric molecules. This absorption is mainly due to vibrational excitation of the H_2O molecules in water vapour, which is abundant in the atmosphere.

Electromagnetic radiation can interact with a molecule in another way. If it has enough energy, the incident photon can break the molecular bonds between atoms. This is called *photo-dissociation* and is illustrated schematically in Figure 4.11 for the example of molecular oxygen. This is a particularly important example because of its role in the formation of ozone in the atmosphere. It takes 5.2 eV of energy to break the molecular bond in oxygen via the reaction

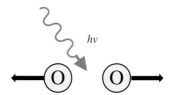

Figure 4.11 Schematic diagram of the photo-dissociation of an O_2 molecule.

$$O_2 + h\nu \rightarrow O + O \qquad (4.34)$$

and so the incident photon must have at least this energy, i.e. a wavelength shorter than 0.24 μm. One of the oxygen atoms produced in the reaction then combines with an O_2 molecule to produce an ozone molecule O_3 via the reaction

$$O_2 + O \rightarrow O_3. \qquad (4.35)$$

The strength of the molecular bond in an O_3 molecule (3.8 eV) is less than for the O_2 molecule and wavelengths up to 0.32 μm can cause it to break:

$$O_3 + h\nu \rightarrow O_2 + O. \qquad (4.36)$$

The oxygen atom formed by Reaction (4.36) can then recombine with another O_2 molecule to form another O_3 molecule. Reactions (4.35) and (4.36) are the basic reactions of the *ozone–oxygen cycle*. If the ozone molecules could be contained in a thin layer of gas at atmospheric pressure, the layer would be just ~3 mm thick. Oxygen absorbs UV radiation between 0.14 and 0.24 μm, while ozone absorbs UV radiation between 0.2 and 0.32 μm. Between them, these molecules absorb most of the UV radiation in the solar spectrum, shielding us from its harmful biological effects.

The Earth's atmosphere is essentially transparent to solar radiation in the visible part of the spectrum (0.4–0.7 μm) and about half the solar energy reaching the Earth's surface arises from this visible region. However, the Earth's atmosphere is opaque in most regions of the electromagnetic spectrum and there are very few *atmospheric windows* that allow us to see the rest of the universe. The main windows are the *optical window*, the one described above, and also the *radio window*, which covers the approximate wavelength range ~1 cm to 20 m. It is upon these two atmospheric windows that ground-based astronomy is based.

Scattering processes

When electromagnetic radiation is incident on a particle of matter, the oscillating electric field of the radiation acts on the electric charges within the particle, causing them to oscillate. The oscillating charged particles then become *radiating dipoles*, emitting *scattered* radiation at the same frequency as the incident radiation. The radiating dipoles emit their radiation over a wide a range of directions and so this scattering process acts to *diffuse* the incident radiation. In the case of solar radiation, molecules, water droplets in clouds, dust and other particles in the atmosphere scatter the Sun's light.

The way light is scattered depends on the size of the scattering particle. If the particle is much smaller than the wavelength of the light, as is the case for atmospheric molecules, we have *Rayleigh scattering.* If, on the other hand, the particle is about the same size as the wavelength or is larger, as is the case for water droplets in a cloud, we have *Mie scattering*. The probability for Rayleigh scattering process depends inversely on the fourth power of the wavelength of the light, $(1/\lambda^4)$. This very strong dependence on wavelength means that shorter (blue) wavelengths are scattered more strongly than longer (red) wavelengths. The effect of this is to reduce the amount of blue light in the solar spectrum reaching the Earth's surface (see Figure 4.9). For example, taking λ_{blue} to be 400 nm and λ_{red} to be 700 nm, we see that blue light is scattered 9.4 times more strongly than red light. When we look at the sky, away from the Sun, we are seeing this predominantly blue scattered light, which is the reason that the sky looks blue. On the other hand, Mie scattering is almost independent of wavelength in the visible region. Consequently, water droplets in a cloud scatter all the wavelengths of white light to the same degree and so the scattered light also appears to be white, giving the white appearance of clouds. Of course, if the clouds are optically thick, little light will penetrate through them and they appear to be grey.

Because of the presence of these scattering processes, the solar radiation arriving on Earth is a mix of *direct* and *diffuse* radiation, as illustrated pictorially in Figure 4.12. Even on a clear day, the proportion of diffusion radiation is ~10% of the total radiation. A significant difference between these two kinds of radiation is that direct radiation can be focused by a lens or mirror, as indicated in Figure 4.12. On the other hand, diffuse light cannot be focused in this way. Consequently, some methods for harvesting solar energy, such as solar power stations that focus or concentrate the Sun's rays using mirrors, need direct solar radiation On the other hand, some other methods, including photovoltaic solar cells, can make use of diffused sunlight.

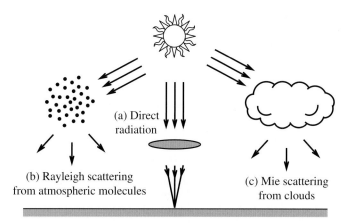

Figure 4.12 Solar radiation may come as *direct* radiation (a) or as *diffuse* radiation due to scattering of light from atmospheric molecules, called Rayleigh scattering (b), or due to scattering of light from larger particles such as water droplets in a cloud, called Mie scattering (c).

Figure 4.13 When the Sun is vertically above the ground on a clear sunny day, with zenith angle $\theta_z = 0$, each square meter of ground receives ~1 kW of solar power. This is reduced when the Sun's position makes a non-zero zenith angle, θ_z, with respect to the vertical. This is because the solar radiation has to pass through a greater depth of the atmosphere and because the cross-sectional area A of the radiation is spread over a larger area on the ground.

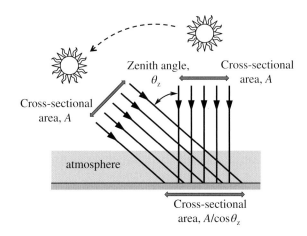

Season, latitude and daily insolation

So far we have considered the situation where the Sun is directly overhead at a given location with the *zenith angle* $\theta_z = 0$, where the zenith angle is the angle between the vertical and the Sun's rays (see Figure 4.13). However, the zenith angle varies constantly, with respect to both the time of day and the time of the year. It also varies with latitude. A non-zero zenith angle reduces the Sun's irradiance at the surface of the Earth, and consequently the Sun's irradiance at a particular location will also vary during the day and during the course of the seasons.

There are two main effects that occur when the zenith angle is not zero. The distance travelled by the direct solar beam through the atmosphere depends on the zenith angle, as illustrated in Figure 4.13. And the greater the distance, the greater will be the atmospheric attenuation. As the atmosphere does not have a well-defined height, it is more convenient in practice to consider the *mass* of the atmosphere when considering atmospheric absorption. For the direct beam at normal incidence ($\theta_z = 0$), passing through the atmosphere at normal pressure, a *standard mass of atmosphere* is encountered by the beam. For a finite zenith angle, the ratio of the increased path length compared with the normal path length is m, where

$$m = \frac{1}{\cos \theta_z} \tag{4.37}$$

(see Figure 4.13). The mass of the increased path length compared with that when the path is normal to the atmosphere is called the *air-mass-ratio* (AM ratio). AM0 refers to zero atmosphere, as experienced by a satellite located above the atmosphere; AM1, refers to $m = 1$, i.e. the Sun directly overhead; AM2 refers to $m = 2$, i.e. corresponding to a zenith angle of 60°, where the solar radiation would pass through a length of the atmosphere equal to approximately twice its height. For the solar spectrum shown in Figure 4.9, $m = 1$.

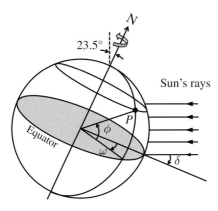

Figure 4.14 The Earth is titled by the angle of 23.5° with respect to its orbital plane about the Sun. The hour angle, ω, is the angular distance between the meridian of an observer at P and the meridian whose plane contains the Sun. It increases by 15° every hour. The angle of declination, δ, is the angle between Earth's equatorial plane and a line drawn from the centre of Earth to the centre of the Sun. ϕ is the latitude of point P.

A second effect is that when the Sun is at a finite zenith angle, a cross-sectional area, A, of the solar radiation now fills an area $A/\cos\theta_z$ (see Figure 4.13), and this reduces the solar irradiance by the factor $\cos\theta_z$. In some systems where mirrors are used to collect sunlight, it is possible to track the position of the Sun by moving these mirrors. Systems that are not moveable, e.g. solar water heaters on roofs, are orientated to maximise the received radiation over the course of a year.

A three-dimensional analysis of the Sun–Earth system gives the following result for the zenith angle θ_z at a particular location P:

$$\cos\theta_z = \sin\phi\sin\delta + \cos\phi\cos\delta\cos\omega, \qquad (4.38)$$

where $\phi = $ latitude, $\delta = $ angle of *declination* and $\omega = $ *hour angle*. These angles are defined by the diagram in Figure 4.14. The hour angle is the angular distance between the meridian of the observer at P and the meridian whose plane contains the Sun, i.e. the angle through which the Earth has rotated since *solar noon*; the hour angle is zero at solar noon and increases by 15° degrees every hour. The angle of declination is the angle between the Earth's equatorial plane and a line drawn from the centre of the Earth to the centre of the Sun. The axis about which the Earth rotates is at an angle of 23.5° with respect to the plane containing the orbit of the Earth. Because of this, the angle of declination varies continuously through the year, and it does so in a sinusoidal manner. Knowing the various parameters in Equation (4.38), the zenith angle at any point on Earth can be determined.

The seasons occur because the Sun's angle of declination changes continuously through the year. Despite the fact that the Earth orbits the Sun, it is simpler to picture the variation in declination angle by imagining, for a moment, that the Sun orbits the Earth. This gives the picture shown in Figure 4.15. The three orientations of the Sun–Earth system correspond, in the northern hemisphere, to (a) the

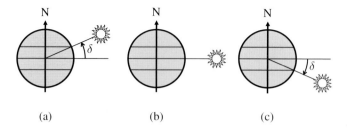

(a)   (b)   (c)

Figure 4.15 Despite the fact that Earth orbits around the Sun, it is simpler to picture the variation in declination angle δ by imagining that the Sun orbits around the Earth. The three orientations of the Sun–Earth system correspond to (in the northern hemisphere): (a) the summer solstice, (b) the autumn solstice, and (c) the winter solstice. At the summer solstice the angle of declination is $+23.5°$ and at the winter solstice it is $-23.5°$.

summer solstice, (b) the autumn solstice and (c) the winter solstice. At the summer solstice ($\sim$21 June), the angle of declination is $+23.5°$ and at the winter solstice ($\sim$21 December 21), it is $-23.5°$. At the autumn equinox ($\sim$23 September), the Sun is directly above the equator, at which point day and night have exactly the same length. Thus, in the summer in the northern hemisphere, the sunlight hits the ground much more directly at say point P than at point P' at the same latitude in the southern hemisphere (see Figure 4.16). Furthermore, a beam of sunlight incident at P' is spread over a larger area than at P, reducing its irradiance. Furthermore, the sunlight reaching P travels a shorter distance in the atmosphere, and the length of daylight will be much longer in the northern hemisphere. Consequently, the total amount of solar energy received per unit per area per day will be much greater in the northern hemisphere. Of course, the situation is reversed when it is winter in the northern hemisphere. The total energy per unit area received in 1 day from the Sun at a particular location is called the *daily insolation*, which, as we can see, varies according to time of the year and to the latitude. Often quoted is the *average annual insolation*, which is the mean of the daily insolation recorded over the course of a year. For example, the average annual insolation for Helsinki, Finland is 2.4 kWh/m^2/day, while for central Australia it is 5.9 kWh/m^2/day.

Figure 4.16 In the summer in the northern hemisphere, the sunlight hits the ground much more directly at point P than at point P' at the same latitude in the southern hemisphere. At P', a beam of sunlight is spread over a larger area than at P, reducing its irradiance. Furthermore, the sunlight reaching P travels the shorter distance in the atmosphere, and the length of daylight will be much longer in the northern hemisphere. Consequently, the total solar energy deposited per unit per area per day is much greater in the northern hemisphere. The situation is reversed when it is winter in the northern hemisphere.

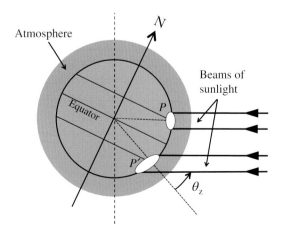

The greenhouse effect

The atmosphere plays a crucial role in maintaining a temperature that is suitable for sustaining life. It acts as a blanket to retain heat from incident solar radiation. As we saw in Section 4.3.2, the atmosphere is essentially transparent to solar radiation in the visible part of the electromagnetic spectrum ($\sim$0.4–0.7 μm) and so a large fraction of this radiation is transmitted by the atmosphere to heat the Earth's surface. On the other hand, Earth, which is at a much lower temperature than the Sun, radiates in the infrared region, in the range $\sim$4–100 μm. Molecules such as H_2O and CO_2 in the atmosphere strongly absorb radiation in the infrared. These molecules absorb the radiation emitted by Earth and then re-radiate this energy, both upwards but also downwards, back to the to the surface of Earth. The radiation that is radiated downwards is absorbed by the Earth's surface instead of escaping into space. In order to maintain thermal equilibrium, the Earth must compensate for this by increasing its temperature, leading to a higher surface temperature than if the atmosphere were absent. This is the *greenhouse effect*. Thus the average surface temperature of the Earth is $\sim$14° C, which compares with our estimate of $-17°$ C for an Earth without an atmosphere (see Section 4.3.1).

The more constituent atoms that a molecule has, the more ways it may vibrate. These are called its *modes of vibration*. In general, the more modes of vibration a molecule has, the greater is its capacity to absorb electromagnetic radiation. Thus, for example, methane CH_4 having five atoms is $\sim$20 times more effective in absorbing the emitted radiation from the surface of Earth than CO_2, which has three atoms. The capacity of a particular molecule to absorb Earth's infrared radiation with respect to that of the CO_2 molecule is called its *global warming potential*.

4.3.3   Penetration of solar energy into the ground

Solar radiation heats the Earth's surface. As the amount of solar energy absorbed each day at a particular location varies over the course of the seasons, so will the surface temperature at that location. Some of the absorbed heat penetrates into the ground by the process of *thermal diffusion*. Because the surface temperature varies, the heat penetrates into the ground somewhat in the manner of a travelling wave, as we shall see. This introduces a time lag so that in winter, the temperature at a depth of, say, $\sim$10 m below the Earth's surface can be significantly higher than the temperature at the surface, while in summer it can be significantly lower. This effect can be put to good use for domestic heating and cooling purposes, as we shall see in Section 4.4.1.

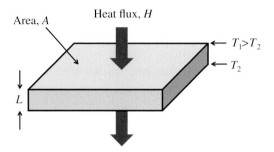

Figure 4.17 Heat conduction across a slab of material of area A, width L and thermal conductivity κ. The faces of the slab are at temperatures T_1 and T_2, with $T_1 > T_2$, and H is the heat flux. The direction of heat flow is from high to low temperature.

Heat will flow across a slab of material when there is a temperature difference between the faces of the slab. This is illustrated in Figure 4.17 where the faces of the slab are at temperatures T_1 and T_2, respectively, and $T_1 > T_2$. For most materials it is found that the *heat flux H*, i.e. the heat conducted per second through the slab, is described by

$$H = \kappa A \frac{(T_1 - T_2)}{L},\qquad(4.39)$$

where κ is the *thermal conductivity* of the material and A and L are the cross-sectional area and thickness of the slab, respectively. The units of H are J/s (W) and the units of κ are W/m K. The heat flows in the direction from higher to lower temperature.

In the case illustrated by Figure 4.17, T_1 and T_2 have fixed values. For the penetration of solar heat, however, the temperature of the Earth's surface changes with time. The result is that the heat flux at a given distance below the surface is a function of the distance and also of time. To deal with this case we use the differential form of Equation (4.39):

$$H = -\kappa A \frac{\mathrm{d}T}{\mathrm{d}z}.\qquad(4.40)$$

This is the one-dimensional form of *Fourier's law of conduction*. The minus sign arises in Equation (4.40) because heat flows in the direction of temperature decreasing. To illustrate the application of this law, we consider a rod of material in which the heat flux is a function of distance z and time t (see Figure 4.18). We assume that there are no heat losses from the sides of the rod, which can be achieved by lagging the rod with thermally insulating material. We consider the heat flux into and out of a small slice of the rod between z and $z + \delta z$. At position z we have

$$H(z, t) = -\kappa A \frac{\partial T(z, t)}{\partial z}$$

and at position $z + \delta z$ we have

$$H(z + \delta z, t) = -\kappa A \frac{\partial T(z + \delta z, t)}{\partial z}.$$

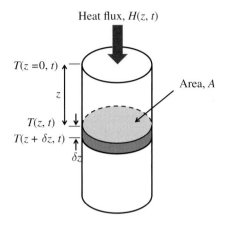

Figure 4.18 The diffusion of heat along a lagged rod of material. The heat flux into and out of a small slice of the rod between z and $z + \delta z$ is deduced in the text.

We use partial derivatives, as the temperature at a particular point is now a function of both z and t. There will be a net heat flux into the slice of material equal to

$$-\kappa A \left(\frac{\partial T(z,t)}{\partial z} \right) + \kappa A \left(\frac{\partial T(z+\delta z,t)}{\partial z} \right). \qquad (4.41)$$

This net heat flux is equal to the rate $\partial Q / \partial t$ at which thermal energy Q increases in the slice. When heat energy Q is added to a mass, the temperature T of the mass increases according to

$$\frac{\mathrm{d}Q}{\mathrm{d}T} = mC, \qquad (4.42)$$

where m is the mass and C is the specific heat. Rearranging this equation and differentiating with respect to t, we obtain for the slice of the rod

$$\frac{\partial Q}{\partial t} = mC \frac{\partial T}{\partial t}. \qquad (4.43)$$

If our slice of thickness δz has density ρ, we obtain

$$\frac{\partial Q}{\partial t} = \rho A \delta z C \frac{\partial T}{\partial t}. \qquad (4.44)$$

Since

$$\frac{\partial Q}{\partial t} = -\kappa A \left(\frac{\partial T(z,t)}{\partial z} \right) + \kappa A \left(\frac{\partial T(z+\delta z,t)}{\partial z} \right),$$

we have

$$\frac{\partial T}{\partial t} = \frac{\kappa}{\rho C} \times \frac{1}{\delta z} \left[\left(\frac{\partial T(z+\delta z,t)}{\partial z} \right) - \left(\frac{\partial T(z,t)}{\partial z} \right) \right].$$

In the limit $\delta z \to 0$,

$$\frac{1}{\delta z} \left[\left(\frac{\partial T(z+\delta z,t)}{\partial z} \right) - \left(\frac{\partial T(z,t)}{\partial z} \right) \right] = \left(\frac{\partial^2 T(z,t)}{\partial z^2} \right).$$

Hence, our final result is

$$\frac{\partial T}{\partial t} = \frac{\kappa}{\rho C} \frac{\partial^2 T}{\partial z^2}. \qquad (4.45)$$

This partial differential equation is called the *one-dimensional heat equation*, and the quantity $\kappa/\rho C$ is called the *diffusivity* of the material. (Equation 4.45 is also known as the *diffusion equation*. More generally, and for the three-dimensional case we write it as

$$\frac{\partial \phi}{\partial t} = \alpha \left(\frac{\partial^2 \phi}{\partial x^2} + \frac{\partial^2 \phi}{\partial y^2} + \frac{\partial^2 \phi}{\partial z^2} \right), \qquad (4.46)$$

where ϕ is the quantity of interest and α is a positive constant. The diffusion equation is of great importance in a wide range of physics, mathematics, engineering and commerce. For example, it describes the diffusion of particles in gases and solids, while it is also used in mathematical models of some transactions in financial markets.)

Worked example

Assume that the temperature at a point on the Earth's surface varies sinusoidally over the course of the seasons and can be described by $T(t) = T_0 + A \sin \omega t$, where T_0 is the average temperature, A is the 'amplitude' of the temperature variation and $\omega = 2\pi/\tau$, where the period τ is 1 year. Show that $T(z,t) = T_0 + Ae^{-z\beta} \sin(\omega t - z\beta)$ is a solution of the one-dimensional heat equation, where z is the depth below the surface, and obtain an expression for β. Taking typical values of ρ, C and κ for Earth to be 1.5×10^3 kg/m^3, 2.0×10^3 J/kg K and 2.5×10^3 W/m K, respectively, find the value of z at which the temperature is hottest in the *winter*.

Solution

We picture a column of rock that lies below ground level, with a geometry just like the rod of material in Figure 4.18. As the depths below ground that we will consider are tiny compared with the Earth's radius, we can assume the ground to be flat and use the one-dimensional heat equation. Note the factor $\sin(\omega t - z\beta)$ in $T(z,t)$ which has the same form as a travelling wave $y = A\sin(\omega t - kz)$, where k is the *wavenumber*.

Differentiating the given solution with respect to t once and with respect to z twice:

$$\frac{\partial T}{\partial t} = \omega Ae^{-z\beta} \cos(\omega t - z\beta),$$

$$\frac{\partial^2 T}{\partial z^2} = 2\beta^2 Ae^{-z\beta} \cos(\omega t - z\beta).$$

Substituting for $\partial T/\partial t$ and $\partial^2 T/\partial z^2$ in Equation (4.45), we obtain

$$\beta = \sqrt{\frac{\omega\rho C}{2\kappa}}.$$

If the value of z is such that $\beta z = \pi$, then $T(z,t) = T_0 + Ae^{-z\beta}\sin(\omega t - \pi)$, which is exactly 'out of phase' with $T(0,t) = T_0 + A\sin(\omega t)$, the corresponding temperature at the surface where $z = 0$. Thus, at $z = \pi/\beta$, the temperature is hottest when it is coldest at the surface. For the given values of κ, ρ and C,

$$\beta = \sqrt{\frac{2\pi \times 1.5 \times 10^3 \times 2.0 \times 10^3}{365 \times 24 \times 60 \times 60 \times 2 \times 2.5}} = 0.35/\text{m},$$

giving $z = \pi/\beta = 9.0$ m.

The 'amplitude' of the temperature variation at $z = \pi/\beta$ is reduced by the factor $e^{-\pi} = 0.043$, and so the temperature, even at $z = 9.0$ m, changes only slightly from T_0 over the whole year. This is why caves are good places to store wine! Typical values of T_0 and A may be 10 and 12°C so that the temperature at $z = 9.0$ m would vary by just $\pm 0.043 \times 12$°C $= \pm 0.52$°C about an average temperature of 10°C. Plots of the temperature variation at the Earth's surface and at a distance $z = 9.0$ m below the surface are shown in Figure 4.19 for the above values. Note the time lag between the two plots.

4.4   Geothermal energy

Geothermal energy is the thermal energy that is generated and stored in the Earth. It is classified as either *shallow* geothermal energy

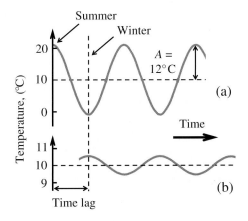

Figure 4.19 (a) A plot of the Earth's surface temperature versus time. (b) A plot of temperature versus time at a distance $z = \pi/\beta = 9.0$ m below the Earth's surface, where the parameter β is defined in the text. Note the time lag between the two plots and the fact that the temperature variation at the given distance below the surface is small.

or *deep* geothermal energy. Shallow geothermal energy is generated and stored within a few tens of metres below the Earth's surface. Deep geothermal energy comes from much lower depths, below, say, 1 km. Shallow and deep geothermal energies are generated in different ways, as described below, but both provide practical sources of energy.

4.4.1   Shallow geothermal energy

We saw in Section 4.3.3 that solar energy penetrates into the ground, heating the surface layer of soil and rock to a depth of a few tens of metres; and that the temperature at a depth of ~10 m is fairly constant, ~10°C in the UK. This shallow geothermal energy is a valuable store of energy and can be used, for example, for space heating in a building. The principle of operation for this is illustrated in Figure 4.20. A long length of copper pipe is buried in the ground at a depth of several metres and the pipe is connected to a *heat pump* (see Section 4.6.5). The action of the heat pump is to extract thermal energy from the water in the pipe and deposit it in the building. This energy is then used to heat radiators or domestic water. This process cools the water, which is then recirculated through the buried pipe, where it absorbs heat from the ground and the cycle repeats. This system can also be used to cool a building in the summer months, when the air temperature is above the temperature at the depth of the buried pipe. In that case, the heat pump is run in the reverse direction so that it extracts thermal energy from the building. It deposits some of the extracted thermal energy in the ground via the copper pipe, but it also uses some of the extracted energy to heat the domestic water.

Figure 4.20 In order to extract shallow geothermal energy, a long length of copper pipe is buried in the ground at a depth of several metres and the pipe is connected to a heat pump. The action of the heat pump is to extract thermal energy from the water in the pipe and deposit it in the building. In the summer months, this process can be reversed so that the air in the building is cooled.

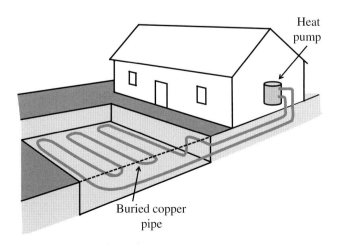

4.4.2 Deep geothermal energy

This form of geothermal energy comes from deep within the Earth's crust. It originates from two sources. It comes from the geological processes that occurred in the original formation of the planet 4.5 billion years ago, and from the energy produced in the decay of radioactive nuclides within the Earth. These radioactive nuclides are concentrated in the Earth's crust and are particularly pronounced in granite. The relative amounts of thermal energy from the two sources are uncertain, but they are comparable to each other.

A sketch illustrating the various regions below the surface of the Earth is shown in Figure 4.21. At the centre of the Earth, the temperature is approximately 4000°C. Heat transfer from the semi-fluid mantle maintains a temperature difference across Earth's relatively thin crust of 1000°C, and a temperature gradient of typically 30°C/km. This results in a flow of heat through the surface of the Earth. The average heat flow due to deep geothermal energy is ~0.1 W/m^2, which is negligible compared with the solar power incident on Earth's surface. Moreover, it is at too low a temperature to be used for heating or for conversion into mechanical energy. However, there are specific locations where deep geothermal energy is brought to the Earth's surface at high temperature by hot springs or steam. These are normally located in regions near tectonic plate boundaries where there is volcanic activity. Indeed, hot springs have been used for bathing since Palaeolithic times and for space heating since ancient Roman times. If the ground releases hot water at temperatures of ~100°C this can be used for space heating or for domestic purposes. If, however, superheated steam at a temperature well above 100°C is released, this can be used to drive steam turbines to produce electricity. The world's largest geothermal field is called the Geysers. It is located approximately 116 km north of San Francisco, California, and contains a complex of 22 geothermal power plants in which superheated steam at ~180°C coming from the ground drives turbines directly. The Geysers facility produces an average power of about 1 GW and is

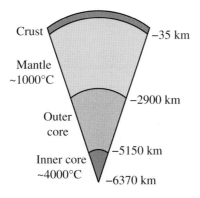

Crust
Mantle
~1000°C
Outer core
Inner core
~4000°C
−35 km
−2900 km
−5150 km
−6370 km

Figure 4.21 The various regions below the surface of Earth. Heat transfer from the mantle maintains a temperature difference across Earth's relatively thin crust of 1000°C, and a temperature gradient of typically 30°C/km. This results in a heat flow through the surface of Earth of ~0.1 W/m^2, called deep thermal energy.

estimated to supply 60% of the power demand for the coastal region between the Golden Gate Bridge and the Oregon state line.

An alternative way to access deep geothermal energy is to drill boreholes into the ground to a depth of ~5 km. Taking the geothermal gradient to be 30°C/km, this gives a temperature at that depth of ~150°C. Cold water is pumped under pressure down one of the boreholes into a natural fracture in the rock which provides a natural heat exchanger. The cold water absorbs thermal energy from the rocks and is pumped back up to the surface at about 150°C. The hot water may then be used for space heating or to produce electricity. This process gradually cools the rocks and, consequently, such a source of deep geothermal energy may have a finite lifetime, say 25 years. After that time, the rocks must be left to recover and regain thermal energy from the mantle below. The Earth's resources of deep geothermal energy are theoretically more than adequate to supply humanity's energy needs and this energy source is environmentally friendly. However, only a very small fraction of it may be profitably exploited becase drilling and exploration for deep resources are very expensive.

4.5 Solar heaters

A solar heater collects the Sun's radiation and transfers the absorbed energy, usually to a fluid. We will describe two different kinds of solar heater which are distinguished by the temperature to which they heat the fluid. The first kind is exemplified by the *solar water heaters* that are seen on rooftops. As we shall see, these can heat water to several tens of degrees above ambient temperature, say to 80°C. Water at this temperature is suitable for domestic purposes like washing or space heating. However, for the conversion of thermal energy into mechanical energy using, say, a steam turbine, much higher temperatures are required (see Section 4.6.5). These higher temperatures are achieved in the second type of solar heaters, called *solar thermal power systems*. The way to achieve such high temperatures is to collect solar radiation over a large area using mirrors and to concentrate the radiation into a much smaller area. We are all familiar with using a magnifying lens to focus the Sun's rays onto a piece of paper and set the paper alight, and indeed, mirrors have been used over several millennia to focus sunlight and produce fire. The Olympic flame is ignited in this way.

4.5.1 Solar water heaters

The flat plate water heater

The principle of operation of a solar water heater is illustrated by the *flat plate water heater*, which is shown schematically in Figures 4.22

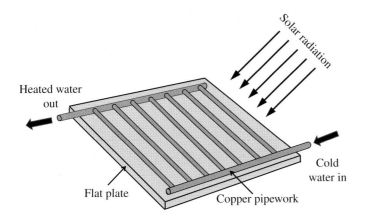

Figure 4.22 Schematic diagram showing the principle of operation of a flat plate water heater. The flat plate collects the Sun's radiation and the absorbed heat is transferred to water in copper pipes that are welded to the plate. The area of the flat plate is typically a few square metres.

and 4.23. The heater has a flat plate with an area of typically a few square metres. This plate collects the solar radiation and transfers the absorbed energy to water that flows through copper pipes welded to the plate. An electric pump (not shown) circulates the water between the pipes and an insulated storage tank. The heater plate is housed in a box that contains a glass cover. This shelters the plate from wind, which would otherwise cool it. Typically, the solar heater is sited on the roof of a building and it is orientated so as to maximise the total amount of solar radiation it receives over the course of a year. Importantly, no focusing action by mirrors is involved and so such water heaters can use both direct and diffuse solar radiation (see Section 4.3.2).

Selective absorption of radiation

The heater plate should absorb as much solar radiation as possible. At the same time it should emit as little of its own thermal radiation as possible as this acts to cool it down. Fortunately, the solar spectrum peaks at a wavelength of ~0.5 μm, while a heater plate emits radiation in the far infrared, peaking at ~10 μm. This is illustrated

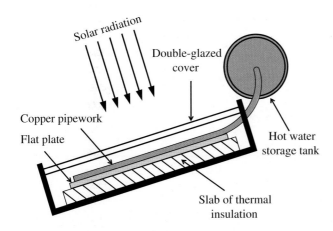

Figure 4.23 Side view of a flat plate water heater. The flat plate is mounted on a slab of thermally insulating material to reduce heat loss due to conduction. A double-glazed cover shelters the flat plate from the wind and also minimises heat loss due to convection. The heated water is stored in a thermally insulated tank.

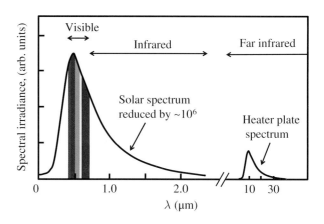

Figure 4.24 Blackbody spectra for temperatures of 5800 and 350 K (77°C), corresponding to the temperature of the Sun and a typical heater plate, respectively. The solar spectrum peaks at about 0.5 μm, while the flat plate spectrum peaks at about 10 μm. To fit on the same scale, the solar spectrum has been reduced in height by a factor of approximately 10^6.

in Figure 4.24, which shows blackbody spectra for 5800 and 350 K, corresponding to the temperature of the Sun and a heater plate at 77°C. This figure is not drawn to scale. The maximum in the Sun's spectrum is about a million times larger than the maximum in the heater plate's spectrum. We see that there is essentially no overlap between the two spectra. Ideally therefore, we would like the surface of the heater plate to have a spectral absorption factor, a_λ, that is close to unity over the region of the solar spectrum, ~0.4–2 μm, but to have a spectral emissivity, e_λ, that is close to zero above 2.5 μm. Some *semiconductors* have absorption/emission characteristics that resemble this ideal behaviour. This is illustrated in Figure 4.25, where the vertical axis corresponds to both e_λ and a_λ. (We recall that the spectral emissivity of a body, at a given wavelength, is equal to its spectral absorption factor at that wavelength.) A semiconductor absorbs only those photons with energies greater than that needed to promote an electron from the *valence band* to the *conduction band*, (see Section 5.4.1). For example, copper oxide is a semiconductor and the minimum photon energy required for electron promotion corresponds to a wavelength of 2 μm. Shorter wavelengths are strongly absorbed, but longer wavelengths are not, which means that

Figure 4.25 Typical absorption and emission coefficients for a metal surface and a semiconductor surface. The vertical axis corresponds to both e_λ and a_λ, as the spectral emissivity of a body, at a given wavelength, is equal to its spectral absorption factor at that wavelength. These surfaces have a high absorption coefficient for solar radiation but a low emission coefficient for the far-infrared radiation emitted by a solar water heater, (~10 μm).

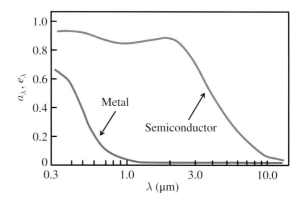

semiconductors are also poor emitters of radiation at longer wave-lengths. Typically, a semiconductor may have a value of a_λ ~0.9 at λ ~0.5 μm and a value of e_λ ~0.1 at λ ~10 μm. Some metals also exhibit an increase in absorption at short wavelengths and low emissivity at long wavelength. For example, copper appears to be reddish in colour because it absorbs more blue light than red light. Moreover metals, unlike semiconductors, are mechanically strong. So a metal surface coated with a semiconductor provides an ideal combination for a selective-absorption material. Copper whose surface has been converted into copper oxide is a good example of this. The oxide layer need only be several microns thick because the high value of a_λ over the solar spectral range means that this radiation is absorbed in a thin layer of material. The heat generated by this semiconductor layer readily passes to the copper substrate. As both the semiconductor layer and the copper substrate have low emissivity at λ ~ 10 μm, heat loss by radiation is minimised. Such a composite material has much lower heat losses than a simple black-painted surface, which has high absorption and emission for both visible and infrared radiation.

The overall efficiency of a flat plate heater for converting solar energy into thermal energy depends on several other factors, including the incident solar power and also heat loss processes, which we will now discuss. However, under typical operating conditions, the overall efficiency is typically between about 40–60%.

4.5.2   Heat transfer processes

Solar energy will raise the temperature of a heater plate above the ambient temperature, and any unwanted heat losses from the plate need to be minimised. The plate will lose thermal energy to its surroundings by the processes of radiation, conduction and convection. We saw earlier that radiation losses are minimised by having a heater plate with a selective absorption surface. Here we consider heat losses by thermal conduction and convection.

Thermal conduction

We saw in Section 4.3.3 that heat will flow across a slab of material when there is a temperature difference between the faces of the slab. In the case of a flat plate water heater, thermal conduction is minimised by mounting the plate on a layer of thermally insulating material such as expanded polystyrene, as shown in Figure 4.23. Expanded polystyrene has the very low value of thermal conductivity κ of 0.03 W/m K. This is because it essentially consists of many small pockets of air. If the width of the expanded polystyrene slab is 75 mm and the temperature of the surface where the copper plate and polystyrene slab join is 30°C above ambient temperature, then the heat loss per unit area due to conduction through the expanded polystyrene would be $0.03 \times 1.0 \times (30/0.075) = 12\,\text{W/m}^2$. This value is small compared with the incident solar power falling on the plate.

Worked example

(a) Calculate the flow of heat through a brick wall of thickness 200 mm and area 6.0 m^2 if the temperatures on either sides of the wall are 20 and 5°C, respectively. Take the thermal conductivity of brick to be 0.60 W/m K. (b) Calculate the heat flow if the wall is lined with a layer of expanded polystyrene of thickness 50 mm and thermal conductivity 0.03 W/m K, as illustrated in Figure 4.26(a).

Solution

(a) From Equation (4.39) the heat flow through the un-insulated brick wall is

$$H_{\text{wall}} = 0.60 \times 6.0 \times \frac{(20 - 5)}{0.20} = 270\,\text{W}.$$

(b) Letting the temperature of the surface where the brick wall and the expanded polystyrene join be T_1, we have for the brick wall

$$H_{\text{wall}} = 0.60 \times 6.0 \times \frac{(T_1 - 5)}{0.20}\,\text{W}.$$

Similarly for the expanded polystyrene, we have

$$H_{\text{poly}} = 0.03 \times 6.0 \times \frac{(20 - T_1)}{0.05}\,\text{W}.$$

The heat flow through the brick wall and the expanded polystyrene must be the same, $H_{\text{wall}} = H_{\text{poly}}$, and so

$$0.60 \times 6.0 \times \frac{(T_1 - 5)}{0.20} = 0.03 \times 6.0 \times \frac{(20 - T_1)}{0.05},$$

Figure 4.26 (a) Cross-section of a brick wall that is thermally insulated by a layer of expanded polystyrene. (b) Analogous electrical circuit. The voltage drop V is analogous to the temperature drop across the insulated wall, while the two resistors are analogous to the thermal resistances of the wall and polysytrene lining.

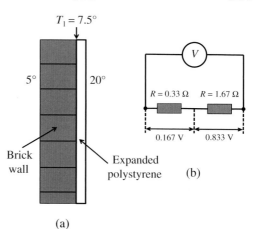

which gives $T_1 = 7.5°C$. Hence the heat flow through the insulated brick wall is

$$0.60 \times 6.0 \times \frac{(7.5 - 5)}{0.20} = 45 \text{ W}.$$

The lining of expanded polystyrene has reduced the heat loss through the wall by a factor of 6.

When designing thermal insulation systems, engineers use the concept of *thermal resistance*, which is denoted by the symbol R. In terms of R, Equation (4.39) becomes

$$H = A\frac{(T_1 - T_2)}{R}, \tag{4.47}$$

where

$$R = \frac{L}{\kappa}. \tag{4.48}$$

Whereas thermal conductivity κ depends only on the type of material, R also depends on the thickness of the material. Consequently, R characterises the effectiveness of a slab of insulating material of a given thickness; the higher the value of R, the better the slab is as an insulator. R has the units of m^2 K/W. Suppose two slabs of material of the same area are joined together *in series*, as in Figure 4.26(a). If the thermal resistances of the two materials are R_1 and R_2, the equivalent thermal resistance R_{equiv} of the combination is given by

$$R_{\text{equiv}} = R_1 + R_2. \tag{4.49}$$

In the above example, the brick wall and the expanded polystyrene have thermal resistances of 0.33 and 1.67 m^2 K/W, respectively. Hence,

$$H = A\frac{(T_1 - T_2)}{R_{\text{equiv}}} = \frac{6 \times 15}{0.33 + 1.67} = 45 \text{ W}.$$

Suppose now that two slabs of insulating material of area A and width L and thermal conductance κ_1 and κ_2, respectively are connected *in parallel*, as shown in Figure 4.27(a). The total heat flow is the sum of the heat flow through each slab of material. Then, as

$$H_1 = \frac{A(T_2 - T_1)}{R_1}; \quad H_2 = \frac{A(T_2 - T_1)}{R_2}$$

$$H_{\text{tot}} = \frac{A(T_2 - T_1)}{R_{\text{equiv}}} = H_1 + H_2 = \frac{A(T_2 - T_1)}{R_1} + \frac{A(T_2 - T_1)}{R_2},$$

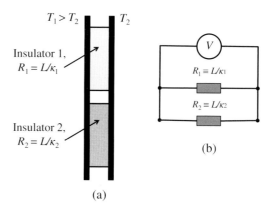

Figure 4.27 (a) Two slabs of insulating material having different values of thermal conductance arranged 'in parallel' between two boundaries at different temperatures. (b) Analogous electrical circuit.

giving

$$\frac{1}{R_{\text{equiv}}} = \frac{1}{R_1} + \frac{1}{R_2}, \tag{4.50}$$

where $R_1 = L/\kappa_1$ and $R_2 = L/\kappa_2$.

In these examples, the analogy between thermal resistance and resistance in an electrical circuit is apparent, especially if we consider the heat flow per unit area through a slab of material:

$$\frac{H}{A} = \frac{(T_1 - T_2)}{R}.$$

We compare this equation with Ohm's law, which gives the current flow I through a resistance R:

$$I = \frac{V}{R}. \tag{4.51}$$

V is the voltage drop across the resistance and corresponds to the temperature drop $(T_1 - T_2)$. The analogous electrical circuit for the example in Figure 4.26a would be a voltage V applied across a *series* combination of two resistors in the ratio 1.67:0.33, as shown in Figure 4.26(b). The analogous electrical circuit for the case where the insulating slabs are connected in parallel Figure 4.27(a) is shown in Figure 4.27(b). This analogy between current flow and heat flow is an example of the similarities that occur in physics, where different physical situations are described by differential equations of the same form.

Worked example

Figure 4.28 shows a hollow cylinder made from an insulating material that has thermal conductivity κ. The radii of the inner and outer surfaces of the cylinder are r_1 and r_2, respectively, and these surfaces are at temperatures T_1 and T_2, respectively, where $T_1 > T_2$. Obtain an expression for $T(r)$, the temperature as a function of radial distance r, within the insulating material.

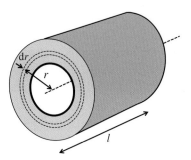

Figure 4.28 A hollow cylinder of insulating material with thermal conductivity κ. The radii of the inner and outer surfaces of the cylinder are r_1 and r_2, respectively, and these surfaces are at temperatures T_1 and T_2, respectively, where $T_1 > T_2$.

Solution

Clearly, the cylinder is a three-dimensional object. However, because of its cylindrical symmetry, the temperature varies only with radial distance r. Hence, we can use the one-dimensional form of Fourier's law (see Equation 4.40):

$$H = -\kappa A \frac{\mathrm{d}T}{\mathrm{d}r},$$

where here the term 'one-dimensional' refers to the number of coordinates needed to describe the distribution of temperature and not to the number of spatial dimensions the cylinder occupies. We consider a cylindrical shell of radius r and width $\mathrm{d}r$ within the insulating material. At equilibrium, the heat flux H passing through any cylindrical surface within the insulating material will be constant. As the area of the shell is $2\pi rl$, we have

$$H = -2\pi rl\kappa \frac{\mathrm{d}T}{\mathrm{d}r}.$$

We integrate this equation from radius r_1, where the temperature is T_1, to some arbitrary radius r, where the temperature is T:

$$\int_{T_1}^{T} \mathrm{d}T = -\frac{H}{2\pi l\kappa} \int_{r_1}^{r} \mathrm{d}r,$$

which gives

$$T(r) = T_1 - \frac{H}{2\pi l\kappa} \ln\left(\frac{r}{r_1}\right).$$

We see that the temperature varies logarithmically with radius r.

Thermal convection

Thermal convection is the transfer of heat by *mass motion* of a fluid from one region of space to another. A familiar example is air rising from a hot radiator. Convection is a complex process and there is no simple equation to describe it. It is found, however, that the heat flow due to convection is proportional to surface area and it is approximately proportional to the temperature difference between the surface and the main body of the fluid. In the case of a flat plate heater, there will be heat loss due to convective heat transfer between the plate and the glass cover of the heater. This heat loss is minimised by double-glazing the glass cover above the heater plate, as shown in Figure 4.23. Then the loss due to convection is similar in magnitude to that due to thermal conduction.

Thermal transmission by glass cover

Glass has a high transmission for visible and near-infrared radiation, but it is essentially opaque for far-infrared radiation. The glass cover thus transmits the solar radiation but absorbs the radiation emitted by the heater plate. Some of the energy absorbed by the glass cover is then re-radiated back to the plate, thus minimising the energy lost. The glass cover does, however, attenuate the solar radiation to some degree; any wave that encounters a boundary, where there is a change in refractive index, is partially reflected at that boundary. When light of normal incidence passes from a medium of refractive index n_1 to a medium of refractive index n_2, the fraction R that is reflected is given by

$$R = \frac{(n_1 - n_2)^2}{(n_1 + n_2)^2}. \tag{4.52}$$

Taking $n_1 = 1$ for air and $n_2 = 1.5$ for the glass, $R = 0.04$. The light passing through a glass window is not only reflected on its front surface, but also on the back surface. In fact, the light may be reflected back and forth several times. Taking this into account, the total reflectance of a glass window is

$$\frac{2R}{(1 + R)} = 0.077.$$

There will also be a small amount of absorption in the glass, $\sim 2\%$, and so the total loss of intensity $\sim 10\%$. If the glass cover is doubly glazed, the reflection losses will be $\sim 20\%$, but this increased loss with double-glazing is usually more than compensated by the reduction in convection losses. Reflection losses at the glass surfaces can be minimised by coating them with a material of appropriate refractive index and thickness, as is done for camera lenses.

Worked example

The average solar power per square metre incident on a south-facing roof in the UK is ~ 120 W and the typical consumption of hot water ($60°C$) is ~ 100 L per household. Making reasonable estimates, work out the required area of a solar water heater to supply this amount of hot water. The specific heat of water is 4.2 kJ/kg K.

Solution

Energy required to heat 100 kg of water through, say, 50°C is

$$100 \times 4.2 \times 50 = 2.1 \times 10^4 \text{ kJ}.$$

Assuming the Sun shines for 6 hours and the efficiency of the solar heater is 50%, the required area of heater is

$$\frac{2.1 \times 10^7}{120 \times 0.5 \times 6 \times 60^2} \sim 16 \, \text{m}^2.$$

Vacuum tube collectors

Heat losses due to convection and conduction that occur in a flat plate heater are eliminated in the *vacuum tube collector*. This collection system has two concentric glass tubes, as illustrated in Figure 4.29. The two glass tubes are held apart and, importantly, the space between them is evacuated. Because the atmospheric gas is removed there is negligible thermal convection or conduction between the two glass tubes. The *outer* surface of the *inner* glass tube is coated with a material with selective absorption to maximise the amount of absorbed radiation. The inner glass tube contains water and is connected to a water storage tank. Typically the water heater would have about 20 tubes, each 1–2 m long. Apart from eliminating convection and conduction losses, the vacuum tube collector design has another important advantage. The selective absorption surface is not exposed to the atmosphere. Thus certain materials that have superior absorption/emissivity characteristics but are not stable in atmosphere can be used. Because of these advantages, the efficiency of a vacuum tube collector is usually higher than that of a flat plate water heater of the same collecting area, and may be ~80%.

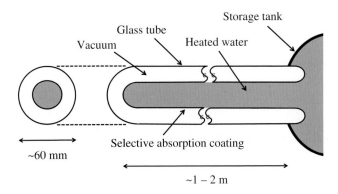

Figure 4.29 Vacuum tube collector. The space between the two glass tubes is evacuated, which eliminates thermal conduction and convection. The *outer* surface of the *inner* tube is coated with a selective absorption material to maximise absorption of solar radiation. A typical water heater would contain about 20 evacuated tubes.

Figure 4.30 (a) Schematic diagram of a parabolic trough concentrator. Solar radiation is concentrated by parabolic mirrors onto a pipe that runs along the focus of the mirrors. The pipe carries synthetic oil which is heated by the solar radiation. Each mirror can be rotated about the axis containing the pipe to track the position of the Sun. (b) The focusing action of the parabolic mirrors.

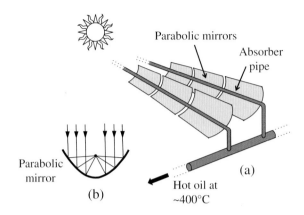

4.5.3 Solar thermal power systems

Higher temperatures than those provided by a flat plate heater or a vacuum tube collector are achieved by collecting solar radiation over a large area and concentrating it with mirrors into a much smaller area. This is done practically on a small scale in the case of solar ovens for cooking food. Such ovens are commonly found, for example, in India. In these ovens, the cooking pot is held at the focus of a reflector and the position of the reflector is adjusted manually to maximise the collection of solar radiation. A demonstration model[2] at the University of Manchester consists of a concave surface with a diameter of 50 cm that is covered with aluminium foil to form a reflector with a focal length of 60 cm. When this reflector is illuminated by direct sunlight it can toast a marshmallow within a matter of seconds! For producing power on an industrial scale, however, mirrors with a total collection area of many square kilometres are required. Two examples of systems that do this are the *parabolic trough concentrator* and the *central receiver with heliostats*, both of which are used for the generation of electricity.

Parabolic trough concentrator

This system has a very large number of reflecting mirrors that have a parabolic cross-section, as illustrated in Figure 4.30. The length of each line of mirrors is several kilometres. Light incident on a reflector is focused onto a *receiver pipe* that lies at the focus of the reflector and runs the length of the mirrors. Because of this focusing action the reflectors can concentrate solar radiation to ~50 times its normal irradiance. Synthetic oil circulates through the pipes, reaching temperatures up to about 400°C. The hot oil is pumped to a heat exchanger to produce steam,

---

[2] This demonstration model was made from a redundant television satellite dish.

which drives a conventional steam turbine. Usually, the long axis of the reflectors is orientated in a north–south direction and the reflectors are mechanically rotated about this axis to track the position of the Sun. One installation, called the Solar Energy Generating System is located in the Mojave Desert, California. The total area of the 232 500 parabolic mirrors is $6.5 \times 10^6 \, \text{m}^2$, and the installation has a total capacity of 354 MW. This generates enough electricity to meet the needs of about half a million people. On cloudy days or at night, alternative heating for the steam turbines is provided by natural gas. In 2015, an announcement was made to build a 1 GW solar energy plant in Oman using parabolic troughs. It is named Miraah and would be one of the largest solar plants in the world.

Central receiver with heliostats

This system consists of a large number of *plane* mirrors, called *heliostats*. These mirrors reflect sunlight onto a central receiver that is placed on top of a tower, as illustrated in Figure 4.31. The mirrors are mechanically rotated about two perpendicular axes to track the position of the Sun. Very high temperatures, ~550°C, can be reached and so molten salt is used as the working fluid. The molten salt is stored in a tank. This stores the thermal energy of the molten salt and it is important to note that *it is much easier to store thermal energy than it is to store electricity*. Salts are an effective storage medium because they have a high specific heat capacity and are low-cost. The molten salt can be used directly to heat water to drive steam turbines. Alternatively, the stored molten salt can be used to drive the turbines at night or during days when the Sun is not shining. The Gemasolar Solar Power Facility, which is situated in Fuentes de Andalucía, Spain, was the first commercial-scale plant in

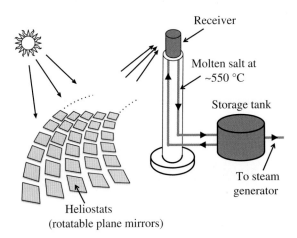

Figure 4.31 Schematic diagram of a solar power plant consisting of a central receiver that is surrounded by plane mirrors (heliostats). The heliostats concentrate solar radiation onto the receiving tower, which contains molten salt. This molten salt is pumped to a storage tank and can then be used to heat water for a steam turbine. The storage tank enables the molten salt to be used when the Sun is not shining.

Figure 4.32 A photograph of the Gemasolar Solar Power Facility, which is situated in Fuentes de Andalucía, Spain. It has 2650 heliostats, which have a total reflective area of 304 750 m² and produces an electrical power of ∼20 MW. The inclusion of the molten salt heat storage system enables independent electrical generation for up to 15 hours without any solar radiation. Source: SENER/ TORRESOL ENERGY.

the world to use a central tower receiver and thermal energy storage technology using molten salt. A photograph of the facility is shown in Figure 4.32. It has 2650 heliostats, which have a total reflective area of 304 750 m², and produces an electrical power of ∼20 MW. The inclusion of the molten salt heat storage system enables independent electrical generation for up to 15 hours without any solar radiation.

4.6   Heat engines: converting heat into work

Solar thermal power systems are sources of *thermal energy* at high temperature, as are nuclear reactors and gas-fired power stations. In our technological world, it is crucially important to be able to convert this thermal energy into mechanical energy. We now turn our attention to how this is achieved. Typically, the thermal energy is used to heat water to produce superheated steam that drives a turbine. The turbine, in turn, drives an electrical generator. A crucially important parameter is the efficiency with which the heat can be converted into work: the ratio of heat input to work output.

Converting work into heat is quite straightforward. We do this when we rub our hands together to warm them in winter. It is the friction that produces the heat. Moreover, we can convert work into heat with almost 100% efficiency. The equivalence of work and heat was established around 1850 by the classic experiments of James Joule, a brewer's son from Manchester. He performed a known amount of work on a mass of water and measured the temperature rise of the water. It takes 4.184 J of work to raise the temperature of 1 g of water by 1°C, a result that is known as the *mechanical equivalent of heat*. On the other hand, and very importantly, *we can never build a machine that converts heat completely into work*. A cyclic machine that partly converts heat into work is called a *heat engine*. Examples of heat engines are the steam turbine and the internal combustion engine. By the beginning of the 18th century, steam engines were being used to pump water out of coalmines. An important question arose: given a source of heat at a certain temperature, e.g. a steam boiler, what is the maximum amount of work that can be extracted from that source, i.e. how efficient can a heat engine be? It was a young engineer called Sadi Carnot who, in 1824, provided the answer. This is a striking example of the discipline of engineering providing a fundamental and far-reaching contribution to physics. Carnot's work was subsequently taken up by Lord Kelvin and others to establish the subject of classical thermodynamics, which is essentially the study of how energy can be transformed into its various forms. Indeed, the first and second laws of thermodynamics are central to a discussion of heat engines. The operation of heat engines usually involves the expansion and compression of a gas and so we will describe the behaviour of gases under various circumstances.

4.6.1   Equation of state of an ideal gas

The state of an *ideal* gas is governed by the following relationship between its pressure, p, volume, V, and temperature, T:

$$pV = NkT, \tag{4.53}$$

where N is the number of molecules in volume V and k is the Boltzmann constant. This equation follows from the empirical laws of Boyle, Charles and Gay-Lussac. Equation (4.53) is called an *equation of state*. Knowing two of the variables p, V and T, the third one fixes. One mole of gas contains Avogadro's number N_A of molecules, and hence we have for 1 mole:

$$pV = N_A kT = RT, \tag{4.54}$$

where $R = N_A k$ is the gas constant. R has the value 8.31 J/mol K, and is the same for *all* gases. If we have n moles of gas the equation of state is

$$pV = nRT. \tag{4.55}$$

This equation is also called the *ideal gas equation*. An ideal gas is one for which Equation (4.55) holds for all pressures and temperatures. This is an idealised model. However, under normal conditions of pressure and temperature, as encountered in our discussions, most gases can be considered to be ideal.

The equation of state strictly applies to a gas in a state of thermal equilibrium. When we expand a gas, say in a cylinder with a moveable piston, the temperature or the pressure, or both, must change since these variables are connected by the equation of state. If we suddenly pull the piston out to expand the gas, the pressure will initially be lower near the piston than far from it. Eventually the gas will settle down to a new equilibrium pressure and temperature, but we cannot determine the pressure or temperature of the entire gas system until equilibrium is restored. If, however, we move the piston *slowly in small steps* and allow equilibrium to be re-established after each step, we can expand the gas in such a way that the gas is never far from an equilibrium state. Such processes in which the system really passes through a succession of equilibrium states are called *quasi-static processes*. For such quasi-static processes, we can always apply the equation of state. In our illustrations of the expansion and compression of gases, we will always assume the processes to be quasi-static.

Worked example

Estimate the average distance between the molecules in a gas at room temperature and atmospheric pressure.

Solution

From the equation of state (Equation 4.53), we have

$$\frac{V}{N} = \frac{kT}{p}.$$

For the purpose of this estimation, we imagine the molecules to be uniformly distributed at a distance d apart from each other in a cube of side length L, with $L^3 = V$. Then

$$\sqrt[3]{N} \times d \approx L, \quad \text{or} \quad d^3 \approx \frac{V}{N}.$$

Hence,

$$d^3 \approx \frac{kT}{p} = \frac{1.38 \times 10^{-23} \times 293}{1.01 \times 10^5} = 4.0 \times 10^{-26} \text{ m},$$

where we have taken $T = 293$ K and $p = 1.01 \times 10^5$ Pa. This gives $d \sim 3 \times 10^{-9}$ m. The diameter of a molecule is $\sim 1 \times 10^{-10}$ m and so this estimate indicates that the molecules are on average about 30 molecular diameters apart from each other. The force between neutral molecules has a short range, i.e. it falls off very rapidly with distance. Consequently, at a distance of ~ 30 molecular diameters, the potential energy of interaction between the molecules is negligible.

4.6.2 Internal energy, work and heat: the first law of thermodynamics

Internal energy, work and heat are different in nature but they are all forms of energy. They are connected through the *first law of thermodynamics*, which simply expresses the fact that *energy is conserved if heat is taken into account.*

Internal energy

Matter consists of particles that have kinetic energy due to their motion and also potential energy due to their interactions with other particles. The *internal energy* of a system is the sum of the kinetic energies of all its particles plus the sum of the potential energies of interaction among these particles. The usual symbol for internal energy is U. The internal energy does *not* include potential energy arising from any interaction between the system and its surroundings. For the case of molecules in an *ideal gas*, the molecules do not interact with each other and so the potential energies of interaction are zero. As we saw in the worked example in the preceding section, gases can be considered to be ideal under normal conditions. The internal energy of an ideal gas is thus due solely to the kinetic energies of the molecules.

We can deduce the internal energy of a gas using the *equipartition theorem*, a result that follows from classical statistical mechanics. This theorem states that when molecules are in thermal equilibrium, the contribution of each *degree of freedom* to the average energy of a molecule is $\frac{1}{2}kT$, where k is the Boltzmann constant and T is temperature. A degree of freedom is a term in the expression for the total energy of the molecule that is quadratic in an independent dynamical variable. We may think of a degree of freedom as an independent way in which a molecule can store energy. For example, the total energy of a molecule in a monatomic gas like helium has the form

$$E = \frac{1}{2}mv_x^2 + \frac{1}{2}mv_y^2 + \frac{1}{2}mv_z^2, \qquad (4.56)$$

where the velocity components v_x, v_y and v_z are independent of each other. This expression has three quadratic terms and so a monatomic gas has three degrees of freedom. Moreover the average energy of each term is $\frac{1}{2}kT$:

$$\left\langle \frac{1}{2}mv_x^2 \right\rangle = \left\langle \frac{1}{2}mv_y^2 \right\rangle = \left\langle \frac{1}{2}mv_z^2 \right\rangle = \frac{1}{2}kT. \qquad (4.57)$$

The total internal energy U of a monatomic molecule is thus $\frac{3}{2}kT$, and the total energy of 1 mole is $\frac{3}{2}RT$.

In the case of a diatomic molecule like nitrogen or oxygen, the molecule also has rotational motion. It can rotate about the two axes that are perpendicular to the molecular bond joining the two atoms; one of the two rotational motions is shown in Figure 4.10(b). Thus, as well as the three terms due to translational motion of its centre of mass in the x, y and z directions, a diatomic molecule has two additional terms due to these rotational motions. The rotational energy can be written as

$$E_{\text{rot}} = \frac{1}{2}I\omega_1^2 + \frac{1}{2}I\omega_2^2, \qquad (4.58)$$

where I is the moment of inertia about the centre of mass of the molecule and ω_1 and ω_2 are the rotational angular frequencies. Each term on the right-hand side of the equation has the quadratic form required for it to contribute $\frac{1}{2}kT$ to the total energy of the molecule. The internal energy U of a diatomic molecule is thus $\frac{3}{2}kT + \frac{2}{2}kT = \frac{5}{2}kT$, or equal to $\frac{5}{2}RT$ for 1 mole of gas.

Diatomic molecules may also vibrate. However, vibrational motion is not usually excited until the gas is heated to a few thousand °C. For example, it takes about 0.3 eV to excite vibrational motion of an N_2 molecule, while the thermal energy ($\sim kT$) available to the molecule at room temperature is ~ 0.025 eV. Hence the molecules cannot store energy in vibrational motion and so in our present discussions we can usually neglect the contribution of vibrational motion to the internal energy of a gas.

We emphasise here that *the internal energy of an ideal gas depends on its temperature T, but not on its pressure or volume*, as illustrated by the above results.

Work

Work is the mechanical energy that we impart to an object when we apply a force to it over a given distance – work is (force) × (distance moved by the force). For example, if we do work by compressing a spring of spring constant k by a distance x, the amount of mechanical energy that we impart to the spring is kx. Imagine that we have a gas that is contained in a cylinder with a moveable and frictionless piston, as illustrated by Figure 4.33(a). The mass and area of the piston are m and A, respectively. The force pushing down on the gas is due to the weight of the piston and the atmospheric pressure, p_{atm}, and is equal to

$$A \times p_{\text{atm}} + mg,$$

where g is the acceleration due to gravity. The upward force exerted by the gas is $p \times A$, where p is the pressure of the gas.

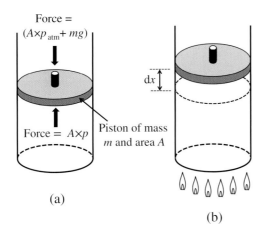

Figure 4.33 Illustration of the
work done when a gas
expands at constant pressure.
(a) Gas at pressure p is
contained in a cylinder with
a moveable and frictionless
piston. At equilibrium, the
downward force due to the
weight of the piston and
atmospheric pressure is
balanced by the upward force
due the pressure of the gas.
(b) If the piston moves an
infinitesimal distance, dx, the
work, dW, done by the gas is
equal to pdV, where
$dV = Adx$ is the infinitesimal
change in volume of the gas.

At the equilibrium position of the piston these forces must balance and so

$$(A \times p_{\mathrm{atm}} + mg) = p \times A.$$

We can expand the gas by gently heating it as depicted in Figure 4.33(b). When the gas expands it does work as it pushes on the face of the piston with a force $F = pA$. If the piston moves an infinitesimal distance dx, the work dW done by the gas is

$$dW = Fdx = pAdx = pdV. \qquad (4.59)$$

where dV is the infinitesimal change in volume of the system. In the expansion of the gas, the force acts in the same direction as the displacement of the piston and we say that the expanding gas does *positive* work. If, on the other hand, the gas is compressed by an external force, the work done by the gas has a negative sign; equivalently, an external force does work *on* the gas. In a finite change of volume from V_1 to V_2, with $V_2 > V_1$, the total work done by the gas is

$$W = \int_{V_1}^{V_2} pdV. \qquad (4.60)$$

In the example shown in Figure 4.33, the pressure p of the gas remains constant during the expansion. In general, the pressure p will vary during the volume change, and to evaluate the integral in Equation (4.60) we have to know how the pressure changes as a function of volume. We can represent this relationship as a plot of p against V and such a p–V *diagram* is shown in Figure 4.34. We identify the elemental area dW indicated on the figure, as the work done in expanding the gas by dV. The total work, W, performed by the gas during an expansion from V_1 to V_2 is represented by the area under the p–V curve between V_1 and V_2. If the gas is compressed, its volume decreases and the work done by the gas is a negative quantity.

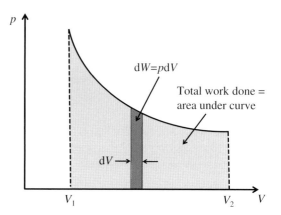

Figure 4.34 A p–V diagram for a gas undergoing an expansion where the pressure p varies with volume V. The elemental area is the work, dW, done when the gas volume changes by dV. The total work performed by the gas is represented by the area under the p–V curve between V_1 and V_2.

We can understand how the gas does work by considering the motion of the molecules in the gas. The pressure of a gas is due to the molecules striking the walls of its container. When a molecule impacts a stationary wall, it exerts a momentary force but it does not do any work because the wall does not move. However the molecule *does* do work if the wall is *moving*. In an expansion of the gas, the piston is moving away from an incident molecule and so the speed of the molecules after the impact will be less than it was before. The decrease in molecular kinetic energy equals the work done on the piston – internal energy of the gas has been converted into work.

Heat

Heat is also a form of energy, but it is not correct to say that a system contains a certain amount of heat. Rather, we can say that a system has a certain amount of internal energy. Heat, denoted by the symbol Q, is the energy that is *transferred* from one system to another because of a difference in temperature; *heat is energy in transition*. (Despite this definition of heat, we are inclined to say that heat *flows* from a body. This is fine so long as we remember that heat is *not* any substance that the body contains.) For example, imagine two identical blocks of copper of mass 1.0 kg; one has a temperature of $10°C$ and the other has a temperature of $20°C$. When brought into thermal contact, the two blocks will eventually come to an equilibrium temperature of $15°C$ and the amount of heat that is transferred between them is

$$Q = C \times m \times (20 - 10)/2 = 386 \times 1.0 \times 5.0 = 1.93\,\text{kJ},$$

where C is the specific heat of copper, equal to 386 J/kg K, and m is the mass of each body. If a system absorbs heat, Q is a positive quantity, and if heat is removed from a system, Q is a negative quantity.

In general, when we add heat to a system, the heat will partly go to increasing the internal energy of the system and partly into doing work (see the worked example in Section 4.6.3). When the transfer of both heat Q and work W occur, conservation of energy requires that

$$Q = \Delta U + W, \qquad (4.61)$$

where ΔU is the total change in the internal energy of the system, or

$$\Delta U = Q - W. \qquad (4.62)$$

This is a statement of the *first law of thermodynamics*. It says that energy is conserved if heat is taken into account.

4.6.3 Specific heats of gases

Suppose that we add an amount of heat to an ideal gas that is confined to a cylinder of fixed volume. This will raise the temperature of the gas. For 1 mole of gas, we define the *specific heat per mole at constant volume C_V* as

$$C_V = \frac{\mathrm{d}Q_V}{\mathrm{d}T}, \qquad (4.63)$$

where $\mathrm{d}T$ is the temperature rise of 1 mole of the gas when an amount of heat $\mathrm{d}Q_V$ is added to it. As the volume does not change, none of the added heat goes into doing work; all of the heat goes into increasing the internal energy of the gas. Hence, from the first law of thermodynamics (Equation 4.62), $\mathrm{d}U = \mathrm{d}Q_V$, giving

$$C_V = \frac{\mathrm{d}U}{\mathrm{d}T}. \qquad (4.64)$$

We see that the specific heat of a gas at constant volume is the differential of internal energy U with respect to temperature T.

We can also heat a quantity of gas while keeping its pressure constant. For 1 mole of gas, we define C_p, the *heat capacity per mole at constant pressure*, as

$$C_p = \frac{\mathrm{d}Q_p}{\mathrm{d}T}, \qquad (4.65)$$

where $\mathrm{d}T$ is the temperature rise of 1 mole of the gas when an amount of heat, $\mathrm{d}Q_p$, is added to it. To maintain a constant pressure p, the gas must expand, say by an amount $\mathrm{d}V$, and so it does work, $\mathrm{d}W$, equal to $p\mathrm{d}V$. Hence from the first law of thermodynamics:

$$\mathrm{d}Q_p = C_p\mathrm{d}T = \mathrm{d}U + p\mathrm{d}V. \qquad (4.66)$$

Raising the temperature of 1 mole of gas by a certain amount takes more energy if we do this at constant pressure than if we do it at constant volume.

We can readily derive a relationship between C_V and C_p for an ideal gas. The essential idea is that the change in the internal energy of an ideal gas is the same for *any* two processes that involve the same change in temperature ΔT. Consider that we raise the temperature of 1 mole of gas by ΔT, at constant volume. Then from Equation (4.64)

$$\Delta U = C_V \Delta T. \tag{4.67}$$

If, on the other hand, we raise the temperature of 1 mole of gas by ΔT, at constant pressure, then from Equation (4.66)

$$\Delta U = C_p \Delta T - p\Delta V. \tag{4.68}$$

As ΔT is the same for both processes, ΔU is also the same for both. Hence,

$$C_p \Delta T - p\Delta V = C_V \Delta T. \tag{4.69}$$

From the ideal gas equation for 1 mole of gas, $pV = RT$, we can say that

$$\frac{\Delta V}{\Delta T} \approx \frac{\mathrm{d}V}{\mathrm{d}T} = \frac{R}{p},$$

Substituting for $p\Delta V = R\Delta T$ in Equation (4.69) we obtain

$$C_p - C_V = R. \tag{4.70}$$

We see that the difference between the two specific heats is equal to R, the gas constant. We have obtained this result for the case of an ideal gas, but it is found experimentally that it is obeyed to within a few percent by many real gases at moderate pressures.

Worked example

1 mole of argon gas at atmospheric pressure is initially at 20°C. (a) Determine its internal energy. (b) Determine the final internal energy and the work done by the gas if 500 J of heat is added to the gas at constant pressure. (c) Determine the same quantities if the heat is added while the volume of the gas is kept constant.

Solution

(a) Argon gas is monatomic and so the internal energy of 1 mole of this gas is

$$U = \frac{3}{2}RT = \frac{3}{2} \times 8.314 \times (273 + 20) = 3.654 \text{ kJ}.$$

(b) $C_V = dU/dT$ and hence for 1 mole of a monatomic gas, $C_V = 3R/2$, giving $C_p = 5R/2$. From $C_p = dQ_p/dT$, the rise in temperature of the gas is

$$\frac{500}{(5/2) \times 8.314} = 24.1°C.$$

Hence the final internal energy is

$$(3/2) \times 8.314 \times (273 + 24.1) = 3.955 \, kJ$$

From the first law of thermodynamics, $\Delta U = Q - W$, the work done by the gas is equal to $500 - (3955 - 3654) = 199$ J.

(c) If the volume is constant, no work is done by the gas and, using $C_V = dQ_V/dT$, the rise in temperature of the gas is

$$\frac{500}{(3/2) \times 8.314} = 40.1°C.$$

Hence, the final temperature of the gas is $60.1°C$, and the change in its internal energy $= 1.0 \times \frac{3}{2} \times 8.314 \times 40.1 = 500$, J, which is equal to the added amount of energy.

4.6.4   Isothermal and adiabatic expansion

Isothermal expansion of a gas occurs when the temperature of the gas is kept constant as it expands. We can arrange this in practice by putting the gas cylinder in thermal contact with a *thermal reservoir*. By thermal reservoir we mean a body that has a sufficiently large heat capacity that it can deliver or absorb heat without incurring any change to its temperature. As the temperature T is constant, the equation of state, $pV = nRT$, gives

$$pV = \text{constant.} \tag{4.71}$$

This is the familiar Boyle's law. The work done by the gas when it expands is provided by the heat absorbed from the reservoir. In effect, the gas serves as a conduit that transfers heat from the reservoir to mechanical energy for the piston, thereby enabling the piston to do work.

Adiabatic expansion occurs when the gas expands with no exchange of heat with its surroundings, so that $dQ = 0$. We can arrange this by lagging the cylinder with thermally insulating material. As $dQ = 0$ for an adiabatic expansion, the first law of thermodynamics gives:

$$\Delta U = -pdV. \tag{4.72}$$

The work done by the gas in an adiabatic expansion is at the expense of the internal energy of the gas and so the temperature of the gas decreases. Equation (4.72) gives the internal energy change, dU, for any process for an ideal gas, whether it is adiabatic or not. So we have $dU = C_V dT$ for 1 mole of gas, where dT is the decrease in temperature. Substituting for $\Delta U = C_V \Delta T$ in Equation (4.72) gives

$$C_V dT = -pdV. \tag{4.73}$$

For a quasi-static expansion of the gas, the equation of state $Pp = RT$ holds, and so for any small changes in p, V and T:

$$pdV + Vdp = RdT. \tag{4.74}$$

Substituting dT from this equation into Equation (4.73) gives

$$C_V(pdV + Vdp) = -RpdV.$$

Then, using $R = (C_p - C_V)$, we obtain

$$Vdp + \gamma pdV = 0, \tag{4.75}$$

where $\gamma = C_p/C_V$ is the *ratio of specific heats*. Integrating:

$$\int \frac{dp}{p} + \gamma \int \frac{dV}{V} = 0, \tag{4.76}$$

and hence for an adiabatic expansion,

$$pV^\gamma = \text{constant}. \tag{4.77}$$

The value of γ is always greater than 1, and so an adiabatic curve is always steeper than an isothermal curve passing through the same point on a P–V diagram.

Worked example

The compression ratio of a diesel engine is 18:1. If the initial pressure and temperature are 1.01×10^5 Pa and 20°C, respectively, calculate the final temperature and pressure after compression, at which point the diesel oil is injected. Take the value of γ for air to be 1.40. Assume that the compression happens so quickly there is no transport of heat to the surroundings and hence the process can be considered to be adiabatic. What would be the final temperature and pressure if the compression process were isothermal?

Solution

As the compression can be considered to be adiabatic, we have, from Equation (4.77),

$$p_2 = p_1 \left(\frac{V_1}{V_2}\right)^\gamma = (1.01 \times 10^5)(18)^{1.4}$$

$$= 57.8 \times 10^5 \, \text{Pa} = 57.2 \, \text{atm}.$$

Substituting for p from the equation of state, we obtain

$$T_2 = T_1 \left(\frac{V_1}{V_2}\right)^{\gamma-1} = 293 \times (18)^{0.4}$$

$$= 931\,\text{K} = 658°\text{C}.$$

This high temperature causes the fuel to ignite spontaneously, without the need for spark plugs. If the compression had been isothermal, the final pressure would be

$$p_2 = p_1 \left(\frac{V_1}{V_2}\right) = 18\,\text{atm}.$$

Because the temperature increases in an adiabatic compression, the final pressure is much greater.

4.6.5   Heat engines and the second law of thermodynamics

A heat engine is any machine that, working in a cycle, transforms heat *partly* into mechanical work. Figure 4.35 illustrates the general arrangement for a heat engine. There is a *hot reservoir* at temperature T_H, which is the source of heat, and a *cold reservoir* at temperature T_C. The heat engine is represented by the circle in the figure. There is also a *working substance* that flows through the heat engine, connecting the hot and cold reservoirs. The working substance absorbs heat from the hot reservoir and the heat engine converts some of this heat into work. However, not all the heat extracted from a hot reservoir can be converted into mechanical work and some of the heat must be discarded to the cold reservoir. This action is repeated in a cyclical manner, with the ongoing conversion of heat into mechanical work. The fact that some heat must be discarded in this process is a consequence of the second law of thermodynamics. There are several statements of this law, but one statement that is particularly appropriate here is:

It is impossible to build a heat engine working in a cycle that absorbs heat from a source of heat and converts the heat completely into mechanical work.

It is the second law of thermodynamics that sets the limit to the highest efficiency that a heat engine can have.

   Referring to Figure 4.35, Q_H is the amount of heat absorbed from the hot reservoir in a complete cycle while Q_C is the amount of heat discarded to the cold reservoir and W is the amount of mechanical work delivered by the heat engine. The arrows indicate the directions of heat flow and work output. In a complete cycle, the working substance and the engine are returned to their original states and so

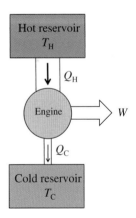

Figure 4.35 Pictorial representation of a heat engine that operates between two thermal reservoirs at temperatures of T_H and T_C, where $T_H > T_C$. A quantity of heat, Q_H, is absorbed from the hot reservoir and a quantity of heat Q_C is discarded to the cold reservoir. The work output of the engine is $W = (Q_H - Q_C)$.

their initial and final internal energies are the same. Hence, from the conservation of energy, i.e. the first law of thermodynamics:

$$W = Q_H - Q_C. \tag{4.78}$$

We define the *thermal efficiency*, ε, of a heat engine as the amount of mechanical work delivered by the engine divided by the amount of heat absorbed from the hot reservoir:

$$\varepsilon = \frac{W}{Q_H}. \tag{4.79}$$

Substituting for W from Equation (4.78) we obtain

$$\varepsilon = 1 - \frac{Q_C}{Q_H}. \tag{4.80}$$

The discarded heat Q_C can never be zero by the second law of thermodynamics and so the efficiency of a heat engine is always less than unity.

We emphasis two essential features of a heat engine:

• A heat engine must operate in a cyclical manner.

• There must be a temperature difference between the two thermal reservoirs.

A heat engine must work in cycles if it is to be useful. We can see this by considering a process that is not cyclical: the isothermal expansion of an ideal gas that is confined in a cylinder with a moveable piston. The gas is held at a constant temperature by being in contact with a thermal reservoir. As the gas expands, it does work. However, this can only be done once. The gas pressure, which is inversely proportional to volume, will decrease until at some point the expanding gas will no longer be able to push the piston and deliver any work.

We can also consider the isothermal expansion and subsequent compression of an ideal gas to understand why a heat engine must operate between reservoirs at different temperatures. As the gas expands in contact with a thermal reservoir at temperature, say, T_0, it does work. Consider now that after it has expanded to a certain volume, we compress the gas to its original volume and we do this while it is still in contact with the reservoir at temperature T_0. Then exactly the same amount of work that was obtained from the expansion of the gas would be needed to compress it to its original volume. No net work has been gained in this two-step cycle. To obtain a net amount of work over a cycle, the gas cylinder must be placed in contact with a second thermal reservoir at a lower temperature when it is recompressed. At a lower temperature, the pressure is lower and so less work is required to compress the gas and so there is a net output of work. We will see how this is achieved in the most advantageous way in the *Carnot cycle*.

The Carnot cycle

Sadi Carnot found that a *reversible engine* is the most efficient heat engine that can operate between two thermal reservoirs. Such a reversible engine is called a *Carnot engine* and its cycle of operation is called a *Carnot cycle*. Carnot reasoned that, if no engine can have a greater efficiency than a Carnot engine, then it follows that all Carnot engines operating between the same two reservoirs have the same efficiency. This efficiency must therefore be independent of the working substance of the engine and so *the efficiency must only depend on the temperatures of the two reservoirs*. Interestingly, Carnot came to this conclusion before the first or second laws of thermodynamics had been established.

A process is reversible if the system *and* its surroundings change in such a way that both can be returned to their original condition by reversing the process. For a process to be reversible we require:

(a) The process must be quasi-static so that the system is always in, or infinitesimally near, an equilibrium state. For example, a gas must not be expanded so fast that eddy currents and turbulence are set up that would dissipate energy.

(b) No mechanical energy is transformed into heat by dissipative forces such as friction. Friction can transform work into heat, but friction can never transform heat into work.

(c) Heat transfer can only occur between bodies at, or infinitesimally near, the same temperature. Heat flows from a hot body to a cold body but never the other way around; the flow of heat from a hot object to a cold object is not reversible. In the Carnot cycle, heat transfer only occurs between bodies at the same temperature, even though the hot and cold reservoirs are at different temperatures.

As the efficiency of a Carnot engine is independent of the working substance, we can choose whichever substance we like to analyse its operation – the results will be valid in general. We therefore chose the ideal gas as the working substance, as Carnot did. A complete cycle, which consists of taking the gas through four steps of reversible expansions and compressions, is shown schematically in Figure 4.36. This figure also represents these four steps on a p–V diagram:

• *Step 1: from a to b on p–V diagram.* A cylinder containing the gas is placed in contact with the hot reservoir and the gas is expanded *isothermally*, at constant temperature T_H. The gas absorbs

Figure 4.36 Illustration of a gas in a cylinder undergoing a Carnot cycle between hot and cold reservoirs, and the corresponding p–V curves. Step 1: isothermal expansion of gas from a to b while in contact with the hot reservoir at temperature T_H; step 2: adiabatic expansion of gas from b to c; step 3: isothermal compression of gas from c to d while in contact with the cold reservoir at temperature T_C; and step 4: adiabatic compression of gas from d to a. Work is done *by* the gas during steps 1 and 2. Work is done *on* the gas during steps 3 and 4. The net work W produced by the gas during the full cycle is represented by the grey area enclosed by the solid curves.

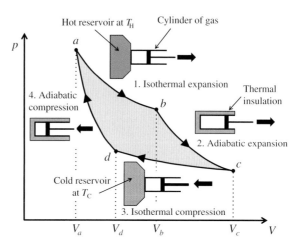

an amount of heat Q_H from the hot reservoir and this heat is transformed into work *by* the expanding gas.

• *Step 2: from b to c.* The gas cylinder is thermally isolated from its surroundings and the gas is expanded *adiabatically*, i.e. without any exchange of heat to its surroundings. During this step, the work done *by* the gas comes from its internal energy, thereby reducing its temperature. This expansion continues until the temperature of the gas reduces to the temperature T_C of the cold reservoir.

• *Step 3: from c to d.* The gas cylinder is placed in contact with the cold reservoir and the gas is compressed *isothermally* at constant temperature T_C. Work is done *on* the gas and this work is converted into an amount of heat Q_C that is deposited in the cold reservoir.

• *Step 4: from d to a.* The gas cylinder is again thermally isolated from its surroundings and the gas is compressed *adiabatically*. During this step, work is again done *on* the gas. As the process is adiabatic, all the work goes into increasing the internal energy of the gas and raising its temperature. The compression continues until the gas reaches the temperature T_H of the hot reservoir. This returns the system to its initial state and completes the cycle.

Work is done *by* or *on* the gas during each step. The net work W produced by the gas during the full cycle is represented by the shaded area on the p–V diagram of Figure 4.36. It is equal to the amount of heat absorbed from the hot reservoir minus the amount of heat discarded to the cold reservoir: $W = Q_H - Q_C$. Note that requirement (c) above is realised. During steps 1 and 3, heat is absorbed or deposited by the gas at constant temperature, while the two adiabatic processes, steps 2 and 4, change the temperature of the gas without adding or taking any heat from it. In this way the gas is always brought to the temperature of the respective reservoir before being in contact with it. This prevents any wasteful flow of heat between bodies at different temperatures.

We have defined the efficiency ε of an ideal heat engine in terms of the amount of mechanical energy it delivers and the amount of heat it absorbs from a hot reservoir. As the efficiency of an ideal heat engine depends solely on the temperatures of the two reservoirs, it is useful to find an expression for ε in terms of these temperatures alone. We do this as follows. During the isothermal expansion in step 1, an amount of heat, Q_H, is absorbed from the hot reservoir. The internal energy of an ideal gas depends only on the temperature of the gas and, as this is constant, the internal energy does not change during this step. Hence, from the conservation of energy, the amount of heat, Q_H, absorbed is equal to the work W_{ab} done by the expanding gas:

$$Q_H = W_{ab} = \int_{V_a}^{V_b} p\,dV = \int_{V_a}^{V_b} \frac{nRT_H}{V}\,dV$$
$$= nRT_H \ln\left(\frac{V_b}{V_a}\right), \tag{4.81}$$

where we have used the equation of state, $pV = nRT$. Similarly, the heat that flows out of the gas during the isothermal compression at the cold reservoir is

$$nRT_H \ln\left(\frac{V_d}{V_c}\right) = -nRT_H \ln\left(\frac{V_c}{V_d}\right). \tag{4.82}$$

This is a negative quantity because heat flows out of the gas. The *amount* of heat, Q_C, deposited in the cold reservoir is the modulus of this quantity, i.e. $nRT_C \ln(V_c/V_d)$. The ratio of the two quantities of heat is then

$$\frac{Q_C}{Q_H} = \frac{T_C \ln(V_c/V_d)}{T_H \ln(V_b/V_a)}. \tag{4.83}$$

We can relate the ratios V_b/V_a and V_c/V_d through the adiabatic processes of steps 2 and 4. For an adiabatic expansion we have $pV^\gamma = \text{constant}$ (Equation 4.77) and using $pV = nRT$, we also have $TV^{\gamma-1} = \text{constant}$. Hence, for the adiabatic expansion of step 2 we have

$$T_H V_b^{\gamma-1} = T_C V_c^{\gamma-1}. \tag{4.84}$$

Similarly for step 4, we have

$$T_H V_a^{\gamma-1} = T_C V_d^{\gamma-1}. \tag{4.85}$$

Dividing Equation (4.84) by Equation (4.85), we obtain

$$\frac{V_b}{V_a} = \frac{V_c}{V_d}. \tag{4.86}$$

Substituting this result into Equation (4.83) gives

$$\frac{Q_C}{Q_H} = \frac{T_C}{T_H}. \tag{4.87}$$

The efficiency of a heat engine is

$$\varepsilon = 1 - \frac{Q_C}{Q_H}, \tag{4.88}$$

and hence

$$\varepsilon = 1 - \frac{T_C}{T_H} = \frac{T_H - T_C}{T_H}. \tag{4.89}$$

This is our main result for a heat engine and one that we have used several times before; the efficiency of a Carnot engine depends only on the temperatures of the two reservoirs, T_C and T_H. It only approaches unity when the temperature of the cold reservoir approaches absolute zero. A plot of ε against T_H is shown in Figure 4.37 for a value of $T_C = 10°C$. This graph demonstrates the need for high source temperatures for the efficient conversion of heat into mechanical work. Hence, for example, jet engines for aircraft are being designed to operate at ever higher temperatures ($>1000°C$), which requires the use of advanced materials to operate at such high temperatures.

Real engines are not ideal, of course. They dissipate energy via friction and they lose appreciable amounts of energy through heat losses to their surroundings. Nevertheless, the ideal Carnot engine is an important practical tool in that it sets an upper limit to the attainable efficiency of any real engine. The efficiency for converting heat from a nuclear reactor into work is typically 40%, while petrol engines in cars achieve efficiencies of typically 20–30%. We know from experience that car engines become very hot and so lose large amounts of energy through conduction, convection and radiation.

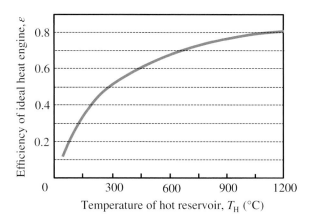

Figure 4.37 The efficiency of an ideal heat engine as a function of the temperature T_H of the hot reservoir. The temperature of the cold reservoir is constant and equal to $10°C$. This graph demonstrates the need for high temperatures of the energy source for the efficient conversion of heat into work.

Worked example

A quantity of 0.5 mole of an ideal diatomic gas is taken through a Carnot cycle as in Figure 4.36. The temperatures of the hot and cold reservoirs are 227 and 27°C, respectively. The pressure p_a at point a in the p–V diagram is 10.0×10^5 Pa. The volume is doubled during the isothermal expansion $a \to b$. (a) Calculate the pressure and volume of the gas at points a, b, c and d. (b) Find Q, W and ΔU for each of the four steps and for the complete cycle. (c) Find the efficiency of the heat engine from your results for Q and W. Compare this value with the value obtained from Equation (4.89).

Solution

For an ideal diatomic gas,

$$\gamma = \frac{C_p}{C_V} = \frac{7/2}{5/2} = 1.4.$$

(a) $V_a = \dfrac{nRT_H}{p_a} = \dfrac{0.5 \times 8.314 \times 500}{10.0 \times 10^5} = 2.08 \times 10^{-3}\ \mathrm{m^3}.$

Volume doubles during step $a \to b$: hence $V_b = 4.16 \times 10^{-3}\ \mathrm{m^3}$.

As expansion is isothermal, $p_a V_a = p_b V_b$, and so

$$p_b = \frac{p_a V_a}{V_b} = 5.0 \times 10^5\ \mathrm{Pa}.$$

For adiabatic expansion $b \to c$, $T_H V_b^{\gamma-1} = T_C V_c^{\gamma-1}$. Hence,

$$V_c = V_b \left(\frac{T_H}{T_C}\right)^{1/(\gamma-1)} = 4.16 \times 10^{-3} \left(\frac{500}{300}\right)^{2.5} = 14.9 \times 10^{-3}\ \mathrm{m^3}.$$

$$p_c = \frac{nRT_C}{V_c} = \frac{0.5 \times 8.314 \times 300}{1.49 \times 10^{-2}} = 0.836 \times 10^5\ \mathrm{Pa}.$$

For adiabatic expansion $d \to a$:

$$V_d = V_a \left(\frac{T_H}{T_C}\right)^{1/(\gamma-1)} = 2.08 \times 10^{-3} \left(\frac{500}{300}\right)^{2.5} = 7.46 \times 10^{-3}\ \mathrm{m^3}.$$

$$p_d = \frac{nRT_C}{V_d} = \frac{0.5 \times 8.314 \times 300}{7.46 \times 10^{-3}} = 1.67 \times 10^5\ \mathrm{Pa}.$$

(b) For isothermal expansion $a \to b$,

$$\Delta U_{ab} = 0 \quad \text{and} \quad W_{ab} = Q_H = nRT_H \ln\left(\frac{V_b}{V_a}\right).$$

Hence,

$$W_{ab} = 0.5 \times 8.314 \times 500 \times \ln 2 = +1441\ \mathrm{J}.$$

This amount of work is done by the expanding gas and is equal to the amount of heat, 1441 J, that is extracted from the hot reservoir.

For adiabatic expansion $b \rightarrow c$, $\Delta Q_{bc} = 0$. From the first law of thermodynamics, $\Delta U_{bc} = Q_{bc} - W_{bc} = -W_{bc}$. Hence, $W_{bc} = -\Delta U_{bc}$, where $\Delta U = nC_V (T_C - T_H)$.

Thus,

$$W_{bc} = 0.5 \times \frac{5}{2} \times 8.314 \times (500 - 300) = +2079 \, \text{J}.$$

In this step work is done by the gas at the expense of its internal energy, which reduces by 2079 J.

For isothermal compression $c \rightarrow d$,

$$\Delta U_{cd} = 0 \quad \text{and} \quad W_{cd} = Q_C = nRT_C \ln \left(\frac{V_d}{V_c} \right).$$

Hence,

$$W_{cd} = 0.5 \times 8.314 \times 300 \times \ln \left(\frac{7.46 \times 10^{-3}}{14.9 \times 10^{-3}} \right) = -863 \, \text{J}.$$

The work is a negative quantity, as the gas is compressed: $V_d < V_c$. 863 J of heat is removed from the gas, to be deposited in the cold reservoir, and so Q_C is also a negative quantity.

For the adiabatic compression $d \rightarrow a$, $\Delta Q_{da} = 0$ and $W_{da} = -\Delta U_{da} = -nC_V (T_H - T_C)$.

Hence,

$$W_{da} = -0.5 \times \frac{5}{2} \times 8.314 \times (500 - 300) = -2079 \, \text{J}.$$

Again the work is a negative quantity in this compression process. Tabulating these results gives:

Step	Q	W	ΔU
$a \rightarrow b$	1441 J	1441 J	0
$b \rightarrow c$	0	2079 J	-2079 J
$c \rightarrow d$	-863 J	-863 J	0
$d \rightarrow a$	0	-2079 J	$+2079$ J
Total	578 J	578 J	0

We see that 1441 J of heat is absorbed from the hot reservoir. Of this, 578 J is delivered as mechanical work and 863 J is discarded to the cold reservoir. Thus efficiency is

$$\varepsilon = \frac{W}{Q_H} = \frac{578}{1414} = 0.4 = 40\%.$$

We can compare this result with

$$\varepsilon = \frac{T_H - T_C}{T_H} = \frac{500 - 300}{500} = 0.4 = 40\%.$$

The steam turbine

The steam turbine is an important example of a heat engine as steam turbines are used in ~90% of the world's production of electricity. The principle of operation of an *impulse* steam turbine is illustrated in Figure 4.38, (see also Section 7.1.3). Superheated steam is passed through a nozzle whose bore reduces to increase the steam velocity. The resulting jet of steam with high velocity is directed at bucket-shaped blades attached to a rotor. The steam jets impinging on the blades exert a tangential force on the rotor, causing it to rotate. In this way, the kinetic energy of the steam is converted into mechanical energy of the rotor. The angle at which the steam jet strikes the blade is chosen to maximise the delivered mechanical energy.

Having just a single rotor does not make efficient use of the available energy of the steam because the steam transfers only a fraction of this energy to the rotor, say 10%. The British engineer Charles Parsons overcame this limitation with his invention in 1884 of the *compound steam turbine*. His idea was to have a large number of nozzle/rotor stages placed one after the other so that maximum mechanical energy could be obtained from the steam's energy. This use of multiple stages results in a closer approach to the ideal reversible expansion process. A schematic diagram of a multi-stage steam turbine is shown in Figure 4.39. The steam boiler is the hot reservoir of this heat engine and water is the working substance. The boiler heats the water to produce superheated steam at a temperature of ~500°C. The turbine has a series of moving rotors that are mounted on a central axle. To these rotors are fixed the bucket-shaped blades. Successive rotors are separated from each other by a set of stationary nozzles that are attached to the casing of the turbine. A detail of this arrangement is shown in Figure 4.40. This arrangement is a key feature of Parson's steam turbine as it reverses the direction of the steam jets between successive rotors. In this

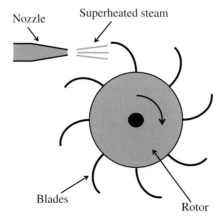

Figure 4.38 The principle of operation of an impulse steam turbine. The nozzle, with a reducing bore, produces a jet of steam with high velocity. This jet is directed at bucket-shaped blades attached to a rotor. The steam jet impinging on the blades exerts a tangential force on the rotor, causing it to rotate. In this way, the kinetic energy of the steam is converted into mechanical energy of the rotor.

Figure 4.39 Schematic diagram of a steam turbine. The turbine has a series of *moving* rotors, which are mounted on a central axle. The rotors are separated from each other by a set of *stationary* nozzles, which are attached to the casing of the turbine. Superheated steam from the boiler passes through the nozzles which accelerate the steam to high velocity, producing jets of steam that are directed at the rotor blades. This produces a tangential force on the rotors, causing them to rotate. The rotary motion of a steam turbine about an axle is an important feature because it is ideally suited to turning the spindle of an electrical generator. The steam is converted back to water in the condenser.

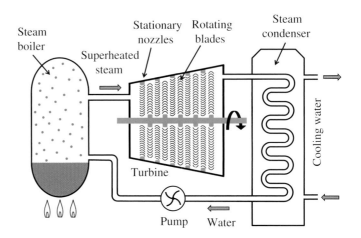

way, a steam jet always strikes a rotor blade at the optimum angle. Steam enters the turbine and expands rapidly, being drawn through the turbine by the low pressure at the turbine exit produced by the steam condenser, which acts as the cold reservoir. The steam passes through the nozzles, which accelerate the steam to high velocity and direct the jets of steam onto the rotor blades. The impinging steam jets exert a force on the rotor blades, which causes each rotor to rotate about the axle, as indicated in Figure 4.40. This converts the kinetic energy of the steam into rotational energy of the rotors. This rotary motion of a steam turbine about an axle is an important feature because it is ideally suited to turning the spindle of an electrical generator. As the steam passes through the turbine from high pressure to low pressure, it occupies increasingly more volume and so the rotor blades steadily increase in diameter to maximise the conversion of steam energy. When the steam emerges from the last rotor, it is cooled and converted back to water in the condenser. The condensed

Figure 4.40 Successive rotors in a steam turbine are separated from each other by a set of stationary nozzles which are attached to the casing of the turbine. This figure illustrates the arrangement of nozzles and rotors. The nozzles change the direction of the steam jets between successive rotors and optimise the angle at which the steam jets strike the blades.

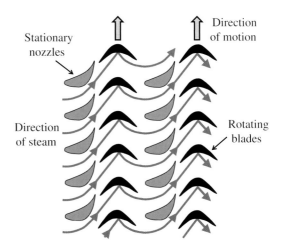

water is then returned to the boiler and the cycle repeats. The action of the steam turbine can be seen to have all the elements of a heat engine, namely hot and cold reservoirs, a working substance that is transferred between the two and the delivery of mechanical energy.

Refrigerators and heat pumps

A Carnot cycle is reversible. Hence, we can run it in reverse so that it absorbs heat from a cold reservoir and deposits heat in a hot reservoir. This is just what a refrigerator does. It absorbs heat from, say, a bottle of milk inside the refrigerator and deposits heat to the air outside. A heat engine delivers a net *output* of mechanical work. On the other hand, a refrigerator requires a net *input* of mechanical work to make the transfer of heat from the cold to the hot reservoir. A schematic diagram showing the flow of heat and work in a refrigerator is given in Figure 4.41. In this case Q_C is a positive quantity and both Q_H and W are negative quantities, as indicated by the figure. However, because we are interested in the *amounts* of energy and work involved, it is convenient to take the modulus of each of them.

The efficiency of a refrigerator is defined as the amount of heat extracted from the cold reservoir Q_C divided by the amount of work W used to extract it. This efficiency is called the *coefficient of performance η*:

$$\eta = \frac{Q_C}{W}. \tag{4.90}$$

The first law of thermodynamics gives $W = Q_H - Q_C$. Hence

$$\eta = \frac{Q_C}{Q_H - Q_C}. \tag{4.91}$$

As a refrigerator operates in a Carnot cycle, we can use the following result:

$$\frac{Q_C}{Q_H} = \frac{T_C}{T_H}. \tag{4.92}$$

Substituting this result into Equation (4.91) gives

$$\eta = \frac{T_C}{T_H - T_C}. \tag{4.93}$$

Again, Equation (4.93) gives the best possible coefficient of performance of a refrigerator for given reservoir temperatures. For example, if $T_C = 5°C$ and $T_H = 20°C$, the best possible value of η is 18.5. Real refrigerators have lower values of η than given by Equation (4.93), with typical values of ~5.

A *heat pump* acts in a same way as a refrigerator and its action can also be represented by Figure 4.41. However, the objective of a heat

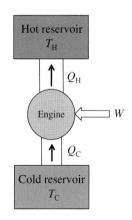

Figure 4.41 Schematic representation of a refrigerator and also of a heat pump. By putting work into the engine, heat can be transferred from a cold reservoir to a hot reservoir.

pump is somewhat different. Rather than taking heat from a cold body, its objective is to deliver heat to a hot body. For example, we saw in Section 4.4.1 that a heat pump can be used to heat a building by taking heat from rocks below the ground. As for the refrigerator, external work must be applied to make this transfer of heat. For a heat pump, the *coefficient of performance* η is defined as the amount of heat deposited in the hot reservoir divided by the amount of work that is put in:

$$\eta = \frac{Q_H}{W} = \frac{Q_H}{Q_H - Q_C} = \frac{T_H}{T_H - T_C},\qquad(4.94)$$

where again we have used the result for a Carnot cycle: $Q_C/Q_H = T_C/T_H$. Real heat pumps have values of η that are lower than predicted by Equation (4.94) and typical vales are ~ 3. Nevertheless, this means that 1 kJ of work provides ~ 3 kJ of heat.

Problems 4

4.1   (a) The remnant radiation from the Big Bang at the beginning of the universe has been observed to be characteristic of a blackbody radiator at the temperature of 2.73 K. At what wavelength does this radiation peak? In what region of the electromagnetic spectrum does this wavelength occur? (b) In Section 4.1.3 we saw that the Sun's core consists of a highly dense plasma at an extremely high temperature. Taking the temperature of the plasma to be 1.5×10^6 K, at what wavelength would the plasma emission peak? In what region of the electromagnetic spectrum does this wavelength occur?

4.2   Following the discussion in Section 4.3.1 regarding the equilibrium temperature of the Earth due to solar heating, we can write the following relationship for the equilibrium temperature T_E: of Earth's surface:

$$\frac{e}{a}T_E^4 = \text{constant},$$

where e is emissivity of Earth and a is the fraction of solar radiation that it absorbs. (a) By what percentage must e change to cause a 1% change in T_E? (1% of T_E is $\sim 3°$C). (b) By what percentage must a change to cause a 1% change in T_E?

4.3   What fraction of solar power is contained in the wavelength range 500.0–505.0 nm. Take the temperature of the Sun to be 5800 K.

4.4   Estimate the net rate of heat loss due to radiation from a naked person in a room at 20°C. Note that the skin temperature of a person is typically 4°C less than the internal body temperature because of the thermal resistance of skin. At what wavelength does the person emit the most power per unit wavelength. Human skin is a near ideal blackbody radiator in the infrared.

4.5   Show how Wien's displacement law can be derived from the Planck radiation formula (Equation 4.17). Evaluate the resultant constant $(hc/4.965k)$ to show that it agrees with the constant in Equation (4.16). The solution of the equation $5 - x = 5e^{-x}$ is $x = 4.965$.

4.6   Show how the Stefan–Boltzmann law can be derived from the Planck radiation formula (Equation 4.17). Evaluate the resultant constant $(2\pi^2 k^4/15h^3c^3)$ to show it agrees with the constant in Equation (4.15).

$$\int_0^\infty \frac{x^3}{(e^x - 1)}\,\mathrm{d}x = \frac{(2\pi)^4}{240}.$$

4.7   A cavity radiator at temperature 5000 K has a hole of 10 mm diameter drilled into its wall. Calculate the power density in the cavity between 600.0 and 602.0 nm. Hence, calculate the power radiated through the hole within that wavelength range.

4.8

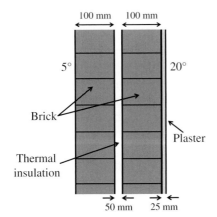

The figure shows the cross-section through the wall of a house. The area of the wall is 6 m². The thermal conductivities of the brick, the insulating layer and the plaster are 0.50, 0.025 and 0.80 W/m² K, respectively, and their widths are as shown in the figure. (a) Calculate the thermal resistance of each of these components. (b) Hence, calculate the flow of heat through the wall when the internal temperature and external temperatures are 20 and 5°C, respectively. (c) Use the thermal resistance of the components to calculate the temperatures at the junctions of the components.

4.9   (a) A hollow cylinder made from an insulating material with thermal conductivity κ has inside radius r_1, outside radius r_2 and length l. The inside surface of the cylinder is held at temperature T_1, and the outside at temperature T_2, with $T_1 > T_2$. Show that the heat flux H through the material is given by

$$H = \frac{2\pi l \kappa (T_1 - T_2)}{\ln(r_2/r_1)}.$$

(b) A thin copper pipe of diameter 25 mm carries water at temperature 100°C. The pipe is insulated by a cylindrical layer of material (see Figure 4.28). The thermal conductivity κ of the insulation material is 0.03 W/m K. If the length of the pipe is 3.0 m, what minimum outer diameter should the insulation have if the heat loss due to conduction must be less than 100 W? Assume an ambient temperature of 20°C.

4.10 A spherical shell of material with thermal conductivity κ has inside radius r_1 and outside radius r_2. The inside surface of the shell is held at temperature T_1, and the outside is at temperature T_2, with $T_1 > T_2$. (a) Show that the temperature $T(r)$ at radial distance r, within the shell is given by

$$T(r) = T_1 - \frac{H}{4\pi\kappa}\left(\frac{1}{r} - \frac{1}{r_1}\right).$$

(b) Obtain an expression for the heat flux H through the shell.

4.11 (a) Show that if the thickness of the cylinder in Question 9(a) is small compared with its inner radius, the equation for heat flux reduces to the linear form given by Equation (4.39). (b) Similarly, show that if the thickness of the spherical shell in Q10 is small compared with its inner radius, the equation for heat flux also reduces to the linear form given by Equation (4.39).

4.12 The Earth's crust has a thickness of approximately 30 km. Heat flows from the hot core of the Earth to the surface, resulting in a temperature drop of approximately 1000°C across the crust. Taking the thermal conductivity of the crust to be 2.0 W/m K, estimate the heat flux per unit area through the Earth's surface due to conduction through the crust. Compare this result with the solar power reaching Earth's surface.

4.13 A solar vacuum tube collector consists of 24 tubes of outer diameter 60 mm, inner diameter 50 mm and length 2000 mm. It is connected to an insulated tank containing 250 L of water. Estimate how long it takes the temperature of the water to rise by 10°C on a sunny day with the solar radiation perpendicular to the plane containing the vacuum tubes. Assume an incident solar power of 500 W/m² and a conversion efficiency of 85%. The specific heat of water is 4200 J/kg K.

4.14 (a) In Section 4.6.3, we had a classical model of a diatomic molecule consisting of two masses connected by a rigid rod. Use this model to calculate the classical rotational frequency of a carbon monoxide molecule that has energy $\frac{1}{2}kT$ with $T = 300$ K. What wavelength does this frequency correspond to in the electromagnetic spectrum? The moment of inertia I is given by $I = \frac{1}{2}\mu r^2$, where r is the separation of the two masses and $\mu = m_1 m_2/m_1 + m_2$ is the reduced mass. The atomic masses of carbon and oxygen are 12 and 16u, respectively, and the value of r for the CO molecule is 0.113 nm. (b) Molecules are never completely rigid, and in a more realistic model of a diatomic molecule, we represent the connection between the two atoms as a Hookes's law spring and not as a rigid rod. The molecules then vibrate about their equilibrium separation along the line joining them. The 'spring constant' for the CO molecule is 1.90×10^3 N/m. Calculate the classical oscillation frequency of the molecule, assuming that it behaves like a simple harmonic oscillator. What wavelength does this frequency correspond to in the electromagnetic spectrum? Again, use the reduced mass in the expression for the classical oscillation frequency.

4.15 (a) One mole of a diatomic gas at temperature of 27°C is expanded *isothermally* to twice its volume. Calculate the work done by the gas, the amount of heat supplied to it and the internal energy of the gas before and after the expansion. (b) One mole of the same gas at initial temperature 27°C is expanded *adiabatically* to twice its volume. Calculate the work done by the gas and any changes in Q and U.

4.16 An ideal heat engine takes 200 J of heat from a hot reservoir at 500°C, does some work and deposits some heat to a cold reservoir at 350°C. Determine how much work is done, how much heat is discarded and the efficiency of the heat engine.

4.17 One mole of an ideal monatomic gas is taken through a Carnot cycle as in Figure 4.36. The temperatures of the hot and cold reservoirs are 450 and 300 K, respectively. The initial pressure of the gas is 10.0×10^5 Pa and, during the initial isothermal expansion stage, the gas is expanded to twice its original volume. (a) Calculate the pressure and volume of the gas at points a, b, c and d on the p–V diagram. (b) Find Q, W and ΔU for each of the four steps of the cycle and for the complete cycle. (c) Find the efficiency of the heat engine from your results for Q and W and compare with the value obtained from Equation (4.89). (d) Using your results from part (b), plot the four sets of (p, V) values to construct the p–V diagram and estimate the area bounded by the four p–V curves. (e) Can more work be delivered in the complete cycle by having a different compression ratio in the initial, isothermal stage of expansion?

4.18 Calculate the efficiencies of the following, assuming they operate as ideal heat engines:

  (i) A steam engine operating between 100 and 10°C.
  (ii) A pressurised water nuclear reactor operating with $T_{core} = 324°C$ and a cold reservoir at 10°C.
  (iii) A sodium-cooled nuclear reactor with $T_{core} = 620°C$ and a cold reservoir at 10°C.

4.19 In Ocean Thermal Energy Conversion (OTEC), the temperature difference between the hot surface of an ocean and the cooler, much deeper, layers below the surface is used to run a heat engine. The idea was first proposed by the French physicist Jacques Arsene d'Arsonval, who also invented the moving-coil galvanometer. Determine the maximum theoretical efficiency of an OTEC power plant if the surface and deep-water temperatures are 30 and 5°C, respectively. If the power plant is to produce 100 kW of power, at what rate must heat be extracted from the warm water? At what rate must heat be deposited in the cold water? The cold water that enters the plant leaves it at a temperature of 10°C. What must be the flow rate of cold water through the system? The specific heat of water is 4200 J/kg K.

4.20 A heat pump is used to extract heat from water at 10°C that flows through a buried pipe, as in Figure 4.20. A power of 5 kW is to be delivered to the building at a temperature of 20°C. Calculate the power to run the heat pump and the required flow rate of water through the pipework if the cooled water leaves the heat pump at a temperature of 5°C. Assume the heat pump to be ideal. A more realistic value of the thermal coefficient of performance of the heat pump is 5. In this case, what is the required power to run the heat pump and the required flow of water? The specific heat of water is 4200 J/kg K.

5

Semiconductor solar cells

In Chapter 4 we described the harvesting of solar energy by solar heaters where we use sunlight to heat a fluid. We saw that if the temperature of the heated fluid is sufficiently high, we can convert its thermal energy into electricity using the combination of a heat engine and an electrical generator. In this chapter we describe harvesting energy from the Sun using *solar cells* that exploit the *photovoltaic effect*. In this case, sunlight that is incident on a solar cell produces an electric current directly, i.e. without involving any thermal energy or indeed any moving parts. This is achieved using a *semiconductor p–n junction*. In this chapter, we describe the properties of semiconductors and the electrical characteristics of a p–n junction and how such a junction is employed in a solar cell.

5.1 Introduction

The discovery of the photovoltaic effect in 1839 is credited to Edmond Becquerel. He was the father of Henri Becquerel, whom we encountered in our discussion of radioactivity in Chapter 2. He observed that when an acid solution containing silver chloride was exposed to light, a current flowed between platinum electrodes inserted into the solution. The conversion of light into electricity using a solid material (selenium) was first demonstrated by William Grylls Adams and Richard Evans Day in 1875. Some years later (1894), Charles Fritts constructed what was probably the first true solar cell. He coated the semiconductor selenium with a thin layer of gold. The efficiency of his solar cell for converting sunlight into electrical energy was only about 1%, so that it could not be used to generate electrical power economically. However, the selenium cell went on to find widespread use as a light meter in photography.

Physics of Energy Sources, First Edition. George C. King.
© 2018 John Wiley & Sons, Ltd. Published 2018 by John Wiley & Sons, Ltd.

The important breakthrough came in the 1950s at the Bell Laboratories, USA. There, Gerald Pearson, Darryl Chapin and Calvin Fuller demonstrated a solar cell using *doped* silicon, which is also a semiconductor. It had an efficiency of 5.7% and made possible the development of the solar cell as a viable source of energy.

We can compare the photovoltaic effect with the more familiar photoelectric effect. In the photoelectric effect, ultraviolet (UV) light that is incident upon a metal surface liberates photoelectrons from the surface. (Photoelectrons are no different from any other electrons but take this name when we talk about their generation by photons.) Einstein received the Nobel Prize in 1923 for his explanation of the photoelectric effect in terms of *quanta* of energy, which we now call photons and which we discussed in Section 4.2.3. Einstein's equation that describes the photoelectric effect is

$$E_{\max} = h\nu - \phi, \tag{5.1}$$

where $h\nu$ is the photon energy, ϕ is the *work function* of the metal and $E_{\max}$ is the maximum energy that a generated photoelectron can have. Clearly, $h\nu$ must be larger than ϕ, which is typically $\sim$5 eV for a metal. For example, the work function of gold is 5.1 eV and incident photons must have a wavelength *shorter* than 243 nm to liberate photoelectrons from a gold surface.

A schematic diagram of an apparatus for observing the photoelectric effect is shown in Figure 5.1. UV light is incident upon the metal surface. An electrode is placed in front of the surface and is held at a positive potential with respect to it. The potential difference produces an electric field that directs the emitted photoelectrons to the positively charged electrode and the electrons flow in the external circuit producing a current i. (Note that *conventional current* flows in the opposite direction to electron flow.) Although this arrangement does not produce electrical power economically, it is widely used to measure the irradiance of UV light in research applications.

Figure 5.1 The photoelectric effect in which UV light is incident upon a metal surface. An electrode is held at a positive potential with respect to the metal surface and collects emitted photoelectrons, which then flow in the external circuit, producing a current i. (Note that conventional current flows in the opposite direction to electron flow.)

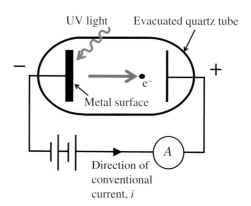

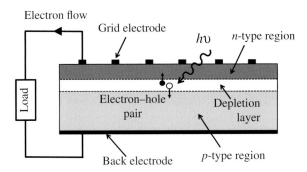

Electron flow

Grid electrode

$h\upsilon$ n-type region

Load

Electron–hole pair

Depletion layer

Back electrode   p-type region

Figure 5.2 Schematic diagram of a solar cell, which is formed from a p–n junction in a semiconductor material. Incident photons cause the generation of electron–hole pairs in or close to the depletion layer that exists at the junction. There is an inbuilt electrical field across the depletion layer that causes the electrons to move into the n-type material and the holes to move into the p-type material. Electrons generated by the incident photons flow around the external circuit from the n-type to the p-type, where they combine with the holes. In this way, electrical power is delivered to an external load.

The photovoltaic effect has some similarities with the photoelectric effect. Now, however, the process occurs in or close to the *depletion layer* that exists at a *p–n* junction in a semiconductor material. Figure 5.2 is a schematic diagram of a solar cell showing the depletion layer between the *p* and *n* regions. Again, there is an incident photon, but now the resulting electron is not ejected from the surface but is promoted to a higher-lying energy level in the semiconductor, which enables it to move freely through the semiconductor material. The excited electron leaves behind a vacancy or *hole* in the semiconductor material, so that the incident photon produces an *electron–hole pair*. Importantly, there is an inbuilt electric field across the depletion region and, consequently, there is no need for an externally applied voltage to cause a current flow. This electric field causes the electron–hole pair to separate and the electron to move to the *n* region and the hole to move to the *p* region. A metallic grid forms one of the electrodes of the cell, and a metallic layer on the back of the solar cell forms the other electrode. If these are connected to an external load, electrons generated by photons incident on the cell flow around the external circuit from the *n* region to the *p* region where they combine with generated holes. In this way, electrical power is delivered to the external load.

An incident photon must have sufficient energy to generate an electron–hole pair. We will see that this means the photon must have an energy, $h\nu$, that is greater than the *band gap*, E_g, of the semiconductor material:

$$h\nu > E_g. \tag{5.2}$$

The band gap of a semiconductor is ~ 1 eV, e.g. $E_g = 1.11$ eV for silicon. We saw in Chapter 4 that the majority of the photons in the solar spectrum have energies >1 eV, and so semiconductor solar cells provide a valuable way to harvest solar energy. To understand the action of a solar cell more fully, we begin with a discussion of semiconductor materials.

5.2 Semiconductors

Conductors and insulators are distinguished by the enormous differences in their electrical resistivities. For example, the resistivity of quartz is $\sim 10^{25}$ times larger than the resistivity of copper. Lying between insulators and conductors with respect to resistivity are semiconductors. For example, the resistivity of the semiconductor silicon is about 10^{10} times larger than that for copper. The reason for such enormous differences is the variation in the number of free electrons that can carry electric current in these kinds of material. These variations arise because in a crystalline solid, such as quartz, copper or silicon, the electrons can only exist in *allowed energy bands*. What is important is the extent to which these bands are occupied by electrons and the energy gaps between adjacent bands.

5.2.1 The band structure of crystalline solids

The energy levels in an isolated atom are discrete and well separated from each other. For example, the energy separation between the two lowest energy levels of hydrogen, with values of principal quantum number n of 1 and 2, respectively, is 3.4 eV. If two identical atoms are brought into close proximity to each other, their energy levels change as their electronic wave functions begin to overlap. As a result, each atomic energy level splits into two levels of slightly different energy for the two-atom system. Similarly, if we bring four atoms into close proximity, a particular energy level splits into four separate energy levels with slightly different energies, which belong to the collection of the four atoms as a whole. When we have N atoms in close proximity, as in a crystal lattice, each energy level of the isolated atom splits into N different energy levels. A macroscopic solid contains $\sim 10^{23}$ atoms/mole and so each particular atomic energy level splits into a very large number of densely packed levels. The energy levels are so numerous and so closely spaced that for all practical purposes they are indistinguishable from a continuous *energy band*. This evolution of energy levels into energy bands is illustrated in Figure 5.3. The energy bands are typically a few eV wide. They may be widely separated in energy, they may be close together or, indeed, they may overlap. It depends on the atoms and the way in which the atoms bond together. The energy gaps between these allowed bands are called *forbidden energy gaps*, since an electron in the solid cannot have an energy that lies within these gaps.

The energy levels in the allowed bands are filled with the atomic electrons according to the *Pauli exclusion principle*, which states that only one electron can occupy a given quantum mechanical *state*. Electrons are *fermions* that can have two values of *spin*

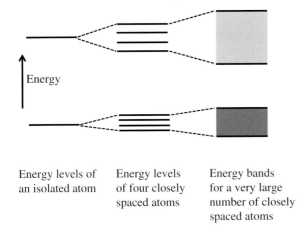

Energy

Energy levels of   Energy levels   Energy bands
an isolated atom   of four closely   for a very large
                    spaced atoms   number of closely
                                    spaced atoms

Figure 5.3 Schematic representation of how energy levels of closely spaced atoms evolve into energy bands. The energy gaps between the allowed bands are called forbidden energy gaps because an electron cannot have an energy that lies within these gaps.

quantum number s. Hence, each discrete energy level can contain two electrons at most. The energy bands are filled, starting with the lowest energy band, then the next higher energy band and so on until all the electrons are accommodated. The electrons in the lowest bands are tightly bound and are generally not important in determining the electrical (or chemical) properties of a solid. The two highest energy bands in a solid are called the *valence band* and the *conduction band.* They are separated in energy by the band gap E_g. In a particular solid, the valence band may be completely filled, or partly filled with electrons, while the conduction band is never more that slightly filled. The extent to which these bands are filled and the size of the band gap, E_g, between them determines whether a solid is a conductor, an insulator or a semiconductor.

When an electron is under the influence of an electric field, it is accelerated by that field and gains energy from it. Classically the electron can gain any amount of energy. It is equal to eV, where V is the voltage drop across the region of acceleration. This would be the case, for example, for an electron being accelerated in an X-ray tube. In a system governed by quantum mechanics, however, an electron can only gain quantised amounts of energy corresponding to differences between the energy levels of the system. When current flows through a solid, the electrons gain energy from the applied electric field and, in so doing, are excited into higher-lying levels in an allowed band. However, current can only flow if energy levels are available for the electrons to be excited into.

Suppose that the valence band in a solid is only partially full when all the electrons have been accommodated, as in Figure 5.4(a). Then there are many empty levels into which an electron can be promoted. Consequently electrons can readily gain energy from an electric field and the result is that the solid has high electrical conductivity. Sodium provides an example of a partially filled valence band. Sodium has 11 electrons and these are arranged in *electronic*

sub-shells according to $1s^2\ 2s^2\ 3p^6\ 3s^1$, which is a shorthand nota-
tion to identify the sub-shells. For example, the 3p sub-shell (with
principal quantum number $n = 3$ and orbital quantum number
$l = 1$) contains six electrons as specified by the numerical super-
script. The single 3s electron is relatively loosely bound in the sodium
atom and it is this electron that takes part in electrical conductivity.
Electrons in the other sub-shells are so tightly bound to the nucleus
that they do not take part. When N atoms of sodium come together
to form a crystalline solid, the 3s energy level splits into N energy
levels, each of which can accommodate two electrons. (The factor of
2 arises because of the two possible spin states of the electrons.) As
the number of 3s valence electrons arising from the N atoms is just N,
it follows that only half of the possible energy levels in the valence
band due to the 3s atomic level will be occupied. Consequently, these
electrons can readily gain energy from an external field and the result
is that sodium has high electrical conductivity. In some other metals, such as magnesium, the valence and conduction bands overlap,
as shown in Figure 5.4(b). Again this provides many energy levels
for electrons to be excited into by an electric field and such metals
also have high electrical conductivity.

   Suppose that in a solid the electrons completely fill all the allowed
energy bands up to and including the valence band and that the
conduction band is empty at absolute zero, $T = 0$ K. This case is
shown in Figure 5.4(c), where the band gap E_g is relatively large.
When we apply an electric field, the electrons cannot be excited into
higher energy levels in the valence band as all the levels in that band
are already occupied. Moreover, electric fields of normal strength do
not provide electrons with enough energy to be promoted into the
conduction band. The electrons have nowhere to be excited to and
the solid is an insulator.

   Now at any temperature T above absolute zero, there is a finite
probability that an electron can be promoted from the valence band

Figure 5.4 Electronic band
structures for: (a) a typical
conductor like sodium in
which the valence band is
partially full; (b) a conductor
like magnesium in which the
allowed energy bands
overlap; (c) a typical
insulator, where there is a
large forbidden energy gap
between the valence and
conduction bands; and (d) a
semiconductor where the
forbidden energy gap is small.

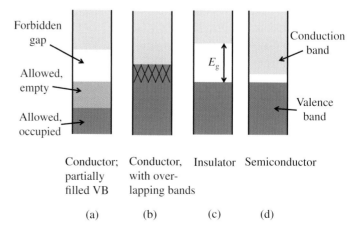

into the conduction band. This is because it has thermal energy $\sim kT$, where k is the Boltzmann constant. The probability is given *approximately* by the *Boltzmann factor* $e^{-E_g/kT}$ (see Section 5.3.3). An example of an electrical insulator is diamond. Its valence band is full and the band gap E_g is 5 eV. We recall that $kT = 1/40$ eV at room temperature. Hence, we find that the probability of an electron being excited into the conduction band at room temperature is about 10^{-87}. This is an extremely small number and, indeed, diamond is a very good insulator.

In the case of a semiconductor at absolute zero, the valence band is full and the conduction band is empty and under this condition, a semiconductor is an insulator. However, the energy gap between the valence and conduction bands is much smaller than for an insulator, as indicated by Figure 5.4(d); for the semiconductor germanium, the band gap E_g is 0.67 eV. Such a band gap is sufficiently small that an appreciable number of electrons do have enough thermal energy to be promoted into the conduction band at room temperature, as illustrated by Figure 5.5. The conduction band contains a very large number of empty levels that these electrons can be excited into by an applied electric field, so that these electrons are free to move through the semiconductor under the influence of the electric field. Hence, germanium becomes reasonably conductive at room temperature, although still far less conductive than, say, copper.

The electrons in a semiconductor that are thermally excited to the conduction band leave behind vacancies or holes in the valence band (see Figure 5.5). As the electrons and holes are produced in pairs, the concentration of electrons in the conduction band and the concentration of holes in the valence band must be equal. As an electron moves through the semiconductor, it may encounter a hole. If it does, the electron and hole recombine and disappear. At equilibrium, the rate of recombination must be equal to the rate of generation of electron-hole pairs.

The hole in a particular atom of the semiconductor may be filled by an electron from a neighbouring atom so that the hole has moved from the first atom to the second. Then an electron from a third

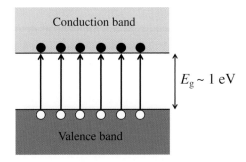

Figure 5.5 Generation of electron–hole pairs by thermal excitation of electrons from the valence band to the conduction band of a semiconductor. The electrons that are excited to the conduction band leave behind holes in the valence band.

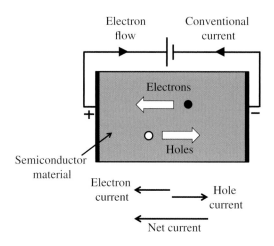

Figure 5.6 If a potential difference is applied across a semiconductor, the free electrons move toward the positive end, while holes appear to move in the opposite direction towards the negative end. Because the holes are positively charged, the flow of holes in one direction is equivalent to a flow of electrons in the opposite direction. Hence, the net current in the semiconductor is the sum of the electron and hole currents.

atom may fill the second hole and the process repeats from atom to atom. If a voltage is applied to the semiconductor, as in Figure 5.6, the electrons will move towards the positive end, while the holes will *appear* to move in the opposite direction, to the negative end. Although only electrons are moving in the crystal, it is convenient to consider the hole as a particle with positive charge $+e$ that is capable of moving in a similar manner to electrons but in the opposite direction. Because the holes are positively charged, the flow of holes in one direction is equivalent to a flow of electrons in the opposite direction. Hence, the net current in the semiconductor is the sum of the electron and hole currents. This is indicated in Figure 5.6, where the direction of the net current points in the direction of the electron flow. (We will consistently use the convention where the net current is in the same direction as the electron current.) Electrons leave the positive end of the semiconductor, travel around the external circuit and re-enter the semiconductor at the negative end, where they recombine with holes.

5.2.2 Intrinsic and extrinsic semiconductors

Our discussion of semiconductors so far has been for pure elements such as pure silicon and pure germanium. These are called *intrinsic semiconductors*. We saw that at temperatures above absolute zero, such intrinsic semiconductors have a finite number of thermally generated electrons in the conduction band and holes in valence band. However, the densities of these charge carriers are small because the thermal energy $\sim kT$ of the electrons is small compared with the band gap energy E_g. The number of charge carriers in a semiconductor can be increased enormously by introducing impurities into it in a process called *doping*. In this process, some of the semiconductor atoms in the crystal lattice are replaced by impurity atoms. By suitable choice of impurity, the number of electrons *or* the number of holes can be increased. A semiconductor in which the number of electrons has been increased is called *n-type*, where *n* stands for negative charge carriers, while a semiconductor where the number of holes has been increased is called *p-type*, where *p* stands for positive charge carriers. Such a doped material is called an *extrinsic semiconductor*. As for the case of intrinsic semiconductors, electron-hole pairs are also created in an extrinsic semiconductor by thermal excitation of electrons across the band gap. However, the concentrations of these thermally generated charge carriers are *extremely small* compared with the concentration of charge carriers due to the added impurities. The more abundant charge carriers in a semiconductor are called *majority carriers*. In *n*-type semiconductors they are electrons and in *p*-type semiconductors, they are holes. The less abundant charge carriers are called *minority carriers*. In *n*-type semiconductors they are holes, and in *p*-type semiconductors, they are electrons.

n-type semiconductors

The silicon atom has 14 electrons and these are arranged in various electronic sub-shells as shown in Table 5.1. Four of these electrons take part in bonding with other silicon atoms: the two 3s and the two 3p electrons. The form of bonding in the crystal lattice is called *covalent bonding* where a silicon atom *shares* one of its electrons with another silicon atom that also contributes an electron to the sharing process. Thus a silicon atom is surrounded by four other silicon atoms, leading to the crystal structure of silicon. A two-dimensional representation of this crystal structure is illustrated in Figure 5.7.

Suppose now that we introduce an atom of phosphorus so that it takes the place of a silicon atom in the lattice, as shown in Figure 5.8. Phosphorus has one more electron than silicon, as shown in Table 5.1; there are two 3s and three 3p electrons. Consequently, one of the electrons from the phosphorus atom will be left over, i.e. it will be spare with no partner to form a covalent bond (see Figure 5.8). This spare electron then becomes only weakly bound to the phosphorus atom and so is easily freed from it. This free electron can then roam through the lattice and conduct electrical current under the influence of an electric field. Impurity atoms that donate spare electrons in

Table 5.1 Electronic configuration of aluminium, silicon and phosphorus

Atom	Electronic configuration
Aluminium, $Z = 13$	$1s^2\ 2s^2\ 2p^6\ 3s^2\ 3p^1$
Silicon, $Z = 14$	$1s^2\ 2s^2\ 2p^6\ 3s^2\ 3p^2$
Phosphorus, $Z = 15$	$1s^2\ 2s^2\ 2p^6\ 3s^2\ 3p^3$

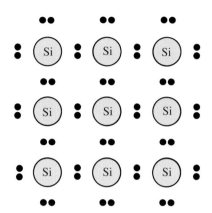

Figure 5.7 A two-dimensional representation of the crystal structure of silicon. Each silicon atom is surrounded by four other silicon atoms and shares one of its electrons with one of its neighbours, which also contributes an electron to this sharing process.

this way are called *donor atoms*. They convert a pure (intrinsic) semiconductor into an *n*-type (extrinsic) semiconductor.

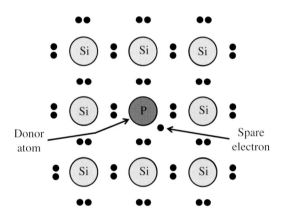

Figure 5.8 Creation of a spare electron in the crystal lattice of silicon by a donor impurity atom of phosphorus, which has one more electron than silicon.

Worked example

Deduce the amount of energy required to remove the spare electron from a phosphorus atom that is embedded in a silicon crystal lattice.

Solution

Removal of the spare electron leaves behind a positive phosphorus *ion* with charge $+e$. Therefore, while it is still bound, we can think of the spare electron that is bound to the phosphorus ion P^+ in rather the same way as an electron that is bound to a proton in a hydrogen atom. According to Bohr theory, the binding energy E_n of an electron in a hydrogen atom is given by

$$E_n = \frac{m_e e^4}{8n^2 h^2 \varepsilon_0^2} = 13.6\frac{1}{n^2} \text{ eV}, \tag{5.3}$$

where n is the principal quantum number of the electron orbital (not to be confused with '*n*' in '*n*-type') and the other symbols have their usual meanings. For example, to remove an electron from the $n = 1$ level of hydrogen, we have to supply 13.6 eV of energy. To use Bohr theory to estimate the binding energy of the spare electron in phosphorus, we must take account of the relative permittivity, ε, of the semiconductor, and we must take account of the *periodic* electrostatic potential produced by the ions in the crystal lattice. This periodic potential modifies the motion of an electron in the lattice and we can take this into account by ascribing an effective mass m_e^* to the spare electron. Taking these two differences into account we obtain from Equation (5.3)

$$E_d = 13.6 \left(\frac{m_e^*}{m_e}\right) \left(\frac{1}{\varepsilon}\right)^2 \text{eV}, \qquad (5.4)$$

where E_d is the binding energy of the spare electron and we have taken the principal quantum number n to be 1. For silicon, $m_e^* = 0.12\ m_e$ and $\varepsilon = 16$, giving $E_d = 6.4$ meV. It takes this amount of energy to remove the weakly bound electron from a phosphorus ion in the lattice. This is much less than the thermal energy kT ($\sim$25 meV) that is available at room temperature. Consequently, virtually all the spare electrons are removed from the phosphorus atoms, and the number of conducting electrons is essentially equal to the number of impurity atoms that are added to the crystal. The P$^+$ ions, of course, remain fixed in the crystal lattice.

E_d is the amount of energy it takes to remove the weakly bound electron from a phosphorus atom and allow it to become a conduction electron. Equivalently, we can say that this is the amount of energy required to lift the electron into the conduction band. Therefore, these weakly bound electrons must reside in energy levels that lie just below the conduction band, by the amount E_d. This, indeed, is the situation at absolute zero ($T = 0$ K) and is illustrated in Figure 5.9(a). The energy levels are called *donor energy levels*. As

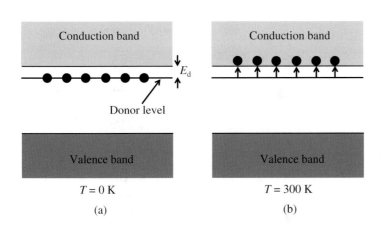

Figure 5.9 (a) In an *n*-type semiconductor at absolute zero ($T = 0$ K), all the electrons that are bound to the donor atoms are in donor levels that lie just below the bottom of the conduction band. (b) At temperatures above absolute zero, these weakly bound electrons can be thermally excited into the conduction band and, at room temperature (300 K), essentially all these electrons are promoted into the conduction band.

noted above, at room temperature ($T \sim 300$ K) essentially all the electrons in the donor levels are promoted into the conduction band, as illustrated in Figure 5.9(b).

p-*type semiconductors*

A complementary situation occurs in a *p*-type semiconductor. Suppose that we introduce an atom of aluminium into the crystal lattice so that it takes the place of a silicon atom, as shown in Figure 5.10. Aluminium has one less electron than silicon, as shown in Table 5.1; it has two 3s electrons and just one 3p electron. Consequently, it has only three electrons to share with its four neighbouring silicon atoms, which means that one of the four valence bonds is incomplete. In order to complete this bond, the aluminium atom *accepts* an electron from one of its silicon neighbours, as illustrated in Figure 5.10. This creates a hole in the neighbouring silicon atom. As we saw previously, such a hole acts as a positive charge carrier that can move freely through the crystal under the influence of an electric field. As the impurity atom has gained an extra electron it becomes a negatively charged ion, Al$^-$, which is fixed in the crystal lattice. Because the impurity atom accepts an electron from a neighbouring silicon atom, it is called an *acceptor atom*.

It needs a small but finite amount of energy E_a for an electron to transfer from a silicon atom to an aluminium atom. Equivalently we can say that E_a is the amount of energy required to generate a hole in the valence band. It follows that the energy level associated with such a transfer must lie at a distance E_g above the top of the valence band, as shown in Figure 5.11(a). And when an electron transfers from a silicon atom to an aluminium atom, it jumps from the valence band into this energy level. As it accepts an electron, it is called an *acceptor level*. At absolute zero, the electrons do not have the necessary thermal energy to make the transfer and remain in the

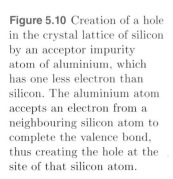

Figure 5.10 Creation of a hole in the crystal lattice of silicon by an acceptor impurity atom of aluminium, which has one less electron than silicon. The aluminium atom accepts an electron from a neighbouring silicon atom to complete the valence bond, thus creating the hole at the site of that silicon atom.

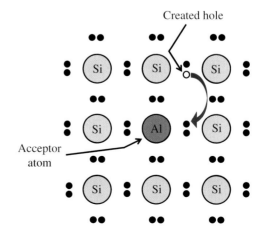

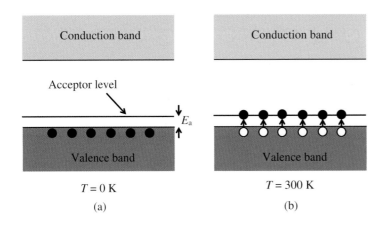

Figure 5.11 (a) In a p-type semiconductor at absolute zero ($T = 0$ K), the electrons remain in the valence band. (b) At temperatures above absolute zero, the electrons can be thermally excited into acceptor levels that lie just above the top of the valence band. This creates holes in the valence band.

valence band as in Figure 5.11(a). By similar reasoning to that used for n-type semiconductors, we find that $E_a \sim 10$ meV. Again, this is substantially smaller than kT at room temperature, and, at that temperature, essentially almost all of the acceptor levels are filled, as indicated in Figure 5.11(b). Then the number of holes is essentially equal to the number of added impurity atoms.

Worked example

The concentration, i.e. number density of free electrons in pure silicon is approximately $10^{16}/\text{m}^3$. If one silicon atom in every million atoms is replaced by a phosphorus atom, by what factor does the concentration of free electrons increase? The density, ρ, of silicon is 2.33×10^3 kg/m^3, and its atomic mass M is 28.1 u.

Solution

The atomic weight of any substance in grams contains Avogadro's number, N_A, of atoms. The volume occupied by M g of the substance $= \frac{M \times 10^{-3}}{\rho}$ m^3. Hence, the number of silicon atoms per cubic metre is given by

$$n_{\text{Si}} = \frac{\rho N_A}{M \times 10^{-3}} = \frac{2.33 \times 10^3 \times 6.02 \times 10^{23}}{28.1 \times 10^{-3}} = 5.0 \times 10^{28} \text{ atoms/m}^3.$$

Hence, the concentration of phosphorus atoms is 5.0×10^{22} atoms/m^3. To a good approximation, each phosphorus atom donates one electron at room temperature. Hence the concentration of free electrons has increased by the factor 5.0×10^6 by doping silicon with just one phosphorus atom per million silicon atoms. The resulting concentration of free electrons is, however, still much smaller than the concentration in conductors. For example, the concentration in copper is $\sim 10^{29}/\text{m}^3$.

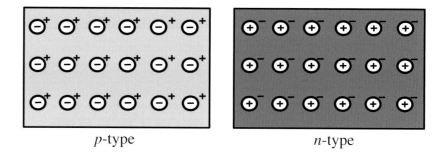

Figure 5.12 The figure shows
isolated pieces of p-type and
n-type semiconductor. For
the sake of clarity, only the
acceptor and donor ions and
the respective majority
carriers are shown. The
acceptor and donor ions are
denoted by the symbols ⊖
and ⊕ respectively. In the
p-type, there is an abundance
of holes, which form the
majority carriers. These are
denoted by the '+' symbol.
Similarly in the n-type, there
is an abundance of electrons,
which are the majority
carriers, denoted by the '−'
symbol. In general, the
concentration of holes in the
p-type will be different from
the concentration of electrons
in the n-type.

p-type

n-type

5.3   The p–n junction

A solar cell is essentially a p–n junction. Indeed, the p–n junction
is at the heart of essentially all semiconductor devices, including
diodes, transistors and light-emitting diodes (LEDs). Such a junction is formed when an n-type semiconductor meets a p-type semiconductor. In this section we describe the movement of electrons and
holes across a p–n junction and the resulting electrical properties of
the junction.

5.3.1   The p–n junction in equilibrium

Consider first two isolated pieces of semiconductor material, a p-type
and an n-type, as shown in Figure 5.12. In this figure, only the acceptor and donor ions and the respective majority carriers are shown.
The acceptor and donor ions are denoted by the symbols ⊖ and ⊕
respectively. In the p-type there is an abundance of holes, which form
the majority carriers. These are denoted by the '+' symbol. There is
charge neutrality because of the presence of the negatively charged
acceptor ions that are embedded uniformly throughout the lattice.
Similarly, in the n-type, there is an abundance of electrons, which
are the majority carriers, denoted by the '−' symbol, and there are
positively charged donor ions that are uniformly embedded in the
lattice; again there is charge neutrality. In general, the concentration of holes in the p-type will be different from the concentration of
electrons in the n-type.

Imagine that we join these two pieces of semiconductor material
together, as shown in Figure 5.13. As there are many more free electrons in the n-type than in the p-type, electrons will *diffuse* from the
n-type into the p-type. Similarly, holes in the p-type will diffuse into
the n-type. An analogy here is a box that is divided into two halves
by a barrier. Imagine that the left side is filled with oxygen and the
right side is filled with nitrogen. If the barrier is permeable, oxygen
diffuses to the right and nitrogen diffuses to the left. The diffusion
of these majority charge carriers across a p–n junction gives rise to
the so-called *diffusion current*, i_{diff}, which is the *net* current due to

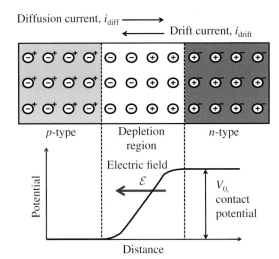

Diffusion current, i_{diff} →

← Drift current, i_{drift}

p-type | Depletion region | n-type

Electric field

$\mathcal{E}$

$V_{0,}$ contact potential

Potential

Distance

Figure 5.13 The p–n junction in equilibrium. Diffusion of electrons and holes across the junction and their subsequent recombination produce a depletion layer that is devoid of mobile charge carriers. When electrons diffuse away from the n-type region, they leave behind positively charged donor ions. Similarly, when holes diffuse away from the p-type region they leave behind negatively charged acceptor ions. This double layer of charge causes an electric field $\mathcal{E}$ to be set up across the junction, producing a difference in potential energy between the n and p regions. This produces a potential barrier of height, V_0, that inhibits holes from the p region diffusing into the n region and similarly inhibits electrons diffusing into the p region. At equilibrium, there must be no net current flow across a p–n junction. Hence, the drift current, i_{drift}, is exactly counterbalanced by the diffusion current i_{diff}, as indicated by their respective arrows on the figure.

the movement of both the electrons and the holes. In Figure 5.13, this diffusion current is represented by an arrow, which points in the same direction as conventional current flow.

As free electrons from the n-type diffuse into the p-type, they recombine with holes in the p-type that are present in high concentration and so move only a short distance before disappearing. Similarly, holes in the p-type diffusing across the junction recombine with electrons in the n-type. As a result of the electrons and holes recombining in this way, there is a *depletion layer* at the p–n junction that is depleted of mobile charge carriers. When the electrons diffuse away from the n-type they leave behind positively charged donor ions that are fixed in the lattice. Thus a layer of positive charge is built up on the n side of the junction. Similarly, a layer of negative charge is built up on the p side of the junction due to the negatively charged acceptor ions; the resulting ions that have lost electrons or holes are said to be *uncovered*. The diffusion of electrons and holes thus creates a double layer of charge. The positive charge layer on the n side repels holes and the negative charge layer on the p side repels electrons, and further diffusion of the electrons and holes is inhibited.

The double-charge layer sets up an electric field $\mathcal{E}$ that points from the n side to the p side of the junction. This is the origin of the inbuilt electric field of a p–n junction. And it leads to a difference in potential energy between the n-type and p-type; the n-type is at the higher potential, as shown in Figure 5.13. We can then picture the resulting situation in terms of a potential barrier of height V_0 that inhibits holes from the p region diffusing into the n region, at higher potential, and also inhibits electrons diffusing into the p region. The height of the barrier V_0 is called the *contact potential*. Its value depends on the levels of doping and the temperature and is typically ~ 0.5 V. We emphasise that this is an *internal barrier*. It is not an externally imposed barrier but is a property of the p–n junction itself.

As well as the majority charge carriers (electrons in the n-type and holes in the p-type), we recall that there are also minority carriers that are produced when thermal excitation generates electron–hole pairs. These are generated in both the n- and p-type regions. In the n-type, the generated holes are the minority carriers, and in the p-type the generated electrons are the minority carriers. Those minority carriers formed close to the p–n junction may migrate towards the junction and will be swept across it by the junction field; the direction of the field causes the generated electrons to drift toward the n side and the holes to drift toward the p side. The current due to these minority carriers is called the *drift current*, i_{drift}, and is the sum of the electron current in one direction and the hole current in the other. In Figure 5.13, this drift current is represented by an arrow, which points in the opposite direction to the diffusion current.

At equilibrium, there must be no net current flow across a p–n junction. Hence, the small i_{drift} due to thermal generation must be exactly counterbalanced by a finite i_{diff}. Thus, although the junction field inhibits the majority carriers from diffusing across the junction, there nevertheless remain a relatively small number of them that must do so. In fact, the contact potential V_0 adjusts until the drift and diffusion currents cancel exactly. Then for a p–n junction in equilibrium, the diffusion and drift currents are equal in magnitude but flow in opposite directions, as indicated by their respective arrows on Figure 5.13; the lengths of the arrows, which represent the magnitudes of the diffusion and drift currents, are equal but the arrows point in opposite directions.

It was convenient to think in terms of bringing together a piece of p-type semiconductor and a piece of n-type semiconductor to form a junction between the two. However we cannot just push pieces of p- and n-type semiconductor together and expect the junction to work properly because of the impossibility of achieving a continuous crystal structure across the junction. Instead, impurity atoms are added to a *single* piece of semiconductor. This can be done by first heating a slice of a pure silicon crystal, ~ 300 μm thick, to a temperature of ~ 1000 °C in a vacuum chamber while passing P_2O_5 gas over one of its surfaces. Phosphorus atoms diffuse into the surface, producing a thin layer of n-type material. The process is continued until the desired thickness of the n-type material is obtained. Then, the p-type layer is made by diffusing aluminium into the other surface of the silicon crystal. This is continued until the desired width of the p-type material is obtained and the p–n junction is formed.

5.3.2   The biased *p–n* junction

The p–n *junction with reverse bias*

Reverse bias is applied to a *p–n* junction, as shown in Figure 5.14, where the negative terminal of a battery is connected to the *p* side and the positive terminal to the *n* side. The effect of this *bias voltage* is to *attract away* from the junction both holes in the *p* region and electrons in the *n* region. This increases the number of donor ions and acceptor ions near the junction that are left uncovered and increases the width of the depletion layer. As the depletion layer is depleted of charge carriers, its resistance is much greater than that of the *p* or *n* regions. Consequently, most of the bias voltage *V* occurs across the depletion layer. The action of this bias voltage is to increase the height of the potential barrier from V_0 to $(V_0 + V)$. This further inhibits the flow of majority carriers across the junction and reduces the diffusion current to essentially zero, as indicated by the reduced length of the diffusion current arrow in Figure 5.14. However, there will still be thermally generated electron–hole pairs in the *n* and *p* regions; the number of these pairs depends on the temperature and the band gap of the semiconductor and these remain the same. And those minority carriers generated close to the junction may migrate towards the junction and will be swept across the depletion layer by the junction field. Hence, the drift current i_{drift} will still exist. The magnitude of this current does not depend appreciably on the applied bias potential, *V*. Hence, the length of the arrow representing the drift current in Figure 5.14 is essentially the same length as for the *p–n* junction in equilibrium, with no bias voltage applied.

Figure 5.14 A reverse-biased *p–n* junction. The effect of the bias voltage *V* is to push the majority charge carriers *away* from the junction, increasing the width of the depletion region. The bias voltage *V* is dropped across the high-resistance depletion layer, increasing the height of the potential barrier from V_0 to $(V_0 + V)$ and reducing the diffusion current to essentially zero. The drift current due to thermally generated electron–hole pairs is essentially insensitive to the bias voltage and remains the same as for the *p–n* junction in equilibrium. Thus the arrow representing the diffusion current is much reduced in length but the length of the drift arrow remains the same.

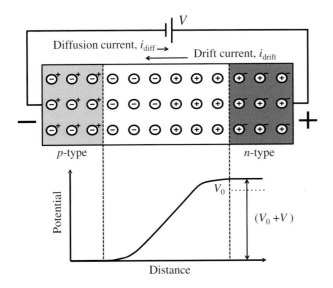

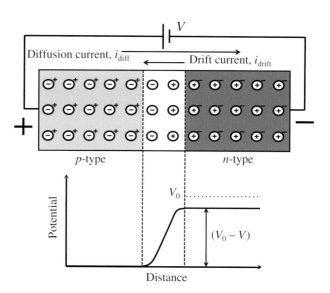

Figure 5.15 A forward-biased *p–n* junction. The effect of the bias voltage *V* is to push the majority carriers toward the junction, reducing the width of the depletion region. The bias voltage is again dropped across the depletion layer and has the effect of reducing the height of the potential barrier to $(V_0 - V)$, so that more majority carriers can surmount the reduced barrier. This increases the diffusion current, which becomes larger than the drift current, as indicated by the length of the respective arrows.

The p–n *junction with forward bias*

Forward bias is applied to a *p–n* junction as shown in Figure 5.15, where the positive terminal of a battery is connected to the *p* side and the negative terminal to the *n* side. The effect of this is to push *towards* the junction the electrons in the *n* region and the holes in the *p* region. This decreases the number of donor ions and acceptor ions near the junction that are left uncovered and reduces the width of the depletion layer. Again, the bias voltage *V* is dropped almost entirely across the high-resistance depletion layer and has the effect of lowering the height of the potential barrier to $(V_0 - V)$. This lowering of the barrier means that more majority carriers can surmount the barrier, i.e. diffuse across the junction. As the applied voltage increases, an increasing number of majority carriers diffuse across the junction and by $V \sim 0.5$ V there is an appreciable diffusion current flowing across the junction. This is indicated in Figure 5.15 by the increased length of the arrow representing the diffusion current. As before, those minority carriers that are formed close to the junction may migrate towards it and be swept across the junction by the inbuilt electric field and so again the drift current does not change appreciably.

The electrons do not travel far into the *p*-type before they recombine with a hole, which are plentiful there. The characteristic distance they travel is called the *diffusion length*. Similarly, the holes have a diffusion length, which is characteristic of the distance they travel in the *n*-type before recombining with an electron. In general the diffusion length for the holes will be different from that for the electrons. In the *p*-type, the resultant flow of holes towards the junction is equivalent to a flow of electrons in the opposite direction.

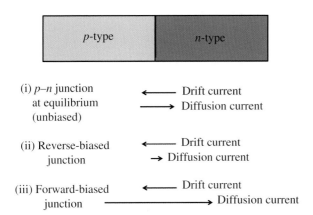

(i) *p–n* junction
at equilibrium
(unbiased)
⟵——— Drift current
———⟶ Diffusion current

(ii) Reverse-biased
junction
⟵——— Drift current
⇢ Diffusion current

(iii) Forward-biased
junction
⟵——— Drift current
———————⟶ Diffusion current

Figure 5.16 A summary of the currents that flow across a *p–n* junction for: (i) *p–n* junction in equilibrium, (ii) reverse-biased junction and (iii) forward-biased junction. The drift current is relatively insensitive to bias voltage and is essentially the same for all three conditions. At equilibrium, when there is no bias, the diffusion and drift currents are equal and opposite. When the *p–n* junction is reverse-biased, the diffusion current is essentially zero. When the *p–n* junction is forward-biased, the diffusion current is much greater than the drift current.

These electrons leave the *p*-type to flow through the external circuit and flow back into the *n*-type to neutralise the holes that have diffused into the *n*-type. When a majority charge carrier crosses the *p–n* junction, it becomes a minority charge carrier. Thus we see that it is the flow of minority carriers (holes in the *n*-type and electrons in the *p*-type) that account for the flow of current in a *p–n* junction.

Figure 5.16 summarises the currents that flow across a *p–n* junction for the three possible bias conditions. These are the diffusion current due to the diffusion of majority carriers across the junction and the drift current due to the minority carriers drifting across the junction. The drift current is fairly insensitive to bias voltage and so is essentially the same for all the three conditions. When there is no bias, the diffusion and drift currents are equal and opposite. When the *p–n* junction is reverse-biased, the diffusion current is essentially zero. When the *p–n* junction is forward-biased, the diffusion current is much greater than the drift current.

We can make an analogy between a *p–n* junction, with its two layers of charge on either side of the junction, and a pair of parallel conducting plates with a voltage difference between them. These plates form an electrical capacitor with a layer of positive charge on one plate and a layer of negative charge on the other. Similarly, the *p–n* junction can also be viewed as a capacitor. As the bias voltage changes, the two layers of charge at the *p–n* junction vary in width and therefore in the amount of charge they contain. Thus the capacitance of the junction can be controlled by the bias voltage. This effect is exploited in a device called the *varactor diode*, which finds application, for example, as the tuning capacitor in a radio receiver.

5.3.3 The current–voltage characteristic of a *p–n* junction

The net current i flowing across a *p–n* junction is the combination of the diffusion and drift currents: $i = i_{\text{diff}} - i_{\text{drift}}$. We have seen (i)

Figure 5.17 The current–voltage characteristic of a p–n junction, where V is the applied voltage and i is the current flowing through the junction. As the voltage V is increased in the forward direction, corresponding to forward bias, the current i increases rapidly. However, when the voltage V is increased in the reverse direction, corresponding to reverse bias, the reverse saturation current i_0 remains essentially constant. The magnitude of i_0 is exaggerated for the sake of clarity.

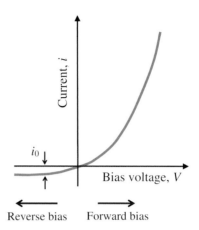

that no net current flows across a p–n junction when the voltage V applied across it is zero; (ii) that current does flow when the junction is forward-biased (V positive) and this current increases rapidly with increasing forward bias voltage; and (iii) that a small but finite current flows across the junction in the opposite direction when the junction is reverse-biased (V negative) and that this current is essentially independent of the value of the bias voltage. This dependence of the current on bias voltage results in the i–V curve shown in Figure 5.17, which is a plot of i against V. As V increases in the forward direction, i increases rapidly, but when V is increased in the reverse direction, i remains constant. The reverse-bias current is called the *reverse saturation current* i_0 and is just the drift current i_{drift}. Its magnitude has been exaggerated in Figure 5.17 for the sake of clarity and for a silicon p–n junction, i_0 typically lies within in the range 10^{-10}–10^{-12} A. The dependence of current i on bias voltage V is described by the equation

$$i = i_0(\mathrm{e}^{eV/kT} - 1), \tag{5.5}$$

where T is the absolute temperature and k is the Boltzmann constant. Obviously the non-linear form of this equation is quite different from the linear form of Ohm's law, $i = V/R$.

It is apparent from Figure 5.17 that a p–n junction essentially allows current to flow in only one direction, apart from the extremely small reverse saturation current. Hence, a p–n junction is a *rectifying* device called a *diode*. Consequently, Equation (5.5), which describes the current–voltage characteristic of a p–n junction, is also known as the *diode equation*. A diode has many important applications. For example, it can be used to convert an alternating current (AC) voltage into a direct current (DC) voltage.

We can use our model of a p–n junction with its potential barrier to arrive at Equation (5.5). For this we need to consider the probability that a majority carrier, electron or hole, can surmount the internal potential barrier of a p–n junction. When a large number of particles is in thermal equilibrium, there is a continual exchange of energy between the particles. At any instant, the energy of an individual particle is not known; it may be larger or smaller than the average. However, there is a well-defined probability distribution for the whole assembly of particles called the *Boltzmann distribution*. It gives the probability of a particle having energy E at temperature T. The Boltzmann distribution is expressed in the form

$$f(E) \propto e^{-E/kT}, \tag{5.6}$$

where T is temperature, k is the Boltzmann constant and the factor $e^{-E/kT}$ is called the *Boltzmann factor*. The precise form of the distribution depends on the system under consideration, but the temperature dependence of the distribution is always governed by a Boltzmann factor. This factor may be multiplied by other energy-dependent terms, but these do not include the temperature. The Boltzmann distribution applies to quantum systems as in the present context but also to classical systems, as in the following example.

Worked example

As we have seen, Earth's atmosphere plays an important role in the interaction of solar radiation with Earth. Use the Boltzmann distribution to deduce the variation in atmospheric pressure with respect to vertical height and to estimate the height of the atmosphere.

Solution

At vertical height h, the gravitational potential energy of a molecule of mass m is mgh. The probability that a molecule at temperature T will have this energy is proportional to the Boltzmann factor, $e^{-mgh/kT}$. The density, and hence pressure, of the atmosphere at height h is directly proportional to the probability of a molecule being at that height. It follows then that the atmospheric pressure $P(h)$ falls exponentially with height according to $P(h) = P_0 e^{-mgh/kT}$, where P_0 is the pressure at Earth's surface. We can use this result to get a rough estimate of the height of the atmosphere. We do this by equating mgh to kT, i.e. the height at which the atmospheric pressure has decreased by a factor of $1/e$. Assuming an average temperature of 250 K for T and a molecular weight of 30 u,

$$h = \frac{kT}{mg} = \frac{1.38 \times 10^{-23} \times 250}{30 \times 1.66 \times 10^{-27} \times 9.8} \approx 7 \text{ km}.$$

Note that it is the thermal energy of the molecules that holds up the atmosphere!

In the case of a p–n junction in equilibrium, the internal barrier of height V_0 inhibits the majority charge carriers diffusing across the junction. According to the Boltzmann distribution, the probability of a charge carrier at temperature T having energy eV_0 is proportional to $e^{-eV_0/kT}$. It follows that the probability that a charge carrier can surmount the barrier and contribute to the diffusion current is also proportional to $e^{-eV_0/kT}$. In the case of the forward-biased p–n junction, the barrier is lowered from V_0 to $(V_0 - V)$. Then the probability that a majority charge carrier has enough energy to surmount this barrier of reduced height is proportional to $e^{-e(V_0-V)/kT}$. The fractional increase in the probability is

$$\frac{e^{-e(V_0-V)/kT}}{e^{-eV_0/kT}} = e^{eV/kT}.$$

The diffusion current, which is proportional to the number of majority carriers surmounting the barrier, increases by the same factor and so

$$i_{\text{diff}} \propto e^{eV/kT} = \text{constant} \times e^{eV/kT}. \qquad (5.7)$$

The net current i flowing through the junction is then

$$i = i_{\text{diff}} - i_{\text{drift}} = \text{constant} \times e^{eV/kT} - i_0,$$

where we have used the fact that i_{drift} is equal to the reverse saturation current i_0. When the junction is at equilibrium, $V = 0$ and $i = 0$ with $e^{eV/kT} = 1$. Hence, the constant is equal to i_0 and $i = i_0(e^{eV/kT} - 1)$ in agreement with Equation (5.5).

5.3.4   Electron and hole concentrations in a semiconductor

In Section 5.2.1 we described the energy bands in a semiconductor and how the atomic electrons fill the available energy levels, starting with the lowest energy band. In order to describe the resulting electron (and hole) densities or *concentrations* in these energy bands, we need to know the following:

(i) the *density of states* in the bands;

(ii) the probability of each of these states being occupied.

The first factor is given by the *density of states function, $Z(E)$,* which may be defined as the number of energy states per unit energy per unit volume. We encountered an analogous function in Section 4.2.3 for the density of standing electromagnetic waves in a blackbody cavity. The form of $Z(E)$ can be obtained in a similar fashion, where geometrical arguments are used to count the number of quantum states that are contained in a given energy range (see Hook and

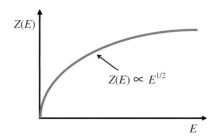

Figure 5.18 The density of states function $Z(E)$ for electrons in the conduction band of a semiconductor.

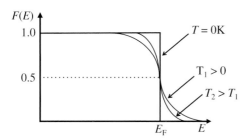

Figure 5.19 The Fermi–Dirac distribution $F(E)$, which gives the probability of a particular state at energy E being occupied by an electron, which is a fermion. The distribution is shown for $T = 0$ K and for increasing temperatures. At any finite temperature, the probability of occupation of an energy level at $E = E_F$ is 0.5, where E_F is defined as the Fermi energy.

Hall,[1] Section 5.3 for further details). For the conduction band, the function $Z(E)$ is given by

$$Z(E) = \frac{4\pi}{h^3}(2m_e^*)^{3/2} E^{1/2}, \qquad (5.8)$$

where h is Planck's constant, the energy E is measured from the bottom of the conduction band, and m_e is the effective mass of the electrons. The form of $Z(E)$ is given in Figure 5.18, which shows its parabolic dependence on energy.

The second factor depends on the fact that electrons are fermions and obey the Pauli exclusion principle. For fermions, the probability of a particular level at energy E being occupied is given by the *Fermi–Dirac distribution*:

$$F(E) = \frac{1}{1 + e^{(E-E_F)/kT}}, \qquad (5.9)$$

where T is the absolute temperature and E_F is a characteristic energy called the *Fermi energy*. We plot this important function in Figure 5.19. At $T = 0$, $F(E)$ is unity for energies less than E_F and zero for energies greater than E_F. At temperatures above absolute zero the distribution changes as also shown in Figure 5.19. Notice that at any finite temperature, the probability of occupation of an energy level at $E = E_F$ is 0.5 and indeed we can use this fact to define the Fermi energy. The symmetry of the Fermi–Dirac distribution about the Fermi energy makes this energy a natural reference point.

[1] J. R. Hook and H. E. Hall, *Solid State Physics*, 2nd edn, John Wiley & Sons Ltd, 1991.

If $(E-E_F) \gg kT$, Equation (5.9) reduces to

$$F(E) \approx e^{-(E-E_F)/kT}, \tag{5.10}$$

which has the same form as the Boltzmann distribution (Equation 5.6). For electrons in the conduction band of a semiconductor, the energy difference $(E - E_F)$ is typically ~ 0.5 eV while the value of kT at room temperature is approximately $1/40$ eV and so this approximation is usually valid. In that case, Equation (5.10) gives us further physical meaning to the Fermi energy. We can use Equation (5.10) just as we would use the Boltzmann distribution, so long as we take the electron to have originated in an energy level corresponding to the Fermi energy. We emphasise, however, that this *Fermi level*, which occurs at the Fermi energy, is not a physical energy level that a particle occupies; rather it is a hypothetical energy level that lies at that energy. In a semiconductor, the Fermi energy lies somewhere between the top of the valence band and the bottom of the conduction band. In particular, the Fermi energy in an *intrinsic* semiconductor lies in the middle of the energy gap.

We will use the case of an *intrinsic* semiconductor to illustrate graphically the resulting concentrations of electrons in the conduction band and holes in the valence band. Figure 5.20(a) shows the density of state function $Z(E)$ for the conduction and valence bands. In the conduction band, $Z(E)$ increases with increasing energy from the bottom of the band. In the valence band, $Z(E)$ has mirror symmetry with respect to $Z(E)$ in the conduction band and decreases as it approaches the top of the band. Figure 5.20(b) shows the Fermi–Dirac distribution. For the case of an intrinsic semiconductor, the Fermi energy is centred on the middle of the band gap. At finite temperatures this Fermi–Dirac distribution has tails that extend into the valence and conduction bands, as shown.

Figure 5.20 (a) The density of state functions $Z(E)$ for the conduction and valence bands of a semiconductor. (b) The Fermi–Dirac distribution in an intrinsic semiconductor, in which the Fermi energy is centred on the middle of the band gap. At finite temperatures this Fermi–Dirac distribution has tails that extend into the valence and conduction bands. (c) The concentration of electrons in the conduction band is obtained by summing the product $Z(E)F(E)$, and is equal to the shaded area as indicated. The concentration of holes in the valence bans is obtained by summing the product $Z(E)[1 - F(E)]$ and is equal to the other shaded area, as indicated. The two shaded areas are the same, reflecting the fact that the number of electrons in the conduction band is equal to the number of holes in the valence band in an intrinsic semiconductor.

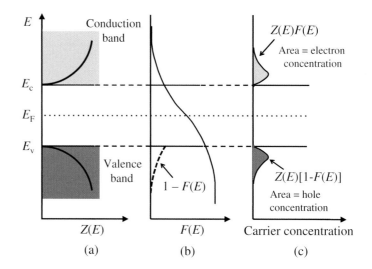

The concentration of electrons in the conduction band is obtained by summing the product $Z(E)F(E)$ of the density of states function and the Fermi–Dirac distribution over the energy range of interest. This product is shown in Figure 5.20(c) and the concentration of electrons is equal to the shaded area indicated in the figure. If the probability of finding an electron with energy E is $F(E)$, the probability of not finding an electron at that energy is $[1 - F(E)]$, which is the probability of finding a hole. Thus the concentration of holes in the valence band is obtained by summing the product $Z(E)[1 - F(E)]$, which is also shown in Figure 5.20(c). The hole concentration is again given by the shaded area indicated in the figure. We expect the two shaded areas to be the same, reflecting the fact that the number of electrons in the conduction band is equal to the number of holes in the valence band in an intrinsic semiconductor. This follows naturally from the symmetry of the density of state function in the two bands and the symmetry of the Fermi–Dirac distribution when the Fermi energy is placed at the middle of the gap. If the Fermi energy were placed at any other energy in the gap, the areas would not be equal. Figure 5.20 shows the situation for an intrinsic semiconductor, but the concentration of charge carriers in any semiconductor, intrinsic or extrinsic, is given by the product of the density of states function and the Fermi–Dirac distribution. We simply need to take the appropriate Fermi energy for the semiconductor.

For the concentration n of electrons in the conduction band, we can write

$$n = \int_{\substack{\text{conduction} \\ \text{band}}} Z(E)F(E)\mathrm{d}E, \tag{5.11}$$

where $Z(E)$ and $F(E)$ are given by Equations (5.8) and (5.9), respectively. Evaluation of the integral gives

$$n = N_\mathrm{c}\mathrm{e}^{-(E_\mathrm{c}-E_\mathrm{F})/kT}, \tag{5.12}$$

where

$$N_\mathrm{c} = 2\left(\frac{m_\mathrm{e}^* kT}{2\pi\hbar^2}\right)^{3/2}, \tag{5.13}$$

and E_c is defined in Figure 5.20. We may think of N_c as a normalisation factor, which describes the total number of electrons that are promoted to the conduction band. In a similar way, to determine the concentration, p, of holes in the valence band, we evaluate the integral

$$n = \int_{\substack{\text{valence} \\ \text{band}}} Z(E)\,[1 - F(E)]\mathrm{d}E, \tag{5.14}$$

which gives

$$p = N_\mathrm{v}\mathrm{e}^{-(E_\mathrm{F} - E_\mathrm{v})/kT}, \tag{5.15}$$

where again, N_v is a constant of normalisation and is given by

$$N_\mathrm{v} = 2\left(\frac{m_\mathrm{h}^* kT}{2\pi\hbar^2}\right)^{3/2}, \tag{5.16}$$

and E_v is again defined by Figure 5.20. m_h^* is the effective mass of the holes, which we have taken to be the same as the effective mass of the electrons. The product of the electron and hole concentrations is given by

$$np = N_\mathrm{c}\mathrm{e}^{-(E_c - E_\mathrm{F})/kT} \times N_\mathrm{v}\mathrm{e}^{-(E_\mathrm{F} - E_v)/kT} \tag{5.17}$$
$$= N_\mathrm{c}N_\mathrm{v}\mathrm{e}^{-E_\mathrm{g}/kT}, \tag{5.18}$$

since $E_g = E_c - E_v$. We see that the product of n and p is fixed at constant temperature. Although we have seen it here for the case of an intrinsic semiconductor, Equation (5.18) also applies to extrinsic semiconductors. It is known as the *law of mass action* for semiconductors. It is a very important relationship as, once we know n or p, we can determine the other. Thus, if we dope an intrinsic semiconductor with acceptor atoms to increase the concentration of holes, the concentration of electrons must decrease accordingly. For the particular case of an intrinsic semiconductor, $n = p = n_i$, where n_i is the *intrinsic carrier concentration*. Hence

$$n_i = n = p = \sqrt{np} = (N_\mathrm{c}N_\mathrm{v})^{1/2}\mathrm{e}^{-E_\mathrm{g}/2kT}. \tag{5.19}$$

We see that the temperature variation of n_i is exponential. i.e. $n_i \propto \mathrm{e}^{-E_\mathrm{g}/2kT}$. Moreover, the product of electron and hole concentrations in any semiconductor can conveniently be written

$$np = n_i^2. \tag{5.20}$$

Worked example

Calculate the intrinsic carrier concentration in silicon at 400 K, given that the electron effective mass is 0.31 m_e, the hole effective mass is 0.56 m_e and the energy gap is 1.11 eV.

Solution

$$n = p = n_i = 2\left(\frac{kT}{2\pi\hbar^2}\right)^{3/2}(m_e^* m_h^*)^{3/4}e^{-E_g/2kT}$$

$$= 2\left[\frac{400 \times 1.38 \times 10^{-23}}{2\pi(1.05 \times 10^{-34})^2}\right]^{3/2} \times [0.31 \times 0.56 \times (0.91 \times 10^{-30})^2]^{3/4}$$

$$\times \; e^{-\left(\frac{1.11 \times 1.6 \times 10^{-19}}{2 \times 400 \times 1.38 \times 10^{-23}}\right)}$$

$$= 2(7.97 \times 10^{46})^{3/2} \times (1.44 \times 10^{-61})^{3/4} \times e^{-(15.96)}$$

$$= 1.2 \times 10^{18} \text{ carriers/m}^3.$$

5.3.5 The Fermi energy in a *p–n* junction

In an *n*-type semiconductor, the electron concentration in the conduction band is high, with the result that the Fermi energy lies just below the bottom of the conduction band. This is illustrated in Figure 5.21(b). On the other hand, in a *p*-type semiconductor, the hole concentration in the valence band is high, with the result that the Fermi energy lies just above the bottom of the valence band, as illustrated in Figure 5.21(a). When, however, we have a junction between a *p*-type semiconductor and an *n*-type semiconductor, *the Fermi energy must be at a constant level throughout the semiconductor*, as shown in Figure 5.22. With this condition, the energy of the system is minimised. We can make an analogy here with temperature in a system consisting of several parts. We know that at equilibrium the temperature must be the same throughout the system. Indeed, both temperature and the Fermi energy, more generally called the *chemical potential*, are thermodynamic quantities.

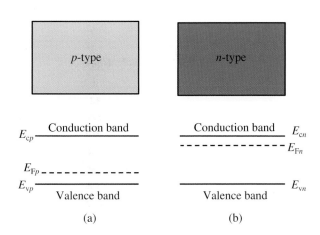

Figure 5.21 (a) The Fermi energy in a *p*-type semiconductor lies just above the top of the valence band. (b) In an *n*-type semiconductor, the Fermi energy lies just below the bottom of the conduction band.

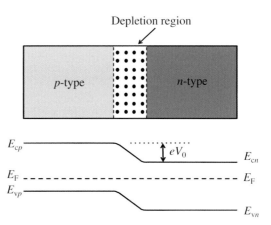

Figure 5.22 The energy levels in an unbiased p–n junction. Note that the Fermi energy is constant throughout the semiconductor. There is a step change, eV_0, in energy between the n and p regions. This inhibits electrons from the n region diffusing into the p region. Similarly, holes are inhibited from diffusing into the n region.

If the p and n regions are made from the same semiconductor, the band gap E_g is the same for both regions. Then, as we can see from Figure 5.22, the requirement for a constant Fermi energy means that the energy levels in the n-type are lowered with respect to those in the p-type. The amount by which they are lowered is eV_0, where V_0 is just the contact potential that we encountered previously. The step in energy inhibits electrons from the n-type diffusing into the p-type, and holes in the p-type are inhibited from diffusing into the n-type, as we also saw in Section 5.3.1. We can change this equilibrium situation by forward-biasing the p–n junction, i.e. by applying a positive voltage, V, to the p-type with respect to the n-type, as shown in Figure 5.23. This raises the energy levels in the n-type relative to the p-type, decreasing the energy difference between them by an amount eV. It then becomes easier for the electrons in the n region to diffuse into the p region and for the holes in the p region to diffuse into the n region. This increases the diffusion current by the Boltzmann factor $e^{eV/kT}$. It is important to note, however, that in a biased p–n junction *the Fermi level is no longer constant throughout the composite*

Figure 5.23 The energy levels in a forward-biased p–n junction. The energy bands in the n-type are raised relative to the bands in the p-type, decreasing the energy difference between them by eV. It then becomes easier for the electrons in the n region to diffuse into the p region and for the holes in the p region to diffuse into the n region. The Fermi level is no longer constant throughout the semiconductor; the Fermi levels on the two sides are displaced by the magnitude of the bias voltage V.

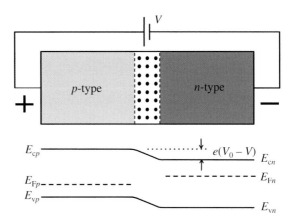

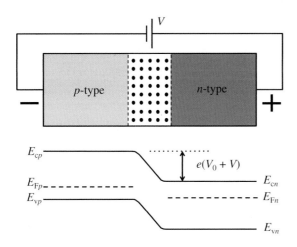

Figure 5.24 The energy levels in a reverse-biased p–n junction. The energy bands in the n-type are lowered relative to the p-type. Electrons from the n region and holes from the p region have an even higher barrier to climb and relatively few of these majority carriers diffuse across the junction. Again, the Fermi level is no longer constant throughout the semiconductor.

semiconductor : the Fermi levels on the two sides are displaced by the magnitude of the bias voltage. When the p–n junction is reverse-biased, as in Figure 5.24, the energy levels in the n-type are *lowered* relative to the p-type. Then the electrons from the n-type and the holes from the p-type have an even higher barrier to climb and relatively few majority carriers diffuse across the junction. Again, the Fermi level is no longer constant throughout the semiconductor.

5.4 Semiconductor solar cells

A semiconductor solar cell is essentially a p–n junction, although one that has a large surface area to collect solar radiation. In this section, we describe what happens when sunlight is incident on a p–n junction and how the electrical power that it delivers to an external load can be maximised. We also describe some ways of increasing the efficiency of solar cells and the use of alternative solar cell materials.

5.4.1 Photon absorption at a p–n junction

When a photon is incident upon a semiconductor such as silicon, an electron may be promoted from the valence to the conduction band if the photon has an energy $h\nu$ that is greater than the band gap E_{g}, i.e.

$$h\nu > E_{\mathrm{g}}, \tag{5.2}$$

or in terms of wavelength λ:

$$\lambda < \frac{hc}{E_{\mathrm{g}}} = \lambda_{\mathrm{c}}, \tag{5.21}$$

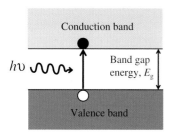

Figure 5.25 When a photon is incident upon a semiconductor material, an electron may be promoted from the valence band to the conduction band if the photon has an energy $h\nu$ that is greater than the band gap E_g.

where λ_c is the *critical wavelength* for the semiconductor. For silicon, $E_g = 1.11$ eV and $\lambda_c = 1.12$ μm. This process produces an electron–hole pair as illustrated in Figure 5.25. Subsequently, the generated electrons and holes recombine with charge carriers of opposite sign. Suppose now that a photon is incident upon the depletion layer of a p–n junction. Again the photon may produce an electron–hole pair, as illustrated in Figure 5.26. But now the inbuilt electric field across the junction causes the electron and hole to separate and drift across the depletion layer. The electrons move into the n region and the holes move into the p region. In terms of the energy level diagram shown in Figure 5.26, an electron is promoted to the conduction band and a hole is left in the valence band. The promoted electron then 'rolls down the hill' into the n region, as the conduction band of the n-type lies energetically below the conduction band of the p-type. Similarly the holes, of opposite charge, move into the valence band of the p-type, which is at a lower energy for them. This movement of electrons and holes across the junction results in a flow of electron current in an external circuit; electrons flow from the n-type around the external circuit and enter the p-type where they recombine with holes. Electron–hole pairs that are generated within a diffusion length or so of the depletion layer may also contribute to the flow of charge carriers across the junction.

Equation (5.2) follows from the conservation of energy. As in any physical reaction, the total momentum of the system must also be conserved. For some semiconductors called *direct band* semiconductors, this happens naturally. Gallium arsenide (GaAs) is an example of a direct band semiconductor. Some semiconductors, however, require a third body to be involved in the reaction to conserve momentum. These are called *indirect band* semiconductors. Silicon (Si), which is the most widely used semiconductor in solar cells, is an example of an indirect band semiconductor. The third body involved is an atom in the crystal lattice of the semiconductor, which absorbs the momentum and becomes vibrationally excited in the process. Such a vibration is called a *phonon*. Importantly, whenever a reaction requires three bodies to be involved, that reaction is less likely to occur than one in which just two bodies are involved.

The difference between direct and indirect band gap semiconductors shows up in their *absorption coefficients* for solar radiation. In Section 3.2.3 we considered the attenuation of a beam of particles as they pass through an absorbing medium and we found that the flux of the beam reduces exponentially with distance travelled in the medium. Similarly, when light passes through a medium, its irradiance I reduces exponentially with distance x according to

$$I(x) = I_0 \exp(-\mu x), \tag{3.61}$$

Direction of electron flow

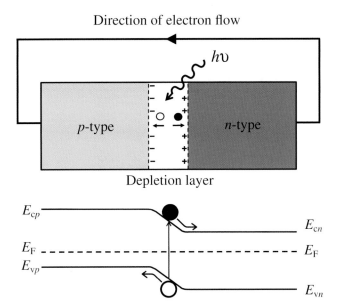

Depletion layer

Figure 5.26 When a photon is incident upon the depletion region of a p–n junction, the photon may produce an electron–hole pair. The inbuilt electric field across the junction causes the electron and hole to separate and drift across the depletion layer. The electron moves into the n region and the hole moves into the p region. In terms of the energy level diagram shown, an electron is promoted to the conduction band and a hole is left in the valence band. The promoted electron then 'rolls down the hill' into the n-type, as the conduction band of the n-type lies energetically below the conduction band of the p-type. Similarly, the holes, of opposite charge, move into the valence band of the p-type, which is at a lower energy for them. This movement of electrons and holes across the junction results in a flow of electron current in an external circuit; electrons flow from the n region around the external circuit and enter the p region where they recombine with holes. Electron–hole pairs that are generated within a diffusion length or so of the depletion layer may also contribute to the flow of charge carriers across the junction.

where μ is the absorption coefficient and I_0 is the initial irradiance. The absorption coefficients for Si and GaAs are plotted against photon energy in Figure 5.27. Each of the curves shows a sharp onset at their respective critical wavelengths, corresponding to their band gap energies; photons of energy less than the band gap of a semiconductor are transmitted with zero or very little absorption. We see that the attenuation coefficient is much lower for the indirect band gap semiconductor Si than the direct band gap semiconductor GaAs over most of the solar wavelength range; the values of μ for Si are $\sim 10^3$/cm and $\sim 10^5$/cm for GaAs. Consequently, a solar cell using Si as the semiconductor has a slice of Si crystal that is ~ 300 μm thick, while a solar cell using GaAs can use a much thinner crystal.

5.4.2 Power generation by a solar cell

As we have described, sunlight that is incident on a solar cell generates electron–hole pairs and these charged carriers drift across the p–n junction to give rise to a photoelectric current that flows through an external circuit. Suppose that the p and n regions of an illuminated solar cell are connected directly together as in Figure 5.28(a). In this case the current flowing in the external circuit is called the *short-circuit current*, i_{sc}. In this figure the arrow points in the direction of *conventional* current flow, (see also Figure 5.29 caption). If, on the other hand, the external circuit is open, as in Figure 5.28(b), the current flowing in the circuit will be zero. In the latter case, the movement of charge carriers across the junction results in an accumulation of charge in both the n and p regions; the p region becomes

Figure 5.27 The attenuation coefficient curves for silicon (Si) and gallium arsenide (GaAs) as a function of photon energy. Each of the curves shows a sharp onset at the respective band gap energies: 1.11 eV for Si and 1.43 eV for GaAs. Photons of energy less than the band gap of a semiconductor are transmitted with zero or very little absorption. The attenuation coefficient is much lower for the indirect band gap semiconductor Si than the direct band gap semiconductor GaAs over most of the solar wavelength range.

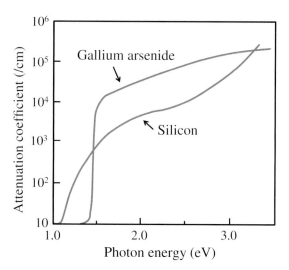

positively charged and the n region becomes negatively charged. This generates a voltage difference, V, between the two regions and, as the p region is at a positive potential with respect to the n region, the p–n junction becomes forward-biased (see Figure 5.15). This reduces the width of the depletion layer so that the diffusion current increases. This continues until V increases to the point where the diffusion current exactly balances the net drift current flowing in the opposite direction. The net drift current is now due to electron–hole pairs that are generated thermally plus those that are generated by the photons incident on the solar cell. The potential difference between the p and n regions when they are not connected externally is called the *open circuit voltage*, V_{oc}.

More generally, an electrical load such as an electrical resistance R_L is connected across a solar cell, as shown in Figure 5.28(c).

Figure 5.28 (a) A solar cell in the short-circuit mode with short-circuit current i_{sc}. (b) A solar cell in the open-circuit mode with open-circuit voltage V_{oc}. (c) A solar cell connected to a load of finite resistance R_L. The current flowing through the load is less than i_{oc} and the voltage across the load is less than V_{oc}.

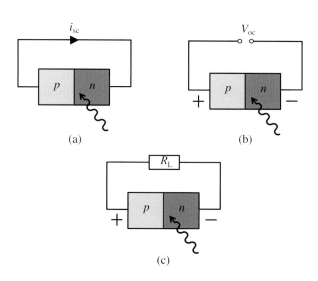

Incident sunlight generates a voltage difference between the p and n regions and a photoelectric current flows in the external circuit. Electrical power is thus delivered to the load. In this case, however, the voltage across the load and the delivered current will be less than the open circuit voltage V_{oc} and the short circuit current i_{sc}, respectively, as we shall see in Section 5.4.3.

Solar cell equation

In our discussions of the p–n junction in Section 5.3, we encountered: (i) the drift current i_{drift} due to minority charge carriers that are thermally generated and which drift across the junction under the influence of the junction field; and (ii) the diffusion current i_{diff} due to the diffusion of majority charge carriers across the junction, which flows in the opposite direction to the drift current. Now for an illuminated p–n junction, we also have a *photocurrent* i_{photo} due to the electron–hole pairs generated by the incident photons. These photon-induced electrons and holes are affected by the junction field in the same way as the electrons and holes that are thermally generated. Thus, the photocurrent flows in the same direction as the drift current and hence the net current i_{net} flowing across the junction is given by

$$i_{net} = i_{diff} - i_{drift} - i_{photo}. \tag{5.22}$$

Following Section 5.3.3, we take $i_{diff} = i_0 e^{eV/kT}$, where now V is the voltage induced across the p–n junction by the incident sunlight, and $i_{drift} = i_0$. This gives

$$i_{net} = i_0(e^{eV/kT} - 1) - i_{photo}. \tag{5.23}$$

In this expression, as i_{photo} increases the magnitude of i_{net} also increases, but i_{net} becomes more negative. A solar cell is conventionally considered to be a battery that delivers a current i_{cell} to an external load that becomes more positive as i_{photo} increases, as illustrated in Figure 5.29. Hence it is convenient to define i_{cell} as $i_{cell} = -i_{net}$ so that

$$i_{cell} = i_{photo} - i_0(e^{eV/kT} - 1). \tag{5.24}$$

where V is now the voltage output of the solar cell and both i_{cell} and V are positive quantities. A plot of solar cell current i_{cell} against cell voltage V is shown as the solid curve in Figure 5.30. The shaded area is the quadrant in which the cell generates electrical power. The figure also illustrates what happens as the illumination of the cell increases; the i_{cell}–V curve moves upward as indicated by the dashed curves. As more current is drawn from the cell, the cell voltage steadily falls, following the solid curve on Figure 5.30.

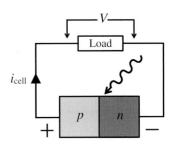

Figure 5.29 A solar cell is conventionally considered to be a battery that delivers a positive current i_{cell} to an external load. V is the voltage output of the cell.

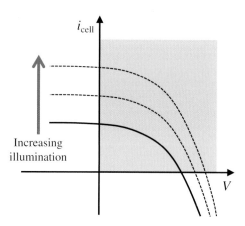

Figure 5.30 The solid curve is a plot of the output current i_{cell} of a solar cell against its output voltage V. The shaded area of the graph is the quadrant in which the cell generates electrical power. The figure also indicates how the i_{cell}–V curve moves upward as the illumination of the cell is increased, as shown by the dashed curves.

We can understand the behaviour of the solar cell voltage in the following way. We recognise the second term in Equation (5.24) from the diode equation:

$$i = i_0(\mathrm{e}^{eV/kT} - 1) \tag{5.5}$$

(see Section 5.3.3). Hence we write Equation (5.24) as

$$i_{\text{cell}} = i_{\text{photo}} - i, \tag{5.25}$$

and we can represent a solar cell by the *equivalent circuit* shown in Figure 5.31. The cell is represented by a current source, delivering current i_{photo}, and this current source is connected in parallel with a diode and with the load. The diode represents the p–n junction of the cell and the voltage across it is the cell voltage V. The arrowheads indicating the direction of current flow point in the direction of conventional current flow. For a given illumination, i_{photo} is fixed. If the load draws more current, i.e. i_{cell} increases, then i must reduce. Then the voltage V must also decrease according to Equation (5.5), giving the shape of the solid i_{cell}–V curve shown in Figure 5.30.

When the p and n regions of a solar cell are connected together as in Figure 5.28(a), the cell voltage V must be zero and the current flowing between the two regions is the short-circuit current i_{sc}. In addition, according to Equation (5.24), with $V = 0$, the cell current flowing in the external circuit is i_{photo}. Hence we take $i_{\text{sc}} = i_{\text{photo}}$, which says that the short-circuit current is equal to the photocurrent

Figure 5.31 A solar cell can be represented by a current source, delivering current i_{photo}, that is connected in parallel with a diode and with the load. The diode represents the p–n junction of the cell. For a given illumination, i_{photo} is fixed. As the load draws more current, i.e. i_{cell} increases, the current, i, flowing through the diode must decrease according to $i_{\text{cell}} = i_{\text{photo}} - i$ and this also reduces the voltage V in accordance with the diode equation, $i = i_0(\mathrm{e}^{eV/kT} - 1)$.

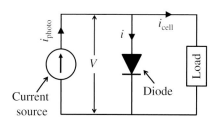

produced by the incident sunlight. Substituting for $i_{\text{photo}} = i_{\text{sc}}$ in Equation (5.24) we obtain

$$i_{\text{cell}} = i_{\text{sc}} - i_0(e^{eV/kT} - 1). \qquad (5.26)$$

This is the solar cell equation.

Worked example

Solar radiation with an irradiance of 500 W/m^2 is incident upon a solar cell of area 100 cm^2. Estimate the short-circuit current of the solar cell if 15% of the incident photons cause electron–hole pairs that lead to this current.

Solution

Taking 2 eV to be the typical energy of a photon in the solar spectrum, the photon flux at the solar cell is $\sim \frac{500}{2 \times 1.602 \times 10^{-19}}$ photons/m^2 s. Therefore the number of electron–hole pairs generated is

$$\sim \frac{500}{2 \times 1.602 \times 10^{-19}} \times (100 \times 10^{-4}) = 1.56 \times 10^{19} \text{ pairs/s.}$$

Hence, the short-circuit current

$$\sim 0.15 \times 2 \text{ (charge carriers/photon)} \times 1.56 \times 10^{19} \times 1.602$$
$$\times 10^{-19} = 0.75 \text{ A.}$$

5.4.3   Maximum power delivery from a solar cell

When the solar cell is left open circuit, as shown in Figure 5.28(b), the output voltage is the open-circuit voltage V_{oc}. In this case, $i_{\text{cell}} = 0$. Then, rearranging Equation (5.26), we obtain

$$\left(\frac{i_{\text{sc}}}{i_0} + 1 \right) = e^{eV_{\text{oc}}/kT}.$$

Usually, $i_{\text{sc}} \gg i_0$ and so

$$V_{\text{oc}} \approx \frac{kT}{e} \ln \left(\frac{i_{\text{sc}}}{i_0} \right). \qquad (5.27)$$

In practice, the value of $V_{\text{oc}} \sim 0.7$ eV.

For a load of finite resistance R_L, as in Figure 5.28(c), the current flowing through the load will be less than i_{sc} and the voltage across the load will be less than V_{oc}. Of course, we would like to deliver as much power to the load as possible and we can achieve this by suitably adjusting the value of R_L. The values of current and voltage that together provide the maximum power are called i_{mp} and V_{mp}, respectively. We can deduce expressions for i_{mp} and V_{mp} as follows. The power P delivered by the solar cell is equal to the product of the cell current i_{cell} and the cell voltage V. Substituting for i_{cell} from Equation (5.26) we have

$$P = V i_{sc} - V i_0 (e^{eV/kT} - 1). \tag{5.28}$$

To find V_{mp}, we differentiate this equation with respect to V and set $dP/dV = 0$. This gives

$$i_{sc} - i_0 (e^{eV/kT} - 1) - \frac{i_0 eV}{kT} e^{eV/kT} = 0.$$

In general, $eV/kT \gg 1$, and we can write

$$i_{sc} - i_0 e^{eV/kT} \left(1 + \frac{eV}{kT} \right) = 0. \tag{5.29}$$

Substituting for $i_{sc} = i_0 e^{eV_{oc}/kT}$ from Equation (5.27), we obtain

$$i_0 e^{eV_{oc}/kT} = i_0 e^{eV/kT} \left(1 + \frac{eV}{kT} \right).$$

Rearranging this equation gives

$$V = V_{oc} - \frac{kT}{e} \ln \left(1 + \frac{eV}{kT} \right). \tag{5.30}$$

The value of V given by Equation (5.30) is just V_{mp}:

$$V_{mp} = V_{oc} - \frac{kT}{e} \ln \left(1 + \frac{eV_{mp}}{kT} \right). \tag{5.31}$$

To make this result more tractable, we note that in practice $V_{mp} \approx V_{oc}$ and, in addition, any logarithmic function $\ln[f(x)]$ varies only slowly with x. Hence we substitute V_{oc} for V_{mp} in the logarithmic term to give our final result for V_{mp}:

$$V_{mp} \approx V_{oc} - \frac{kT}{e} \ln \left(1 + \frac{eV_{oc}}{kT} \right). \tag{5.32}$$

In a similar fashion, we find that the current for maximum power is given by

$$i_{mp} \approx i_{sc} \left(1 - \frac{kT}{eV_{mp}} \right). \tag{5.33}$$

(a)

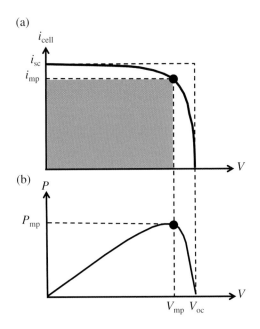

(b)

Figure 5.32 (a) The $i_{\text{cell}}-V$ curve for a solar cell in the power generation quadrant. Maximum power P_{mp} occurs for $V = V_{\text{mp}}$, $i_{\text{cell}} = i_{\text{mp}}$ and is given by the area of the shaded rectangle. The figure also illustrates that, in practice, $V_{\text{mp}} \approx V_{\text{oc}}$, and $i_{\text{mp}} \approx i_{\text{sc}}$. The ratio of the maximum power, P_{mp}, and the product, $i_{\text{sc}} V_{\text{oc}}$, is the fill factor. (b) For $V = 0$ (closed circuit), $i_{\text{cell}} = i_{\text{sc}}$, and the output power P is zero. For $V = V_{\text{oc}}$ (open circuit), $i_{\text{cell}} = 0$ and P is also zero.

For strong illumination, $kT/eV_{\text{mp}} \ll 1$, in which case $i_{\text{mp}} \approx i_{\text{sc}}$. Also under this condition, the second term in Equation (5.32) for V_{mp} is small compared with V_{oc} and hence $V_{\text{mp}} \approx V_{\text{oc}}$.

The above results are presented graphically in Figure 5.32(a), which shows the $i_{\text{cell}} - V$ curve for a solar cell in the power generation quadrant. For $V = 0$ (closed circuit), $i_{\text{cell}} = i_{\text{sc}}$, and the output power P is zero, as we see in Figure 5.32(b). For $V = V_{\text{oc}}$ (open circuit), $i_{\text{cell}} = 0$ and P is also zero. Maximum power, P_{mp}, occurs for $V = V_{\text{mp}}$ and $i_{\text{cell}} = i_{\text{mp}}$ and is given by the area of the shaded rectangle in Figure 5.32(a). The figure also illustrates that, in practice, $V_{\text{mp}} \approx V_{\text{oc}}$, and $i_{\text{mp}} \approx i_{\text{sc}}$. The ratio of the maximum power P_{mp} and the product $i_{\text{sc}} V_{\text{oc}}$ is called the *fill factor, FF*:

$$FF = \frac{P_{\text{mp}}}{i_{\text{sc}} V_{\text{oc}}} = \frac{i_{\text{mp}} V_{\text{mp}}}{i_{\text{sc}} V_{\text{oc}}}. \tag{5.34}$$

The area corresponding to $i_{\text{sc}} V_{\text{oc}}$ is also indicated in Figure 5.32(a). In practice, fill factors are $\sim 80\%$.

Perhaps the most useful figure of merit for a solar cell is its *power conversion efficiency η*, which is defined as the ratio of the maximum electrical power P_{mp} to the incident solar power P_{S}:

$$\eta = \frac{P_{\text{mp}}}{P_{\text{S}}} = \frac{FF \times i_{\text{sc}} V_{\text{oc}}}{P_{\text{S}}}. \tag{5.35}$$

By convention, solar cell efficiencies are measured under standard test conditions. These are an incidence irradiance of 1000 W/m^2 with an AM1.5 air mass spectrum to simulate atmospheric absorption. For commercial solar cells, $\eta \sim 20\%$.

5.4.4   The Shockley–Queisser limit

We saw in Chapter 4 that the Sun can be considered to be a black-body at a temperature of ~5800 K and that the solar spectrum contains a continuous range of photon wavelengths. We will now see that not all the sunlight incident on a solar cell can be converted into electrical power. In 1961, William Shockley and Hans Queisser made a detailed analysis of the p–n junction to determine the maximum possible efficiency of a semiconductor solar cell; the efficiency of converting solar power into electrical power. The important result of their theoretical considerations is that the maximum possible efficiency of a solar cell is about 34%, when atmospheric absorption is taken into account. So for example, if the solar irradiance is 1000 W/m², the maximum electrical power that can be obtained is 340 W/m² of the solar cell. The analysis of Shockley and Queisser is based on several factors. These include: (i) not all solar photons have enough energy to promote an electron to the conduction band of the semiconductor material; (ii) a fraction of the photon energy is dissipated as heat in the crystal lattice; and (iii) a fraction of the photon-generated electrons and holes recombine before they can contribute to the photocurrent.

Dissipation of energy in the crystal lattice

To illustrate this effect, we consider, as Shockley and Queisser did, a blackbody spectrum for a temperature of 5800 K to represent the solar spectrum. This is shown in Figure 5.33, which plots spectral irradiance against wavelength. The total area under the curve gives the total power in the spectrum. Also indicated in the figure is the critical wavelength for silicon, i.e. the maximum wavelength that can promote an electron into the conduction band. The figure shows

Figure 5.33 A blackbody spectrum for a temperature of 5800 K. Also indicated is the critical wavelength (1.12 μm) for silicon. Photons lying in region A do not have enough energy to cause promotion of an electron across the band gap. For photon energies greater than the band gap, a fraction of their energy is dissipated as heat in the crystal lattice. The area of region C represents the power that is lost in this way. The area of region B represents the only part of the solar power that can, in principle, be obtained from a solar cell. The area of region B compared with the total area under the blackbody curve is about 45%.

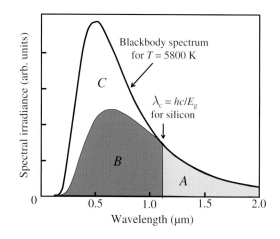

three different regions of the 'solar' spectrum, A, B and C. Photons lying in region A do not have enough energy to cause promotion of an electron across the band gap, i.e. $hc/\lambda < E_g$, and so they do not produce any electrical power; they may simply heat the crystal lattice. For photon energies greater than the band gap, i.e. $hc/\lambda > E_g$, the excited electron goes into an energy level that lies above the bottom of the conduction band, as shown in Figure 5.34. This electron very quickly drops down to the bottom of the conduction band and its excess energy $(hc/\lambda - E_g)$ is dissipated as heat in the crystal lattice. The greater the energy of the photon, the greater is the fraction of its energy that is lost as heat. The area of region C in Figure 5.33 represents the power that is dissipated in this way. Thus the area of region B in the figure represents the only solar power that can, in principle, be obtained from a solar cell. The relative size of the area of region B depends on the value of the semiconductor band gap. For the case of silicon, the area of region B compared with the total area of the spectrum is about 45%. If a semiconductor with a larger band gap is chosen, fewer photons at the infrared end of the solar spectrum are able to promote electrons across the band gap, but visible photons will do so more efficiently; less of their energy will be lost as heat to the lattice. On the other hand, if the band gap is made smaller, more photons at the infrared end of the solar spectrum are able to promote electrons but visible photons will do so less efficiently. This correctly suggests that there is an optimum band gap for maximum solar cell efficiency, as we shall see.

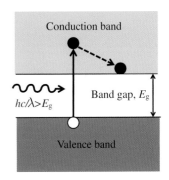

Figure 5.34 For photon energies greater than the band gap, the excited electron goes into an energy level that lies above the bottom of the conduction band. This electron very quickly drops down to the bottom of the conduction band and its excess energy $(hc/\lambda - E_g)$ is dissipated as heat in the crystal lattice.

Electron–hole recombination

Electrons and holes that are generated by incident photons have a tendency to recombine with charge carriers of opposite sign. If recombination occurs before the electron–hole pair can contribute to the photocurrent, the efficiency of the cell is reduced. The fundamental recombination process is *radiative recombination*. This process is illustrated in Figure 5.35. An electron recombines with a hole and the result is that a photon is emitted with energy roughly equal to the band gap. Shockley and Queisser took account of radiative recombination using a *detailed-balance principle*, based on thermodynamic considerations. This principle essentially says that when the solar cell is not illuminated, but is in thermodynamic equilibrium with its surroundings, the rate of generation of electron–hole pairs must be equal to the recombination rate. Then when the cell is illuminated by sunlight, the rate of electron–hole pair generation is equal to the rate of radiative recombination plus the rate at which electrons are consumed in the delivery of current to the external load. Using the detailed balance principle, Shockley and Queisser deduced that radiative recombination reduces solar cell efficiency by about 10%.

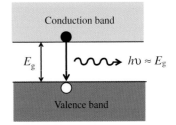

Figure 5.35 The process of radiative recombination. An electron recombines with a hole and the result is that a photon is emitted with an energy that is roughly equal to the band gap.

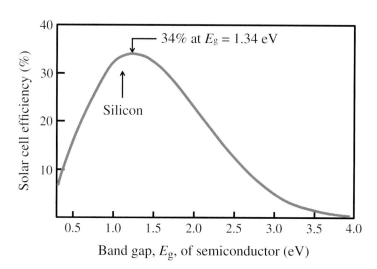

Figure 5.36 The efficiency of a solar cell as a function of band gap energy, E_g, when the cell is illuminated by solar radiation and atmospheric effects are taken into account. This curve is deduced from the theoretical considerations of Shockley and Queisser. The maximum efficiency is 34% and this is achieved for a band gap of 1.34 eV.

From their theoretical treatment of the p–n junction, Shockley and Queisser obtained a curve for solar cell efficiency as a function of band gap energy. The form of this curve when atmospheric effects are taken into account is shown in Figure 5.36. The maximum efficiency is 34% and this is achieved for a band gap of 1.34 eV. Silicon has a band gap that is reasonably close to this optimum value, and for its band gap of 1.11 eV the efficiency is 29%. The efficiency of modern commercial silicon solar cells is about 22%. The difference is partly due to practical effects such as reflection of light from the front surface of the cell and light blockage from the thin connecting wires on its surface. And in practical cells, not all of the available charge carriers are collected by the electric field of the junction. An efficiency of 22% may seem modest but we recall that the efficiency of say a petrol engine has a similar value. The Shockley and Queisser limit applies to a *single p–n* junction. We will see in Section 5.4.6 that the limit can be circumvented by using in tandem two or more p–n junctions that have different band gaps.

5.4.5   Solar cell construction

The first practical solar cells used crystalline silicon as the semiconductor material and the majority of solar cells still use this material. Use of silicon as the semiconductor material has a number of advantages: silicon is one of the most abundant elements on Earth; its band gap is close to being ideally matched to the solar spectrum; it is chemically stable; and it is non-toxic. And because of the microelectronics industry, the production and processing of extremely pure crystalline silicon are well developed. The disadvantage of silicon is

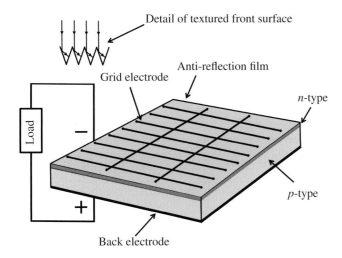

Detail of textured front surface

Grid electrode

Anti-reflection film

n-type

Load

−

+

p-type

Back electrode

Figure 5.37 The construction of a typical solar cell. The solar cell is based on a thin single crystal of silicon, the kind used in microelectronic devices. A *p–n* junction is formed within the silicon layer. The thickness (∼1 μm) of the *n*-type region is much less than the thickness (∼300 μm) of the *p*-type region, but the *n*-type is more heavily doped. This allows most of the absorption of light to occur in the *p* region. Typically a cell may be ∼10 cm × 10 cm in size. A grid of wires is placed between an anti-reflectance coating and the semiconductor surface to provide electrical contact to the cell. There is also an electrode at the back of the cell to complete the circuit. Light collection can be enhanced by texturing the front surface of the solar cell by a chemical etching process.

that it has a relatively low attenuation coefficient for solar radiation because it is an indirect band gap semiconductor (see Section 5.4.1).

Figure 5.37 illustrates the construction of a typical solar cell based on a single crystal of silicon. A *p–n* junction is formed within the silicon crystal by diffusion of the appropriate impurities into the crystal. It is favourable to have the thickness (∼1 μm) of the *n*-type region much less than the thickness (∼300 μm) of the *p*-type region, but for the *n*-type to be more heavily doped. This allows most of the absorption of light to occur in the *p* region. The diffusion length of the charge carriers in the *p* region is ∼100 μm and, as noted previously, charge carriers produced within this length may migrate to the junction and contribute to the photocurrent. A grid of wires is placed between an anti-reflectance coating and the semiconductor surface to provide electrical contact to the cell. Semiconductor materials have relatively high values of electrical resistivity and the grid is arranged to minimise resistance losses in the surface of the material. There is also an electrode at the back of the cell to complete the electrical circuit. Typically, a cell may be ∼10 cm × 10 cm in size and a number of these individual cells can be connected to form a *module*. For example, typically 36 cells are connected in series to provide an overvoltage to charge 12 V batteries. Under clear sky conditions, the output current of the cell is ∼300 A/m^2.

As we saw in Section 4.5.2, the fraction R of light reflected at a surface is given by

$$R = \frac{(n_0 - n_1)^2}{(n_0 + n_1)^2}, \qquad (4.52)$$

where, in the present case, n_0 is 1.0 for air and n_1 is the refractive index of the semiconductor. Semiconductors have high values of refractive index; typically ~ 3.5. This gives a reflectance of 0.31, which would mean that nearly 70% of the radiation would be lost through reflection. This reflection is minimised, however, by coating the front of the solar cell with a thin film of appropriate thickness and refractive index, just as camera lenses are similarly coated. The thickness t of the film should be equal to one-quarter of the wavelength of the incident radiation. Under this condition the waves reflected off the front of the film and those reflected off the semiconductor surface differ in phase by π radians and interfere destructively. Moreover, the refractive index of the thin film should be equal to $\sqrt{n_0 n_1}$ so that these two reflected waves are equal in intensity. Of course, this only works for one wavelength. Nevertheless, taking $\sqrt{n_0 n_1}$ to be 1.9 for silicon and $t = 0.08$ μm, the broadband reflectance is much reduced, to $\sim 6\%$. The use of multiple thin layers can further reduce the reflectance to below 3%. Light collection can also be enhanced by chemically etching the front surface of the solar cell to give it a textured finish, as shown in Figure 5.37.

Instead of using single crystals of silicon, it is also possible to use silicon in polycrystalline form. Polycrystalline crystals have the same valence-conduction band structure and they have the advantage that they are much cheaper to produce than single crystals. The disadvantage is that the presence of boundaries between the crystal grains increases recombination of electron–hole pairs. Consequently, polycrystalline solar cells have lower efficiencies compared with single-crystal cells, $\sim 15\%$ compared with $\sim 20\%$. Nevertheless, because of their lower cost, the polycrystalline type account for about half the commercial production of solar cells.

Solar panels are now frequently seen on the rooftops of buildings to supply electricity. For example, the Alan Turing Building at the University of Manchester is fitted with photovoltaic cells that provide an average power output of 38 kWh/day. As described in Section 4.3.2, the daily insolation, i.e. the total amount of solar energy per unit area received in 1 day from the Sun, varies according to time of the year and according to latitude. Consequently, fixed solar panels are positioned so as to maximise the solar energy they receive, just as for solar flat plate heaters. There are also much larger solar power plants that provide electrical power on an industrial scale. Figure 5.38 shows an array of silicon solar cells at the Nellis Solar Power Plant located within the Nellis Air Force Base in Clark County, Nevada, USA. The plant has nearly 6000 solar panels and these can be rotated about a single axis to track the Sun. The peak power capacity of the plant is approximately 14 MW. In one striking application of solar cells, in 2016, Bertrand Piccard and Andre Borschberg

Figure 5.38 An array of silicon solar cells at the Nellis Solar Power Plant located within the Nellis Air Force Base in Clark County, Nevada, USA. The plant has nearly 6000 solar panels and these can be rotated about a single axis to track the Sun. The peak power capacity of the plant is approximately 14 MW. Courtesy of United States Air Force. http://www.nellis.af.mil/photos/media_search.asp?q=solar&btnG.x=0&btnG.y=0

circumnavigated the globe in a plane powered only by solar cells; a journey of 40 000 km.

5.4.6 Increasing the efficiency of solar cells and alternative solar cell materials

There is much ongoing research into maximising solar cell efficiency. Here we give examples of techniques that increase this efficiency and describe alternative solar cell materials.

Multi-junction solar cells

We saw in Section 5.4.4 that any energy an incident photon has in excess of the semiconductor band gap is lost as heat to the crystal lattice. So the most efficient use of solar energy comes from those photons that have an energy just above the band gap energy. As there is a range of photon energies in the solar spectrum, this suggests stacking together two or more *p–n* junctions to form a *tandem* or *multi-junction cell* in which the band gaps of the junctions are different. And, indeed, this is a way to circumvent the Shockley and Queisser limit.

The principle of the multi-junction solar cell is illustrated schematically in Figure 5.39(a). This shows a stack of three sub-cells. Each sub-cell consists of a *p–n* junction that is made from a different semiconductor from the other two. The three sub-cells are separated from each other by a *tunnel junction*. Its purpose is to provide a low electrical resistance connection between adjacent sub-cells and also

Figure 5.39 (a) The principle of operation of the multi-junction solar cell, which consists of a stack of three sub-cells. The topmost sub-cell is made from a semiconductor that has a large band gap and absorbs photons at the blue end of the solar spectrum. The transmitted photons encounter the middle sub-cell that is made from a semiconductor that has a medium band gap and absorbs lower-energy, green photons. The bottom sub-cell is made from a semiconductor with a low band gap that absorbs the lowest-energy photons, at the red end of the spectrum. The *tunnel junctions* provide low-loss electrical and optical connections between adjacent sub-cells. (b) The three regions of the solar spectrum that are absorbed by the three different semiconductor materials, each contributing to the photocurrent. The shaded areas represent the electrical power obtained.

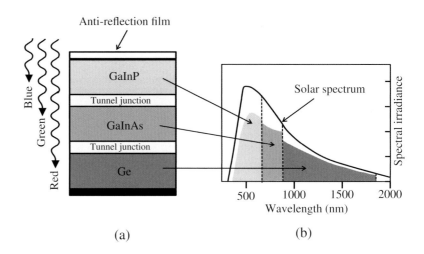

(a)   (b)

a low loss optical connection. Electrical contacts at the top and bottom of the device deliver the photocurrent to an external circuit. The topmost sub-cell is made from a semiconductor that has a large band gap. This absorbs photons at the blue end of the solar spectrum and these are converted into a photocurrent. However, the semiconductor is transparent to photons of lower energy than the band gap. These transmitted photons then pass to the middle sub-cell that is made from a semiconductor that has a medium band gap. Lower-energy (green) photons are absorbed by this sub-cell, but photons of energy lower than the band gap are transmitted to the bottom sub-cell. This is made from a semiconductor with a small band gap and absorbs the lowest energy photons at the red end of the spectrum. Three semiconductor materials that are typically used are gallium indium phosphide (GaInP), gallium indium arsenide (GaInAs) and germanium (Ge). These have band gaps of 1.89, 1.42 and 0.67 eV, respectively. By using these three materials, photons across a wide range of the solar spectrum are efficiently absorbed and contribute to the photocurrent. Figure 5.39(b) shows the three regions of the solar spectrum that are absorbed by the three different semiconductor materials. The vertical dashed lines indicated the positions of the critical wavelengths of the three semiconductors and the shaded areas represent the electrical power obtained. Multi-junction solar cells are more costly to produce than single-junction cells. Consequently, they are usually used in more specialised applications where conversion efficiency outweighs cost, as in the use of solar cells to power satellites.

Light concentrators

One way to increase the output power of a solar cell is simply to collect sunlight over a large area and concentrate it onto the cell with a lens or mirror. The *concentration ratio* X is the ratio of the

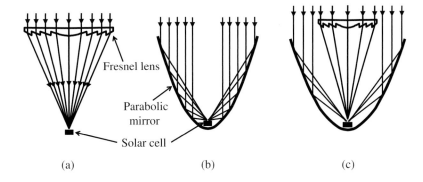

Fresnel lens

Parabolic
mirror

Solar cell

(a)   (b)   (c)

Figure 5.40 Various ways to concentrate solar radiation onto a solar cell: (a) a Fresnel lens, (b) a parabolic mirror, and (c) a combination of the two.

concentrator input aperture to the surface area of the cell. This is an effective technique, as the active semiconductor material is usually the most expensive part of the cell; this is especially the case for multi-junction cells. Concentration of incident light can be obtained with a Fresnel lens or a parabolic mirror, as shown in Figure 5.40. Indeed, a combination of the two may be used, as also shown. Concentration ratios of 1000 or more can be achieved, increasing the light irradiance by the same factor. Such power densities are sufficient to heat up a solar cell substantially and high-capacity heat sinks are necessary to avoid large temperature rises that reduce the efficiency of the cell.

If sunlight is concentrated by a factor of X, it is reasonable to assume that the photocurrent i_{photo}, and hence the short circuit current i_{sc}, is multiplied by the same factor. In addition, however, increasing the incident light irradiance also increases the efficiency of the solar cell. This is because increasing i_{sc} also increases the open-circuit voltage V_{oc}. In turn, this increases the fill factor and the power conversion efficiency of the cell. We can see how this increase in V_{oc} arises from Equation (5.27):

$$V_{\mathrm{oc}} \approx \frac{kT}{e} \ln\left(\frac{i_{\mathrm{sc}}}{i_0}\right). \tag{5.27}$$

If we concentrate the light by a factor X, the resulting open-circuit voltage, V'_{oc}, is given by

$$V'_{\mathrm{oc}} \approx \frac{kT}{e} \ln\left(\frac{X i_{\mathrm{sc}}}{i_0}\right)$$
$$= V_{\mathrm{oc}} + \frac{kT}{e} \ln(X).$$

If we take $V_{\mathrm{oc}} = 0.72$ V, $kT/e = 1/40$ V and $X = 1000$, we see that V'_{oc} is 24% greater than V_{oc} and the efficiency of the solar cell will increase by a similar factor. The efficiency of multi-junction solar cells using concentrators is now above 40%.

Quantum dot solar cells

Quantum dots are tiny crystals of semiconductor material; the quantum dots used in solar cells have dimensions of about 10 nm. They have electrical properties that make them useful for solar cells, as well as for a wide range of other applications. In particular, the separation of the energy levels in a quantum dot can be controlled. This means that the band gap of a quantum dot may be 'tuned' to a particular wavelength range. This is in sharp contrast to semiconductor materials of macroscopic size, which have a fixed band gap. Controlling the band gap offers several possibilities. For example, it can be optimised to match the solar spectrum according to the Shockley-Queisser limit, or a multi-junction cell can be made from stacked layers of quantum dots of steadily reducing band gap.

The band gap can be varied because the allowed energy levels in a quantum dot are governed by the quantum mechanics of particles that are confined to microscopic dimensions. We described in Section 2.2.3 a one-dimensional quantum well. We saw that the allowed energy levels in such a well depend on the width of the well. By changing the width of the well, we change the allowed energy levels and the energy separations between them. The energies and their separations in a one-dimensional well scale as $1/L^2$, where L characterises the width of the well. In a quantum dot, electrons are confined in three dimensions but quantised energy levels are again obtained. Moreover, their energy levels and separations again scale as $1/L^2$, where L characterises the size of the quantum dot. To change the band gap in a quantum dot, we simply need to change its size. Quantum dots can be cheaply produced as colloids in a suitable solvent by wet-chemical techniques, and the size of the dot can be readily controlled.

There are a number of solar cell designs that use quantum dots. A schematic diagram of one kind is shown in Figure 5.41. The cell consists of quantum dots made from the semiconductor cadmium

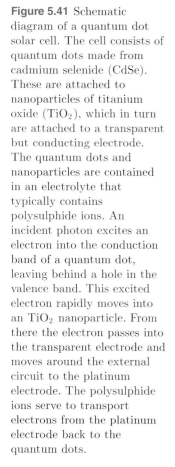

Figure 5.41 Schematic diagram of a quantum dot solar cell. The cell consists of quantum dots made from cadmium selenide (CdSe). These are attached to nanoparticles of titanium oxide (TiO$_2$), which in turn are attached to a transparent but conducting electrode. The quantum dots and nanoparticles are contained in an electrolyte that typically contains polysulphide ions. An incident photon excites an electron into the conduction band of a quantum dot, leaving behind a hole in the valence band. This excited electron rapidly moves into an TiO$_2$ nanoparticle. From there the electron passes into the transparent electrode and moves around the external circuit to the platinum electrode. The polysulphide ions serve to transport electrons from the platinum electrode back to the quantum dots.

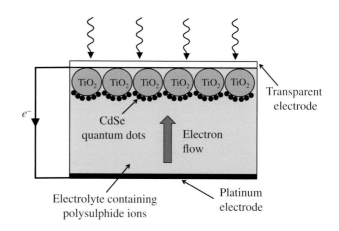

selenide (CdSe). The quantum dots are attached to *nanoparticles* of titanium oxide (TiO_2), which in turn are attached to a transparent but conducting electrode. The quantum dots and nanoparticles are contained in an electrolyte that typically contains polysulphide ions. These are compounds consisting of chains of n negatively charged sulphur ions with the symbol S_n^{2-}, with $n = 2$–5. There is also a second electrode made of platinum at the back of the cell.

The operation of the cell is as follows. An incident photon excites an electron into the conduction band of a quantum dot, leaving behind a hole in the valence band. This excited electron lies energetically above the conduction band of the TiO_2 nanoparticle and the electron rapidly moves into the nanoparticle. From there the electron passes into the transparent electrode and moves around the external circuit to the platinum electrode. An S_n^{2-} ion from the electrolyte accepts electrons from the platinum electrode via the reaction

$$S_n^{2-} + 2e^- \rightarrow S_{n-1}^{2-} + S^{2-}. \tag{5.36}$$

The S^{2-} ions diffuse through the electrolyte and migrate to the quantum dots. There they deliver electrons to fill the holes generated by the photo-excited electrons according to the reactions

$$S_n^{2-} + 2h^+ \rightarrow S$$
$$S + S_{n-1}^{2-} \rightarrow S_n^{2-},$$

where the symbol h^+ denotes a hole. This returns S_n^{2-} ions to the electrolyte and the cycle continues. In this way, electrons are transported from the platinum electrode back to the quantum dots.

In another kind of solar cell, called a *dye-sensitised solar cell*, the quantum dots are replaced by dye molecules. The mechanism in a dye-sensitised solar cell and the structure of the cell are similar to those of a quantum dot cell. Photo-excitation of the sensitised dye results in the injection of an electron into the conduction band of a TiO_2 nanoparticle. The electron then moves to a transparent electrode and passes around an external circuit to an electrode at the back of the cell. As before, there is an electrolyte that serves to transport electrons from the electrode back to the dye molecules. Interestingly, research on these dye-based systems began as a way of understanding the mechanism of photosynthesis, Nature's way of harvesting solar energy.

Quantum dot solar cells and dye-sensitised solar cells are presently not as efficient as a silicon solar cell, although their efficiencies are expected to increase with further research and development. However, they do have a distinct advantage. Quantum dots and dye molecules can be readily deposited onto a plastic substrate by spraying or printing. This dramatically reduces module construction costs, especially for large surface area solar cells.

Problems 5

5.1 (a) Three semiconductor materials that are commonly used in multi-junction solar cells are gallium indium phosphide, gallium indium arsenide and germanium. These have band gaps of 1.89, 1.42 and 0.67 eV, respectively. Calculate the critical wavelengths for these three semiconductors. (b) At what wavelength would you expect silicon to become transparent to solar radiation? The band gap of silicon is 1.11 eV.

5.2 The work function of copper is 4.7 eV. (a) What is the maximum wavelength that an incident photon can have if it is to produce a photoelectron? (b) One type of particle accelerator that is used to produce a beam of high-energy electrons utilises the photoelectric effect for its source of electrons. In this source, a beam of UV radiation ($\lambda = 226$ nm) from a pulsed laser is incident on a copper surface. There are 500 laser pulses/s and each pulse contains 5.0×10^{-4} J of energy. Given that the *quantum efficiency*, defined as the number of photoelectrons produced per incident photon, of the copper surface is 1.5×10^{-3}, calculate the mean current of the electron beam in the accelerator.

5.3 Light of wavelength 550 nm is incident upon a slice of silicon. What are the frequency and wavelength of the light within the slice? The refractive index of silicon is 3.42.

5.4 A phosphorus atom replaces an atom of silicon in the lattice of a silicon crystal, giving rise to a loosely bound electron. Use the modified Bohr model of the hydrogen atom to calculate the Bohr radius of this electron. The relative permittivity, ε, of silicon is 11.8 and the effective mass, m_e^*, of the electron is $0.26\ m_e$. The Bohr radius, a_0, for the hydrogen atom is 0.53×10^{-10} m.

5.5 The impurity concentration in a semiconductor has to be relatively low; at low concentration, the impurity atoms produce discrete energy levels, but if the impurity atoms are sufficiently close, their energy levels spread into bands. Use the modified Bohr theory of the hydrogen atom to estimate the maximum concentration of arsenic impurities in germanium that avoids overlap between adjacent impurity atoms. The relative permittivity, ε, of germanium is 15.8 and the effective mass, m_e^*, of the electron is $0.55\ m_e$. Germanium has a density of 5.32×10^3 kg/m^3 and its atomic weight is 72.6 u.

5.6 (a) The band gap of gallium arsenide is 1.43 eV and the effective masses of the electrons and holes are $0.067\ m_e$ and $0.48\ m_e$, respectively. Find the intrinsic concentrations of the electrons and holes. (b) Compare your value with the number of free electrons in copper, assuming each copper atom gives rise to a free electron. The density of copper is 8.96×10^3 kg/m^3 and its atomic weight is 63.55 u.

5.7 The effective masses of the electrons and holes in germanium are $0.55\ m_e$ and $0.37\ m_e$, respectively, and the band gap is 0.67 eV. A sample of germanium is doped with donor impurity atoms at the concentration 1.4×10^{23} atoms/m^3. What is the hole concentration when the sample is at equilibrium at $T = 300$ K?

5.8   Intrinsic semiconductor gallium arsenide has a band gap of 1.43 eV. Use the Fermi–Dirac distribution to calculate the probability that a state at the bottom of the conduction band is occupied by an electron at a temperature of 300 K. By how much does the probability increase if the temperature is raised by 10 K?

5.9   The effective mass of a charge carrier in a semiconductor can be obtained from a measurement of its cyclotron frequency when a magnetic field is applied to the semiconductor. When a magnetic field of 1.5 T is applied to a sample of indium arsenide, the cyclotron frequency of the donor electrons is found to be 1.32×10^{13} rad/s. What is the effective mass of these electrons?

5.10 A crystal of pure germanium can be used to measure the energy of γ-rays. The principle of operation is that the γ-ray produces electron–hole pairs as it passes through the crystal. This gives rise to a current pulse that is measured by a *charge sensitive preamplifier* in an external circuit, as illustrated in the figure. Suppose that a 660 keV γ-ray emitted by a ^{137}Cs nucleus loses all its energy in passing through a slab of germanium by producing electron–hole pairs. Calculate the number, N, of electron–hole pairs that can be generated. According to Poisson statistics, the number of electron–hole pairs generated will have statistical fluctuations given by $N \pm \sqrt{N}$. Hence, estimate the energy resolution of a germanium detector at 660 keV. Explain why is it important to maintain a germanium detector at the temperature of liquid nitrogen and why the energy resolution of a germanium detector is better than that of a silicon detector.

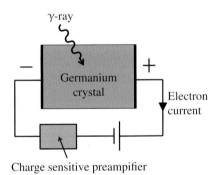

5.11 A light-emitting diode (LED) is essentially a forward-biased p–n junction. As electrons cross from the n-region to the p-region, they combine with holes generating light. (a) Given that the band gap of gallium arsenide is 1.43 eV, what is the wavelength of the emitted light and in which region of the solar spectrum does it lie? (b) If a battery provides a current through a gallium arsenide LED of 5 mA, what will be the total power of the emitted light, assuming that 10% of the electrons generate a photon? (c) What band gap would be required for a semiconductor to produce violet light? Suggest a suitable semiconductor. Note: Isamu Akasaki, Hiroshi Amano and Shuji Nakamura were jointly awarded the Nobel Prize in Physics, 2014 'for the invention of efficient blue light-emitting diodes. . . .'

5.12 Silicon has a linear attenuation coefficient of 3080/cm at wavelength 653 nm. (a) What thickness of silicon crystal is necessary for 99% of incident photons of this wavelength to be absorbed? What thickness is necessary for a crystal of gallium arsenide, which has an attenuation coefficient of 34 468/cm at this wavelength?
(b) Assuming that the absorption of a photon leads to the generation of an electron–hole pair, how many electron–hole pairs are produced per second in the first 1.0 μm of a silicon crystal of area 100 cm^2, when 25 W of radiation of wavelength 653 nm is incident upon the crystal? What is the mean rate of generation per unit volume?

5.13 In a quantum dot, the electrons are confined within three dimensions. The allowed energy levels of the electrons have the form

$$E = \frac{h^2 \pi^2 n_1^2}{2m_e^* L_1^2} + \frac{h^2 \pi^2 n_2^2}{2m_e^* L_2^2} + \frac{h^2 \pi^2 n_3^2}{2m_e^* L_3^2}.$$

where m_e^* is the effective mass of an electron and is equal to 0.65 m_e. Suppose that a particular quantum dot has dimensions 8 nm × 8 nm × 6 nm. (a) Calculate the energies of the first three levels. (b) What is the wavelength of the photon emitted in a transition between the lowest two levels?

5.14 The rated power of a solar cell is its maximum power output with an incident solar irradiance of 1000 W/m^2. A particular solar cell has a power rating of 50 W. What maximum power would the module produce at a manned station on Mars? The mean distances of the Earth and Mars from the Sun are 1.5×10^{11} m and 2.28×10^{11} m, respectively. Note that Mars has no significant atmosphere.

5.15 A particular solar cell has a reverse saturation current, i_0, of 1.0×10^{-10} A. The photocurrent i_{photo} when the solar cell is irradiated by sunlight is 2.0 A. Taking the temperature to be $T = 300$ K: (a) determine the short circuit current i_{sc} and the open circuit voltage V_{oc}; (b) calculate V_{mp} and the i_{mp}, the values of voltage and current for maximum power; (c) make a plot of the solar cell equation

$$i_{cell} = i_{sc} - i_0(e^{eV/kT} - 1)$$

over the range of i_{cell} from $i_{cell} = 0$ to $i_{cell} = i_{sc}$, to show how the cell voltage V varies with load current i_{cell}. Mark on the plot i_{sc}, i_{mp}, V_{oc} and V_{mp}. Calculate the fill factor for the solar cell. Determine the number of solar cells required for a module that provides a nominal voltage of 12 V. What would be the optimum load resistance for this module?

5.16 Making reasonable assumptions, estimate the area of the solar cells that are needed on a south-facing roof to provide an electrical power of 3 kW to a household.

5.17 Compare and contrast solar water heaters and photovoltaic cells for the purpose of harvesting solar energy.

6

Wind power

We have already seen that the Sun delivers a huge amount of energy to the Earth's surface. Approximately 2% of this is converted into wind energy. This is a small fraction but still amounts to a global wind power of $\sim 2 \times 10^{12}$ kW. The purpose of wind turbines is to harvest some of this power. A wind turbine as described in this chapter is a machine that converts wind power into electricity, usually to supply electricity to a national grid. This is in contrast to a windmill, which converts wind power into mechanical power. Wind power has important advantages. It is a renewable energy source and it is also a clean source; it does not release harmful gases into the environment. Moreover, wind turbines can deliver power to remote areas that are not connected to a national grid. The main disadvantage is, of course, that the wind is variable and so is an intermittent source of energy. This means that wind turbines must be used in conjunction with other energy sources and/or energy storage systems. Nevertheless, because of its advantages, it is anticipated that wind power will in the future contribute at least 20% of global energy requirements.

In this chapter we describe the origin and directions of the wind and how a wind turbine works. As we are dealing with the flow of air, a fluid, through the blades of the turbine, it is also appropriate to introduce the physics of *fluid flow*. However, we first give a brief history of wind power.

6.1 A brief history of wind power

The use of wind power has a long history that goes back thousands of years. The first known historical reference is from the Greek mathematician and engineer Hero of Alexandria (*c.* 10–70 AD). He described a windmill that provided air to drive a musical organ.

Physics of Energy Sources, First Edition. George C. King.
© 2018 John Wiley & Sons, Ltd. Published 2018 by John Wiley & Sons, Ltd.

(Hero of Alexandria is also credited with the invention of the first steam engine.) In the following centuries the main use of wind power was to perform mechanical work. Holland, a flat and windy country, is famous for its use of windmills. They were used to pump water out of low lands and back into rivers beyond dikes so that the land could be farmed. The windmills were also used for many other purposes, including grinding grain for food and sawing logs. Similarly, the plains of America were made fertile by the use of windmills, which were used to pump underground water from beneath the plains to irrigate the fields. Wind continued to be a major source of energy up until the Industrial Revolution, but then began to recede in importance. Coal-driven steam engines had important advantages that the wind did not posses, such as portability and reliability of fuel supply.

Then, in 1831, the British scientist Michael Faraday succeeded in producing a current by moving a wire in a magnetic field. Faraday's law states that the induced electromotive force (emf), ε, in a closed circuit is equal to minus the rate of change of the magnetic flux ϕ through the circuit:

$$\varepsilon = -\frac{\mathrm{d}\phi}{\mathrm{d}t}. \tag{6.1}$$

This discovery of electromagnetic induction was the beginning of electrical engineering. It led to the development of electrical generators that used electromagnetic induction to produce both DC and AC electrical power. These generators enabled the rotating mechanical motion of a wind turbine to be turned into electrical energy. In 1888, the American inventor Charles Brush built the first automatically operated turbine for the production of electricity. Bush's turbine had a *rotor* consisting of 144 blades of cedar wood with a diameter of 17 m. The turbine produced 12 kW of electrical power and Bush used the electricity to charge batteries that, in turn, powered the lights in his house. So he also took into account the fact that wind turbines, which depend on the availability of wind, require the storage of part of the energy produced. Bush's wind turbine system embodied many of the features of the future generations of wind turbines.

The latter part of the 19th century and the first half of the 20th century saw the construction of a number of larger wind turbines, which greatly influenced the development of today's technology. Probably the most important developments occurred in Denmark. In particular, Poul la Cour (1846–1908) was a Danish scientist and pioneer of wind power. He built the first wind tunnels for the purpose of aerodynamic tests to identify the best shape for turbines blades and he also investigated ways in which the energy from wind turbines could be stored. By the late 1960s people were becoming more aware of the environmental consequences of industrial development and this was followed by the mid-1970s' oil crisis. This gave a new impetus to the development of wind power so that, by the end of 2015, the total installed capacity of wind power in the world was greater than 400 GW. Commercial wind turbines now have output powers in the range 1.5–8 MW.

6.2   Origin and directions of the wind

Wind is the movement of air from a region of high pressure to a region of low pressure; the larger the difference between the high and low pressures, the greater the wind speed. Pressure differences arise due to the rising and sinking of air in the atmosphere. Where air is rising, we get lower pressure at the Earth's surface, and where it is sinking we get higher pressure. The regions around the equator are heated by the Sun more than those regions around the poles for the geometric reasons we described in Section 4.3.2. Hot air is lighter than cold air and at the equator the air rises high into the atmosphere, producing a region of low pressure. This process pushes the air at the top of the atmosphere towards the north and south poles. As this air reaches colder regions, it starts to sink back to Earth where the pressure increases. If the Earth did not rotate, we would expect that the air would simply arrive at the north and the south poles, sink down and return to the equator, which is at a lower pressure, as illustrated in Figure 6.1. This would generate winds near the Earth's surface that travel from the poles back to the equator. In practice, however, this does not happen. This is because the direction of the wind is strongly influenced by the Earth's rotational motion. This rotational motion causes winds in the northern hemisphere to be deflected towards the right of their direction of travel, while winds in the southern hemisphere are deflected towards their left. This effect is described in terms of the so-called *Coriolis force*.

6.2.1   The Coriolis force

Imagine that we launch a rocket from the north pole towards the equator and view its trajectory from outer space. Imagine also for a moment that planet Earth does not rotate about its axis. The rocket's trajectory that we would see is shown in Figure 6.2(a) and the rocket would land at point A on the equator. But Earth does rotate. So while the rocket is flying through the air, the point A on Earth is also moving; in an easterly direction at 465 m/s. Thus when the rocket lands, it does so at a point that is to the west of point A, as shown in Figure 6.2(b). For an observer located close to A, the rocket has apparently been deflected towards the west, i.e. towards the right

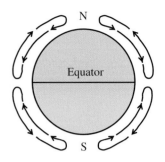

Figure 6.1 Circulation of the Earth's atmosphere in the absence of the Coriolis force. The air rises at the equator, producing low pressure there, and sinks at the colder poles, giving high pressure there.

Figure 6.2 (a) The trajectory of a rocket in the northern hemisphere if the Earth did not rotate. (b) The trajectory of the rocket as the Earth does rotate. For an observer at point A, the rocket is deflected to the right of its direction of travel because of Earth's rotation.

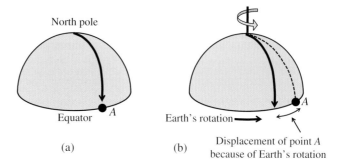

(a)   (b)

with respect to its direction of travel. The apparent bending force is called the Coriolis force. The Coriolis force is not a true force like the force of gravity. It simply arises when an object moves in a rotating frame of reference, and is sometimes described as a *pseudo force*. It is named after the French scientist, Gaspard-Gustave Coriolis, who studied the transfer of energy in rotating systems.

An object moving in a northerly direction in the northern hemisphere is also deflected towards the right because of the Earth's rotation. We can see this as follows. Suppose that we now launch a rocket from the equator towards the north pole, as illustrated by Figure 6.3(a). While it is on the ground, the rocket is moving along with the Earth in an easterly direction at a speed u_{east} of 465 m/s. At its launch, the rocket is given a component of velocity u_{north} in the northerly direction, but it maintains its velocity component u_{east} in the easterly direction. The angular velocity, Ω, of the Earth about its axis is the same for all latitudes, of course. However, the linear velocity $(r \times \Omega)$ of a given point on the surface of the Earth reduces as latitude increases, r being the perpendicular distance to the axis of rotation. The linear velocity is largest at the equator. So as the rocket travels to higher latitude, its component of velocity u_{east} parallel to the equator becomes greater than the speed of the ground beneath it. Hence, looking at Figure 6.3(b), we see that while point A, at a particular latitude, has moved a certain distance east since the rocket launch, the rocket has moved a greater distance to the

Figure 6.3 (a) The velocity components of a rocket that is fired in a northerly direction. (b) The rocket is deflected to right of its direction of travel according to an observer at point A because of the Earth's rotation.

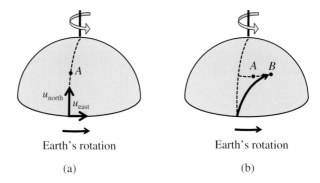

(a)   (b)

east, given by point B. Again, to an observer close to point A it would appear that the rocket is deflected to the right. Similar considerations show that an object travelling in the southern hemisphere is deflected towards the left of its direction of travel.

The Coriolis force similarly affects the apparent direction of the wind. In the northern hemisphere, winds are deflected towards the right and in the southern hemisphere they are directed towards the left. This has profound consequences. The air flow that starts above the equator and travels at high altitude towards the north pole is deflected to the right by the Coriolis force. By the latitude of approximately 30°, the Coriolis force prevents the air from moving much further north and the air sinks to the Earth's surface. This leads to high pressure at that latitude; higher than the pressure at the equator. This sets up a *cell* of circulating air called the *Hadley cell*, as shown in Figure 6.4, where the arrows indicate the directions of the circulating air. This cell is named after George Hadley, an English lawyer and amateur meteorologist who proposed the mechanism by which winds are sustained. There are also two other such cells, called the *Ferrel* and the *polar* cell, as shown in Figure 6.4. These circulating currents of air produce the winds in the lower atmosphere. These winds are also affected by the Coriolis force; winds moving a northerly direction are deflected to the right and those moving a southerly direction are deflected to the left. This gives rise to the *global wind pattern* illustrated in Figure 6.4, which shows the well-know *trade winds* and *westerlies*. This pattern explains why, for example, the prevailing winds across the UK are west to southwesterly. At the junction of two cells, the horizontal wind speed is relatively small; most of the movement of the air is in the vertical direction. Thus, for example, in the region around the equator where the two Hadley cells meet, there is a region called the *doldrums*. This

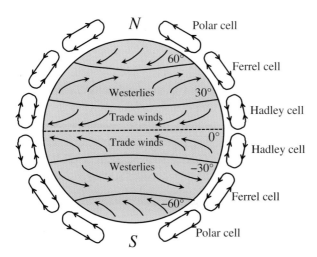

Figure 6.4 The cells of circulation in the atmosphere caused by the Earth's rotation, and the directions of the resulting global winds.

region was infamous among 18th-century sailors because of its calm wind conditions. The heights of the atmosphere in Figures 6.1 and 6.4 are greatly exaggerated. We saw in Section 5.3.3 that the height of Earth's atmosphere is ~10 km. If we were to imagine Earth to be the size of an orange, the extent of the atmosphere would be less than 0.1 mm.

Although these global winds are of much importance in determining the prevailing winds at a particular location, local climatic conditions and landscape also influence wind direction. Again this involves the rising and sinking of air in the atmosphere. One of the simplest examples of a local wind is a sea breeze, which occurs where the land meets the sea. On sunny days, solar radiation heats the land quite quickly. By contrast, the sea is slower to warm up because it has a higher heat capacity. This leads to a temperature difference between the warm land and the cooler sea. As the land heats up, it warms the air above it, leading to lower pressure over the land. The air over the sea remains cooler and denser and so the pressure is higher than inland. This pressure difference produces a sea breeze blowing from the sea to the land. Clearly, the directions of the prevailing wind and local winds form a major consideration in the siting of wind turbines.

6.3   The flow of ideal fluids

The flow of a fluid can be extremely complex, as is evident in the rapids of a fast-flowing river. However, some systems of fluid flow can be represented by a relatively simple, idealised model. Moreover, in many situations we can often treat the fluid as an *ideal fluid*. This is a fluid that is (i) incompressible and (ii) has no viscosity, which means no internal friction. Liquids are essentially incompressible, but we can also treat gases as incompressible if the pressure difference from one region to another is not too great. And in many cases the frictional forces in the fluid can be neglected compared with forces arising from pressure differences or gravity.

A practical way to visualise the flow of a fluid is to inject dye into the fluid through a series of narrow, equally spaced channels so that the dye enters parallel to the direction of flow; if the fluid is air, smoke can be used instead. Figure 6.5 illustrates schematically what would be observed for the flow of a fluid past a cylinder. If the overall pattern of lines does not change with time, as in Figure 6.5(a), the die traces out a set of smooth and continuous lines. This is called *steady flow* and the lines are called *streamlines*. A streamline can be thought of as the path that a particle would take if placed in a flowing fluid. Thus a streamline is a curve whose tangent at any point is in the direction of the fluid velocity at that point. Note that

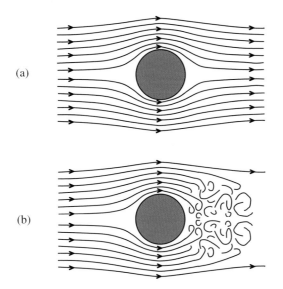

Figure 6.5 Schematic for the flow of a fluid past a cylinder. (a) Steady or laminar flow, where the overall pattern of streamlines does not change with time; the streamlines curve around the cylinder becoming closer together at the top and bottom of the cylinder. (b) Turbulent flow where the flow pattern changes continuously, making it irregular and chaotic.

all particles that pass through the same point follow the same path, i.e. the same streamline. Streamlines never intersect; no fluid particle can move across from one streamline to another. At the far left of Figure 6.5(a) the streamlines are parallel and horizontal. As they approach the cylinder the streamlines curve around the cylinder, getting closer together at the top and bottom of the cylinder. Then they spread out again before becoming parallel again at the far right. The flow pattern shown in Figure 6.5(a) is typical of *laminar flow*, where adjacent layers of fluid slide smoothly past each other. At sufficiently high flow rates or when obstacles cause abrupt changes in fluid velocity, the flow can become irregular and chaotic. This is called *turbulent flow* and the flow pattern changes continuously, as illustrated by Figure 6.5(b). In turbulent flow, the kinetic energy of the fluid is dissipated as thermal energy. Clearly turbulent flow has to be minimized as far as possible in the harvesting of energy from the wind.

6.3.1   The continuity equation

Imagine now that a fluid travels down a pipe of decreasing cross section as shown in Figure 6.6, where the fluid flows from left to right. The shaded volume on the left depicts the volume ΔV of fluid that flows through area A_1 in time Δt. If the speed of the fluid is u_1, the volume, ΔV, is given by:

$$\Delta V = A_1 u_1 \Delta t. \tag{6.2}$$

We assume that the fluid is incompressible so that an equal volume of fluid must flow past any point along the pipe during the same time, Δt. The volume that flows through area A_2 is depicted by the

Figure 6.6 The figure shows a
fluid that flows from left to
right down a pipe of
decreasing cross-section. The
shaded volume on the left
depicts the volume of fluid
that flows through area A_1 in
time Δt. The volume that
flows through area A_2 in time
Δt is depicted by the shaded
area on the right. These
volumes must be equal
according to the continuity
equation.

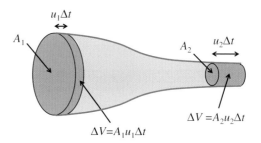

shaded area on the right of Figure 6.6. If the speed of the fluid is
u_2 at A_2 this volume of fluid is equal to $A_2 u_2 \Delta t$. As these volumes
must be equal we have

$$A_1 u_1 \Delta t = A_2 u_2 \Delta t, \tag{6.3}$$

so that

$$A_1 u_1 = A_2 u_2. \tag{6.4}$$

This is a *continuity equation*, where the quantity (Au) is called the
volume flow rate. We see that Au is a conserved quantity. We also
have $\Delta m = \rho \Delta V$, where ρ is the density of the fluid and Δm is the
mass of fluid in volume ΔV. Hence we can write

$$\Delta m = \rho A u \Delta t, \tag{6.5}$$

giving

$$\frac{\Delta m}{\Delta t} \approx \frac{\mathrm{d}m}{\mathrm{d}t} = \dot{m} = \rho A u, \tag{6.6}$$

where we denote $\mathrm{d}m/\mathrm{d}t$ as $\dot{m}$. As Au is a conserved quantity, $\dot{m}$ is
also a conserved quantity for constant density ρ; the mass flow $\dot{m}$ is
the same across any cross-sectional area of the pipe.

6.3.2   Bernoulli's equation

A fluid will flow between two different regions if there is a pressure
difference between them. The resulting force on the fluid due to
this pressure difference causes the fluid velocity to increase. It is
Bernoulli's equation that relates the pressure and velocity of a fluid
at a particular point.

   Imagine a particular streamline in a steady flow situation as
shown in Figure 6.7. And imagine a very small cube of fluid mov-
ing along this streamline with velocity $u = \mathrm{d}s/\mathrm{d}t$, where distance s
is measured along the streamline. The inset on the figure shows an
enlarged view of the front face of this cube, which has length l. The
pressure p will, in general, vary along the streamline, as also will the
velocity u of the cube. If p is the pressure at the centre of the cube,
the pressure acting on the left face of the cube can be expanded as

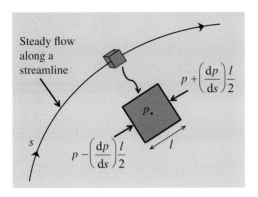

Figure 6.7 A very small cube of fluid moving along a particular streamline with velocity $u = \mathrm{d}s/\mathrm{d}t$. The inset shows an enlarged view of the front face of this cube, which has length l. The pressure p will, in general, vary along the streamline, as also will the velocity of the cube.

$[p - (\mathrm{d}p/\mathrm{d}s)\,l/2]$ when l is small. Similarly, the pressure acting on the right face of the cube can be expanded as $[p + (\mathrm{d}p/\mathrm{d}s)\,l/2]$.

Hence, the net force acting on the cube of fluid along the s-direction is

$$[p - (\mathrm{d}p/\mathrm{d}s)\,l/2]\,l^2 - [p + (\mathrm{d}p/\mathrm{d}s)\,l/2]\,l^2.$$

Then, applying Newton's second law, we obtain

$$[p - (\mathrm{d}p/\mathrm{d}s)\,l/2]\,l^2 - [p + (\mathrm{d}p/\mathrm{d}s)\,l/2]\,l^2 = m\frac{\mathrm{d}u}{\mathrm{d}t} = \rho l^3 \frac{\mathrm{d}u}{\mathrm{d}t}, \quad (6.7)$$

where ρ is the fluid density. Simplifying this equation gives

$$-\frac{\mathrm{d}p}{\mathrm{d}s} = \rho\frac{\mathrm{d}u}{\mathrm{d}t}. \quad (6.8)$$

As velocity u is a function of s, we have

$$-\frac{\mathrm{d}p}{\mathrm{d}s} = \rho\frac{\mathrm{d}u}{\mathrm{d}s}\frac{\mathrm{d}s}{\mathrm{d}t} = \rho\frac{\mathrm{d}u}{\mathrm{d}s}u.$$

This gives

$$-\mathrm{d}p = \rho u\,\mathrm{d}u = \frac{\rho}{2}\mathrm{d}(u^2),$$

where we have used $\mathrm{d}(u^2) = 2u\,\mathrm{d}u$. Then, integrating along the streamline from pressure p_1 to p_2, we have

$$-\int_{p_1}^{p_2}\mathrm{d}p = \frac{1}{2}\rho\int_{u_1}^{u_2}\mathrm{d}(u^2).$$

Hence,

$$(p_1 - p_2) = \frac{1}{2}\rho\left(u_2^2 - u_1^2\right), \quad (6.9)$$

which we can also write as

$$\frac{p}{\rho} + \frac{1}{2}u^2 = \text{constant}. \quad (6.10)$$

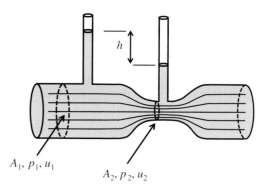

Figure 6.8 The principle of operation of a *Venturi meter*, which is used to measure fluid velocity in a pipe.

A_1, p_1, u_1

A_2, p_2, u_2

Equations (6.9), and (6.10) are expressions of Bernoulli's equation. Notice, in particular, that Bernoulli's equation says that as the pressure p reduces, the velocity u must increase. This result was first expressed in words by the Swiss mathematician Daniel Bernoulli in 1738 and was later derived in equation form by his associate Leonard Euler in 1755. Bernoulli's equation is much used in situations of fluid flow where compressibility and viscosity can be ignored.

Pressure can be thought of as potential energy per unit volume, as we will see further in Section 8.2.2. This is evident from the dimensions of both these quantities: $[\mathrm{ML}^{-1}\mathrm{T}^{-2}]$. Hence the quantity p/ρ is potential energy per unit mass. And we can recognise the quantity $u^2/2$ as kinetic energy per unit mass. Hence, Bernoulli's equation is simply a statement of the conservation of energy.

Worked example

Figure 6.8 shows a *Venturi meter* that is used to measure fluid velocity in a pipe. Obtain an expression for the fluid velocity u_1 in terms of the cross-sectional areas A_1 and A_2 and the difference in height, h, of the liquid levels in the two vertical tubes.

Solution

Applying Bernoulli's equation to the wide and narrow pipes respectively, we obtain

$$\frac{p_1}{\rho} + \frac{1}{2}u_1^2 = \frac{p_2}{\rho} + \frac{1}{2}u_2^2,$$

where the symbols have their usual meanings. We also have the continuity equation: $A_1 u_1 = A_2 u_2$ (Equation 6.4). Substituting for u_2 we obtain

$$p_1 - p_2 = \frac{1}{2}\rho u_1^2 \left(\frac{A_1^2}{A_2^2} - 1\right).$$

Since $A_1 > A_2$, we have $u_1 < u_2$ and hence $p_2 < p_1$. The pressure difference $(p_1 - p_2) = \rho g h$, where ρ is the fluid density and H is the difference between the liquid levels. Hence we obtain

$$u_1 = \sqrt{\frac{2gh}{(A_1/A_2)^2 - 1}}$$

From Figure 6.8, we see that when streamlines get closer together the pressure goes down and the fluid velocity goes up. This is a general result.

As we have seen, Bernoulli's equation is a statement of the conservation of energy. If gravitation forces are also involved, i.e. the fluid changes its vertical height, then we also include its gravitational potential energy, mgh, as for example in a hydroelectric power plant (see Section 7.1). (In the previous example, the change in height h was sufficiently small that we could neglect any change in gravitational potential energy.) Recalling that each term in Equation (6.10) represents *energy per unit mass*, we just need to add a term gh to Equation (6.10), We thus obtain:

$$\frac{p}{\rho} + \frac{1}{2}u^2 + gh = \text{constant.} \tag{6.11}$$

This is the general form of Bernoulli's equation.

Worked example

Figure 6.9 shows a container of water with a small hole in its side that produces a jet of water. H is the depth of the container and h is the distance of the hole below the surface of the water. The jet of water strikes the floor at a distance x from the side of the container. (a) Determine x if $H = 1.6$ m and $h = 0.40$ m. (b) At what value of h should a second hole be drilled to give the same value of x? (c) What value of h gives the maximum value of x?

Solution

The size of the hole is small so that we can assume that the velocity at which the water level in the container falls is negligible. Also the pressure at the water surface and at the small hole are the same and equal to atmospheric pressure.

(a) Taking the bottom of the container as the reference level for gravitational potential energy, we have from Bernoulli's equation

$$\frac{p_{\text{atmos}}}{\rho} + \frac{1}{2}(0)^2 + gH = \frac{p_{\text{atmos}}}{\rho} + \frac{1}{2}u^2 + g\left(H - h\right),$$

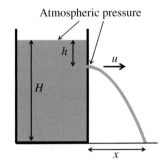

Figure 6.9 A container of water with a small hole in its side that produces a jet of water. H is the depth of the container and h is the distance of the hole below the surface of the water. The jet of water strikes the floor at a distance x from the side of the container.

where u is the velocity of the water jet. This gives $u = \sqrt{2gh}$. Interestingly, this is the same result for the final velocity of a body that falls under gravity through a distance h. The time, t, before the water jet strikes the ground is given by the expression:

$$\frac{1}{2}gt^2 = (H - h)$$

and hence $x = 2\sqrt{h\,(H - h)}$. For the given values of H and h, $x = 1.4$ m.

(b) Let height h' give the same value of x as $h = 0.40$ m. Then, using $x = 2\sqrt{h\,(H - h)}$, we obtain:

$$2\sqrt{0.40\,(1.6 - 0.40)} = 2\sqrt{h'\,(1.6 - h')}.$$

This simplifies to

$$h'^2 - 1.6h' + 0.48 = 0,$$

from which $h' = 1.2$ m or 0.40 m.

(c) The distance x will be maximised when $\dfrac{\mathrm{d}}{\mathrm{d}h}\left[h\,(H - h)\right] = 0$. This is when $h = H/2$, giving $h = 0.80$ m.

Bernoulli's equation explains qualitatively the lift provided by an aeroplane wing. A section of a wing that moves through the air is shown in Figure 6.10(a). The streamlines crowd together above the wing corresponding to increased flow speed and reduced pressure, just as in the smaller diameter tube of the Venturi meter. Hence, the downward force due to the air on the top side of the wing is less than the upward force of the air on the underside of the wing, and

Figure 6.10 (a) When the air flows past an aeroplane wing, the streamlines crowd together above the wing, producing a lower pressure in that region. The upward force on the wing is then greater than the downward force on the wing and there is a net upward or lift force, F_L. The wing is also subject to a drag force F_D, which is due to frictional forces between the wing and the flow of air. (b) F_L can be increased by tilting the blade away from the incident wind direction within a certain range; the angle through which it is tilted is called the *angle of attack*, α. However, this also increases F_D.

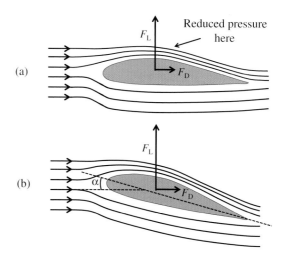

there is a net upward or *lift force*, F_L. This force is perpendicular to the direction of the incoming air flow as shown in the figure.

In addition to the lift force, the wing is also subject to a drag force, F_D. The drag force is due to frictional forces between the wing and the flowing air. As illustrated in Figure 6.10(a), the drag force is parallel to the direction of incoming air flow. Both of these forces can be readily experienced by holding your arm out of a moving car and shaping your hand as the section of an aeroplane wing. The lift force, F_L, can be increased by tilting the blade away from the incident wind direction within a certain range, as in Figure 6.10(b). The angle through which it is tilted is called the *angle of attack*, α. However, this also increases the drag force F_D. We will see in Section 6.4 that the blades of a wind turbine experience the same forces as an aeroplane wing; the lift force and the drag force. In that case it is usually desirable to have the lift force to be as large as possible large compared with the drag force, as the drag force detracts from the desired action of the turbine. In a wind turbine, the angle of attack is controlled by twisting the blades about their long axis.

6.4   Extraction of wind power by a turbine

Before considering the extraction of wind power by a turbine, we consider the somewhat similar situation of a hosepipe directing a jet of water at a wall. A section of the hosepipe, which has uniform cross-sectional area A, is shown in Figure 6.11.

Water travels through the pipe at velocity u_0 from left to right. In time Δt the volume ΔV of water that passes through area A is $Au_0\Delta t$. The mass Δm of this volume is $\rho Au_0\Delta t$, where ρ is the water density and the momentum Δp is $\rho Au_0^2\Delta t$. Hence the rate at

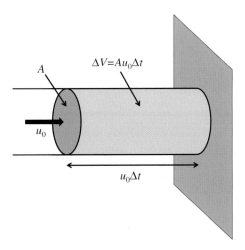

Figure 6.11 A section of a hosepipe with a uniform cross-sectional area A that carries water at velocity u_0 from left to right. In time Δt the volume of water that passes through area A is $Au_0\Delta t$.

which momentum passes through area A, i.e. the momentum flow $\dot{p}$ is given by

$$\frac{\Delta p}{\Delta t} \approx \frac{\mathrm{d}p}{\mathrm{d}t} = \dot{p} = \frac{\rho A u_0^2 \Delta t}{\Delta t} = \rho A u_0^2. \tag{6.12}$$

When the water eventually strikes the wall it exerts a force $F = \dot{p}$ on the wall. Assuming that the water velocity reduces to zero at the wall, this force is then given by

$$F = \dot{p} = \rho A u_0^2. \tag{6.13}$$

The power P delivered to the wall is the rate at which the kinetic energy of the water is deposited at the wall. The volume ΔV of water has kinetic energy ΔK, which is given by

$$\Delta K = \frac{1}{2}(\Delta m)u_0^2 = \frac{1}{2}\left(\rho A u_0 \Delta t\right)u_0^2 = \frac{1}{2}\rho A u_0^3 \Delta t. \tag{6.14}$$

Hence,

$$P = \frac{\mathrm{d}K}{\mathrm{d}t} \cong \frac{\Delta K}{\Delta t} = \frac{1}{2}\rho A u_0^3. \tag{6.15}$$

Importantly, the delivered power P *is proportional to the third power of the water velocity.* Using $\dot{m} = \rho A u_0$ (see Equation 6.6) we can also write the power as

$$P = \frac{1}{2}\dot{m}u_0^2. \tag{6.16}$$

We see that the flow of water exerts a force on the wall and delivers power to the wall. We can make a similar analysis for a wind turbine with air as the flowing fluid, as we will do below. However, there is a significant difference. The air that is incident upon the turbine cannot come to a sudden halt. It has to pass through the turbine in order for the blades to rotate and this must be taken into account. Albert Betz, a German physicist, obtained an expression for the maximum power, P_{max}, that can be extracted from the wind by a turbine. We can write his result as

$$P_{\mathrm{max}} = 0.59 \times \frac{1}{2}\rho A_{\mathrm{T}} u_0^3, \tag{6.17}$$

where A_{T} is the area swept out by the turbine blades and u_0 is the wind speed. This equation is a statement of the *Betz criterion*. No matter what the design of the turbine or the shape of the blades, Equation (6.17) gives the maximum power that a turbine can extract from the wind. It has the same form as Equation (6.15) apart from the addition of the constant 0.59. In particular, the power is proportional to the third power of the wind speed. Of course, the density of air is much smaller than that of, water. However, this is compensated by the fact that commercial wind turbines sweep out a large area.

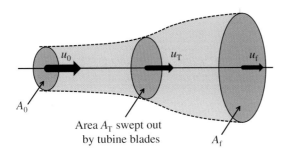

Figure 6.12 A column of air that is incident upon a wind turbine. The wind speed steadily decreases as the air approaches and passes through the turbine blades, which explains the evolution of the cross-sectional area of the air column. The area A_T represents the area swept out by the blades of the turbine. The wind exerts a force on the rotor blades and power is extracted from the wind.

6.4.1   The Betz criterion

To obtain an expression for the Betz criterion, we consider a column of air of initial cross-sectional area A_0, as depicted on the left side of Figure 6.12. The initial wind speed is u_0. As the air approaches the turbine and passes through it, the wind speed reduces and hence, according to Equation (6.4), the cross-sectional area of the air column increases, as shown in Figure (6.12). In this figure the area A_T represents the area swept out by the turbine blades, at which point the wind speed is u_T, which is less than u_0. We do not need to include the details of the turbine blades. We simply need to know that the wind exerts a force on them and that power is extracted from the wind. Finally, downstream of the turbine, the wind speed is reduced to u_f and the final cross-sectional area becomes A_f.

In this process, the air mass flow $\dot{m}$ through any cross-sectional area of the column is conserved and hence the mass flow at the three areas, A_0, A_T, and A_f, is the same:

$$\dot{m}_0 = \dot{m}_T = \dot{m}_f = \dot{m}. \qquad (6.18)$$

We can find the power, P_T, extracted from the wind by considering the rate at which the wind loses its kinetic energy K. Recalling Equations (6.15) and (6.16) we have

$$P = \frac{\mathrm{d}K}{\mathrm{d}t} = \frac{1}{2}\dot{m}u_0^2.$$

Thus, at initial area A_0, the power in the wind is $\frac{1}{2}\left(\dot{m}u_0^2\right)$ and at final area A_f it is $\frac{1}{2}\left(\dot{m}u_f^2\right)$. The power lost by the wind is equal to the power P_T extracted by the turbine, which is therefore given by

$$P_T = \frac{1}{2}\dot{m}\left(u_0^2 - u_f^2\right). \qquad (6.19)$$

We can also find the power that is extracted from the wind by considering the change in momentum flow $\dot{p}$ that the wind suffers in passing through the turbine. The momentum flow $\dot{p}$ across initial area A_0 is, from Equations (6.12) and (6.6),

$$\dot{p}_0 = \rho A_0 u_0^2 = \dot{m}_0 u_0.$$

Similarly, the momentum flow across the final area A_f is

$$\dot{p}_f = \rho A_f u_f^2 = \dot{m}_f u_f.$$

The difference between the initial and final momentum flow $(\dot{p}_0 - \dot{p}_f)$ is just the force F acting on the turbine. Thus we have

$$F = (\dot{p}_0 - \dot{p}_f) = \dot{m}_0 u_0 - \dot{m}_f u_f = \dot{m}(u_0 - u_f). \qquad (6.20)$$

We recall that mechanical work is force times distance and that power is the rate of doing work, i.e. force times velocity. Hence, the power P_T extracted by the turbine is given by

$$P_T = F \times u_T = \dot{m}(u_0 - u_f) u_T, \qquad (6.21)$$

where u_T is the wind speed at the turbine. Equating Equations (6.19) and (6.21) we obtain

$$\dot{m}(u_0 - u_f) u_T = \frac{1}{2}\dot{m}\left(u_0^2 - u_f^2\right)$$

which gives

$$u_T = \frac{1}{2}(u_0 + u_f). \qquad (6.22)$$

This result says that the wind speed at the turbine is the mean of the initial and final wind speeds. We have $\dot{m} = \rho A_T u_T$ (see Equation 6.6), and substituting for $\dot{m}$ in Equation (6.21) gives

$$P_T = \rho A_T u_T^2 (u_0 - u_f). \qquad (6.23)$$

Then using Equation (6.22) to eliminate u_f we obtain

$$P_T = 2\rho A_T u_T^2 (u_0 - u_T). \qquad (6.24)$$

We define the parameter a as the fractional decrease in wind speed:

$$a = \frac{(u_0 - u_T)}{u_0}, \qquad (6.25)$$

giving

$$u_T = (1 - a) u_0. \qquad (6.26)$$

Substituting for u_T from Equation (6.26) into Equation (6.24) we obtain

$$P_T = 2\rho A_T (1-a)^2 u_0^2 [u_0 - (1-a) u_0]$$
$$= [4a(1-a)^2]\left(\frac{1}{2}\rho A_T u_0^3\right). \qquad (6.27)$$

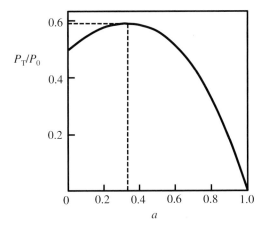

The power P_0 in an unobstructed column of air of area A_T travelling with wind speed u_0 is given by

$$P_0 = \frac{1}{2}\rho A_T u_0^3 \qquad (6.28)$$

(see Equation 6.15). Thus the ratio $P_T/P_0 = 4a(1-a)^2$ is a measure of the fraction of wind power extracted by a turbine and is called the *power coefficient* C_P. The power coefficient is plotted against a in Figure 6.13. The maximum value of C_P occurs when $a = 1/3$, when it has the value 0.59. From this value of a it follows from Equations (6.22) and (6.26) that for maximum power extraction the final wind speed u_f should be one-third of the initial wind speed u_0.

In this analysis, we have assumed that the flow is laminar and there is no turbulence, i.e. that the air passes smoothly through the turbine. In practice, the motion of the turbine blades does produce some turbulence and, consequently, practical turbine efficiencies are less than the Betz limit. However, commercial wind turbines typically achieve 75–80% of the Betz limit. Note that the Betz limit has nothing to do with thermodynamic efficiency, but is instead a mechanical limit.

We can use Equation (6.28) to obtain a value for the typical power of unobstructed wind. Taking a wind speed $u_0 = 10$ m/s and an air density $\rho = 1.2$ kg/m^3, we obtain a value for P_0 of 600 W/m^2. This is comparable to the power density of solar radiation at the surface of Earth.

Worked example

Estimate the number of wind turbines that would produce the same amount of power as a typical nuclear power station. Assume a blade length of 50 m and a wind speed of 10 m/s and that the turbine produces 80% of the Betz limit. Take the density of air to be 1.2 kg/m^3.

Figure 6.14 (a) One of the blades of a three-blade rotor as viewed from the front of the rotor. The blades rotate at angular velocity Ω in the plane that is perpendicular to the wind direction. An element of the blade at radial distance r from the rotor hub is indicated by the shaded area. It has linear velocity $u_{\text{lin}} = r\Omega$. The blade can be twisted about its long axis to control the angle of attack. (b) The cross-section of the shaded element as viewed from the blade tip. When a wind flows across a turbine blade, a pressure difference develops between the two surfaces of the blade and a lifting force is produced. As the blade is attached to the hub of the rotor, the lifting force causes the blade to rotate about the axis of the rotor.

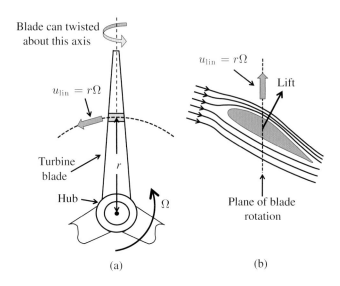

(a)　　　　　　　　(b)

Solution

The Betz criterion gives a value of P_{max} of

$$0.59 \times \frac{1}{2}\rho A_{\text{T}} u_0^3 = 0.59 \times \frac{1.2 \times \pi \times 50^2 \times 10^3}{2} = 4.7 \text{ MW}.$$

With a factor of 0.8 for overall efficiency, this reduces to 3.8 MW. Taking the output of a typical nuclear power station to be $\sim$1 GW, (see Section 3.4.5) this indicates that it takes $\sim$260 wind turbines to produce the same amount of power.

6.4.2　Action of wind turbine blades

The action of the blades of a wind turbine is to convert the linear motion of the incident wind into rotational motion of the rotor. Figure 6.14(a) shows one of the blades of a three-blade rotor as viewed from the front of the rotor. An element of the blade is indicated by the blue shaded area. It is at distance r from the axis of the rotor and rotates about it with angular velocity Ω and linear velocity $u_{\text{lin}} = r\Omega$. Figure 6.14(b) shows the cross-section of this element as viewed from the blade tip and also indicates the plane of rotation of the blade. The turbine blade is similar in cross-section to the wing of an aeroplane (see Figure 6.10) and indeed the basic aerodynamics are the same for both. In the case of an aeroplane, the engines propel the wing through the air and a pressure difference develops between the top and bottom surfaces of the wing, thus producing a lift force, as we saw in Section 6.3. Similarly when the air flows across a turbine blade, a pressure difference develops between the two surfaces of the blade and a lift force is produced, as illustrated in Figure 6.14(b). The similarity between a turbine blade and an aeroplane wing can

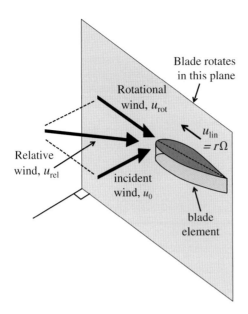

Figure 6.15 A turbine blade experiences the incident wind with velocity u_0 coming directly towards it. As the blade is rotating, it also experiences air moving towards it in its plane of rotation with velocity $u_{rot} = u_{lin} = r\Omega$. These two winds combine vectorially to produce a resultant wind with velocity u_{rel}.

be seen by turning the page so that the direction of the incoming air flow in Figure 6.14(b) is horizontal. However, the blade is attached to the hub of the rotor and the net effect is that the blade rotates about the axis of the rotor.

There is one particular difference between the action of a turbine blade and an aeroplane wing. The aeroplane wing experiences the wind coming directly towards it (see Figure 6.10). Similarly, a turbine blade experiences the incident wind with velocity u_0 coming directly towards the turbine. However, as the blade is rotating, it also experiences air coming towards it in its plane of rotation. This is called the *rotational wind*. The velocity of this rotational wind is just the linear velocity, $u_{lin} = r\Omega$, of the blade. These two winds combine vectorially to produce the resultant or *relative wind*. It is the air flow due to the relative wind that is shown in Figure 6.14(b) and now we can see why its direction is not perpendicular to the blade's plane of rotation. The directions of the incident wind, the rotational wind and the relative wind with respect to the rotational plane of the rotor blades are illustrated in Figure 6.15. The velocities of these winds are u_0, u_{rot} and u_{rel}, respectively. Taking a value of u_0 of 12 m/s with $r = 25$ m, and a rotational frequency of the rotor of 15 rpm, we obtain $u_{lin}/u_0 = (2\pi \times 25)/(4 \times 12) = 3.3$. We see that the tips of the blades move much faster than the incident wind speed. This is because the blades experience the lift force.

In most modern wind turbines, the angle of attack can be adjusted by changing the *pitch* of the blade, i.e. by twisting the blade about its long axis, as indicated in Figure 6.14. It is also the case that the optimum angle of attack depends on the ratio of the velocities

Figure 6.16 (a) The forces acting on a turbine blade are the lift force F_{L} and the drag force F_{D}. The lift force is perpendicular to the direction of the incoming air flow and is the consequence of the unequal pressure on the upper and lower surfaces of the blade. The drag force is parallel to the direction of incoming air flow, and is due to frictional forces between the wing and the flowing air. The lift force, F_{lift}, and drag force, F_{drag}, combine to produce the total force, F_{tot}, acting on the element. This total force has a component F_{pow} in the plane of the rotation of the blades and it is this component that produces a torque on the rotor and hence the production of power by the turbine. The magnitude of F_{pow}, and hence the amount of power absorbed from the wind, is controlled by adjusting the angle of attack α.

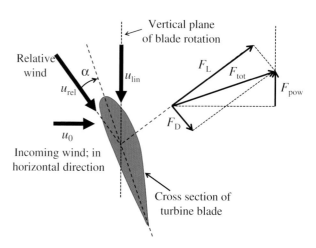

u_{rot}/u_0 and clearly this ratio varies along the length of the turbine blade. This is taken into account by constructing the blade to have an inbuilt twist along its length.

Figure 6.16 illustrates the forces acting on an element of a turbine blade and the relevant wind velocities. The lift force, F_{lift}, and drag force, F_{drag}, combine to produce the total force, F_{tot}, acting on the element. This total force has a component F_{pow} in the plane of the rotation of the blades and it is this component that produces a torque on the rotor and hence the production of power by the turbine. The magnitude of F_{pow}, and hence the amount of power absorbed from the wind, is controlled by adjusting the angle of attack α, i.e. by adjusting the pitch of the blades.

Commercial wind turbines that are used to generate electricity exploit the lift force F_{L} to turn the turbine rotor, as we have just described. But it is also possible to use the drag force F_{D} to cause a rotor to rotate. Indeed, the wind turbines that are seen, for example, on the plains of America exploit the drag force. Such turbines are recognisable by the many blades that their rotors have. They have characteristics that make them suitable for tasks such as pumping water from underground reserves.

Optimum rotational speed of turbine blades

The rotational speed of a wind turbine is very important in maximising its power output. If the rotational speed is too low, some wind will pass through without interacting with the blades. If it is too high, there will be excess wind turbulence, which decreases the efficiency of the turbine. In order to optimise the angular velocity of a rotor with respect to the incident wind speed, we compare the time t_{b} that it takes one rotor blade to move into the position occupied by the previous blade. If n is the number of blades and their angular velocity is Ω rads/s, t_b is given by

$$t_b = \frac{2\pi}{n\Omega}. \qquad (6.29)$$

Suppose that the turbulence created by a turbine blade lasts time t_w. We need to wait at least that time before the following blade moves into its position. If d is the length of the wind that is perturbed by a rotating blade and u_0 is the wind speed,

$$t_w = \frac{d}{u_0}. \qquad (6.30)$$

It follows that for maximum power extraction we require t_b and t_w to be equal. This gives

$$\frac{2\pi}{n\Omega} = \frac{d}{u_0}$$

or

$$\frac{2\pi R}{nd} = \frac{R\Omega}{u_0}, \qquad (6.31)$$

where R is the radius of the blades. Defining the *tip-speed ratio* λ as

$$\lambda = \frac{\text{speed of rotor blade tip}}{\text{speed of incoming wind}} = \frac{R\Omega}{u_0}, \qquad (6.32)$$

there is maximum power extraction when $\lambda = 2\pi R/nd$. Experimental measurements show that $d \approx R/2$, and hence $\lambda \approx 4\pi/n$. This suggests that a three-blade rotor, say, has an optimum tip-speed ratio of ~ 4. By designing turbine blades with highly aerodynamic efficiency, the effects of turbulence can be minimised and this can increase the value of the tip-speed ratio to about 6 or 7. This leads to higher rotational speeds of the rotor and hence higher power output. Tip-speed ratio is probably the most important parameter of a wind turbine, as it is a function of the three most important variables: blade radius R, wind speed u_0 and rotor angular velocity Ω. Being dimensionless, it is an essential scaling factor in wind turbine design.

6.5   Wind turbine design and operation

There are various types of wind turbine. These can be divided into those that rotate about a horizontal axis and those that rotate about a vertical axis. The horizontal axis type is the most common today and is our primary focus. These typically have three blades, as illustrated in Figure 6.17(a). Turbines that rotate about a vertical axis are exemplified by the *Darrieus turbine*, which is illustrated in Figure 6.17(b). Vertical axis turbines have some advantages. They work

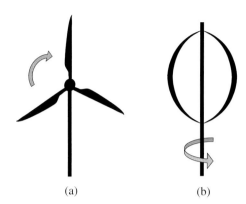

Figure 6.17 (a) The horizontal axis wind turbine is the most common type today and typically has three blades. (b) The Darrieus wind turbine exemplifies those types that rotate about a vertical axis.

(a)   (b)

in any wind direction. Moreover, as their axis of rotation is vertical, they do not suffer from the gravity-induced stress/strain cycles that occur in a horizontal axis turbine when the blades rotate. Disadvantages of a vertical axis turbine are that their overall efficiencies are generally less than for horizontal axis turbines and they not self-starting; they must be given an impulse to start rotating.

The main components of a horizontal axis wind turbine are shown in Figure 6.18. These include the rotor, the gearbox and the electrical generator. The rotor consists of the turbine blades and a supporting hub that connects the blades to the main shaft of the turbine. The blades are typically made from composites, primarily fibreglass or plastics reinforced with carbon fibres. As noted previously, in modern turbines, the pitch of the blades can be varied to control the power output from the turbine. The housing that contains the mechanical and electrical components of the turbine and protects them from the weather is called the *nacelle*. The whole assembly is

Figure 6.18 The main components of a horizontal axis wind turbine. These include the rotor, the gearbox and the electrical generator. The rotor consists of the turbine blades and a supporting hub that connects the blades to the main shaft of the turbine. The housing that contains the mechanical and electrical components of the turbine and protects them from the weather is called the nacelle. The whole assembly is mounted on a support tower. The yaw mechanism consists of a large bearing that connects the body of the turbine to its support tower and is used to turn the turbine into the wind. The purpose of the gearbox is to match the rate of rotation of the rotor (5–20 rpm) to that of the generator (~1000 rpm). Wind turbines also incorporate a brake that can be engaged to close down the turbine for maintenance or for when the wind speed is so high that it could damage the turbine.

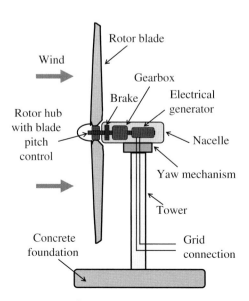

mounted on a support tower. The rotor may be in front (upwind) of the support tower or behind it (downwind). Wind veers frequently in direction and horizontal axis turbines must be turned into the wind for maximum power output. Downwind turbines are, in principle, self-orientating. However, the rotor is then in the shadow of the tower, and this produces extra turbulence. Moreover, as the rotor rotates, there will be times during the cycle when one of the blades falls within the wind shadow and this leads to uneven loading across the rotor and mechanical stress. Upwind turbines, as in Figure 6.18, avoid these disadvantages but they must be driven to face the incident wind. This is done under computer control by the *yaw mechanism*, which consists of a large bearing that connects the body of the turbine to its support tower; this mechanism gets its control signal from a wind direction sensor mounted on the nacelle. Most commercial wind turbines are of the upwind variety. Wind turbines also incorporate a brake that can be engaged to close down the turbine for maintenance or for when the wind speed is so high that it could damage the turbine.

Most commercial wind turbines have three blades. This number, like many design considerations, is a compromise between competing requirements. These include aerodynamic efficiency, mechanical stability, manufacturing cost, and environmental and aesthetic issues. The aerodynamic efficiency of a rotor does increase with the number of blades but with diminishing returns. For example, the improvement in going from two to three blades is ∼5% and in going from three to four blades is ∼1%. As the number of blades increases, the balance and hence the stability of the rotor increases, which means less mechanical stress on the turbine as it rotates. Thus the lack of balance in a one-blade rotor can cause can cause high levels of mechanical stress in the rotor, even though the blade is balanced by a counterweight. Twin-blade rotors are better balanced. However, yawing operations, where the nacelle and rotor turn around a vertical axis, have to take place slowly to limit fluctuating dynamic loads during the operation; when the blades are vertical the forces required to yaw the rotor are low, but when the blades are horizontal, the forces are much higher. These cyclic forces impose significant stresses on the turbine. Three-blade rotors are more balanced than one- or two-blade rotors. Moreover, the fluctuating dynamic loads are much lower when a three-blade machine is yawed, as the asymmetric forces encountered as the rotor rotates are smaller. Four-blade rotors are better balanced still. There are, however, some drawbacks that arise with an increase in the number of blades. As the number of blades increases, so too do the manufacturing costs. Moreover, the *solidity* of a rotor must be kept to a reasonably low value, where solidity is the ratio of the total area of the blades and the area swept out by the blades. Thus, as the number of blades increases, they must be thinner. And then it becomes difficult to build blades that are sufficiently strong and more expensive materials are required. Taking all considerations into account, three-blade turbines appear to provide the best compromise, as evidenced by the fact that most commercial turbines have three blades, and it is generally accepted that three-bladed turbines are less noisy and less visually disturbing than other possible designs.

The rotor of a wind turbine usually rotates within the range 5–20 rpm. On the other hand, the drive shaft of a conventional electrical generator rotates at ~ 1000 rpm. The purpose of the gearbox is to match the rotational speeds of the rotor and the generator. A typical gearing ratio is 1:50. On large wind turbines, the generated electricity is usually 690 V *three-phase* AC.

A wind turbine may be designed to operate at fixed rotor speed, i.e. at fixed angular velocity or with variable rotor speed. Most commercial wind turbines are connected, via a transformer, to a national electricity grid. As these operate at a fixed frequency, usually 50 or 60 Hz, it follows that the electrical generator must also operate at a constant or nearly constant frequency. In a fixed-speed turbine, this determines the angular velocity, Ω, of the rotor. The disadvantage of this fixed-speed mode of operation is that the optimum tip speed ratio:

$$\lambda = \frac{\text{speed of rotor blade tip}}{\text{speed of incoming wind}} = \frac{R\Omega}{u_0}.$$

is not maintained if Ω is kept constant but wind speed u_0 changes, although in practice the reduction in overall efficiency is usually not substantial. Most modern turbines have variable-speed operation. This has the advantage that the tip-speed ratio can be kept constant at its optimum value over a wide range of wind speeds, thus operating the turbine at maximum efficiency. However, the varying speed of the rotor clearly complicates the generation of AC electricity at constant frequency. There are a variety of ways to allow variable-speed operation of the turbine rotor while keeping the generating frequency constant. These may be mechanical or electrical, although most approaches to variable-speed operation in use today are electrical in nature. These involve solid-state power convertors, which can change electrical power from one frequency to another.

The operating regions for a wind turbine are illustrated by the example of a *wind power curve* shown in Figure 6.19. At very low wind speeds, there is insufficient torque exerted on the turbine blades to make them rotate. However, as the speed increases, the turbine begins to rotate and to generate electrical power. The minimum speed at which the turbine blades overcome friction and start to rotate is called the *cut-in speed* and is typically between 3 and 4 m/s. As the wind speed rises above the cut-in speed, the output power

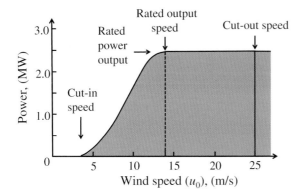

Figure 6.19 The operating
regions of a wind turbine. At
very low wind speeds, there
is insufficient torque exerted
on the turbine blades to
make them rotate. However,
as the speed increases, the
turbine begins to rotate and
to generate electrical power.
The minimum speed at which
the turbine blades rotate is
called the cut-in speed. As
the wind speed rises, the
output power rises rapidly;
the blue shaded area in the
figure is where the turbine
produces output power.
When the wind speed
becomes greater than
~ 14 m/s, the absorbed wind
power is limited to a constant
value to avoid exceeding safe
electrical and mechanical
loading limits. This limit is
called the rated power output
and the speed at which it is
reached is called the rated
output speed. At wind speeds
above the cut-out speed, the
turbine rotor is stopped from
turning to avoid excessive
operating loads.

rises rapidly; we recall that the power in the wind increases as the third power of its speed. The range in which the turbine produces output power is indicated by the blue shaded area in Figure 6.19. When the wind speed becomes greater than ~ 14 m/s, the absorbed wind power is limited to a constant value to avoid exceeding safe electrical and mechanical loading limits. This limit is called the *rated power output* and the speed at which it is reached is called the *rated output speed*. In large turbines, the power is typically limited by controlling the pitch of the rotor blades. As the wind speed increases above the rated output wind speed, the forces on the turbine structure continue to rise and, at some point, there is a risk of damage to the rotor. As a result, the braking system is employed to bring the rotor to a standstill. This is called the *cut-out speed* and is usually ~ 25 m/s.

Typical values for a commercial wind turbine that is connected to a national grid are as follows: a rotor diameter of ~ 100 m; a hub height of ~ 80 m; an operational range of wind speed of 4–25 m/s; and a nominal output power of 3.6 MW. This is sufficient for more than 3000 average households. Currently, the largest wind turbines have a rated capacity of 8 MW and a rotor diameter of 164 m.

A possible electrical arrangement for a variable-speed wind turbine that is located in a remote region and not connected to a national grid is shown schematically in Figure 6.20. The generator produces an AC voltage of variable frequency and variable amplitude. Some electrical loads, such as resistive elements to heat water, do not need constant AC frequency or supply voltage and so the output from the generator can be applied directly to them. Applications that are sensitive to frequency and amplitude can be supplied by a separate voltage line, as outlined in Figure 6.20. The AC voltage from the generator is converted to a smoothed DC voltage by a diode bridge/capacitor circuit. This voltage is then regulated in amplitude and the regulated voltage is connected to a battery that provides energy storage. This regulated voltage also drives an oscillator that

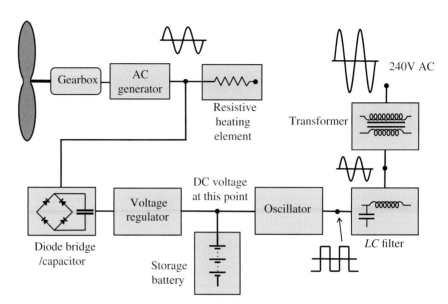

Figure 6.20 A possible electrical arrangement for a variable-speed wind turbine that is not connected to a national grid. This arrangement provides electrical power for applications, such as resistive elements to heat water that do not need constant AC frequency or supply voltage. It also provides a separate 240 V, 50 Hz AC voltage line for applications that are sensitive to frequency and amplitude.

operates at the required frequency, say 50 Hz. Typically the oscillator gives an output waveform that is not a pure sine wave; it may give a square waveform as shown in the figure. The square waves can, however, be converted into a waveform that is closer in shape to a sine wave by a *low-pass filter*. This filter consists of a combination of inductances and capacitors that reduce the harmonics of the square wave that lie above the fundamental frequency. The filtered voltage is applied to a transformer that raises the voltage amplitude to the required value, say 240 V.

Worked example

Figure 6.21(a) illustrates a simple version of an alternator that converts mechanical energy to electrical energy in the form of alternating current. The wire loop of area A rotates with constant angular frequency ω about the axis shown. The magnetic field B is uniform and

Figure 6.21 (a) A simple version of an alternator that converts mechanical energy to electrical energy in the form of alternating current. The wire loop of area A rotates with constant angular frequency ω about the axis shown. The magnetic field B is uniform and constant. (b) Side view of the alternator.

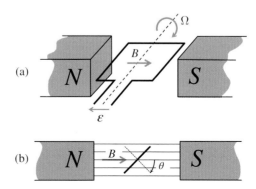

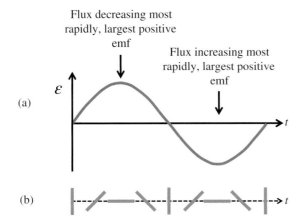

Flux decreasing most rapidly, largest positive emf

Flux increasing most rapidly, largest positive emf

(a)

(b)

Figure 6.22 (a) The variation of the induced emf, ε, with time for the alternator in Figure 6.22; (b) the corresponding orientations of the loop.

constant. Obtain an expression for the induced emf ε as a function of time t. At time $t = 0, \theta = 0$.

Solution

When the loop makes angle θ with respect to the direction of the magnetic field B, the magnetic flux ϕ through the loop is $\phi = BA\cos\theta$ (see Figure 6.21b). Hence by Faraday's law (Equation 6.1), the induced emf is

$$\varepsilon = -\frac{\mathrm{d}\phi}{\mathrm{d}t} = -\omega BA\sin\omega t.$$

Figure 6.22(a) shows the variation of ε with time and also the variation of magnetic flux through the loop. Figure 6.22(b) shows the corresponding orientations of the coil. The physical principle outlined in this example is the basis for the operation of electrical generators, in this case for an AC current generator.

6.6 Siting of a wind turbine

The siting of a wind turbine is of paramount importance in maximizing the amount and speed of the wind it receives. We recall that the power harvested by a wind turbine varies rapidly as the third power of the wind speed and, in practice, the wind must have a minimum average speed of ~5 m/s for a turbine to function effectively. Factors that influence turbine siting include: (i) the directions of prevailing and local winds; (ii) the local terrain such as hills and mountains; (iii) any local obstacles such as houses and trees; (iv) environmental issues such as noise and visual impact; and (v) what is called the *roughness* of the terrain. For example, a water surface or a flat, open landscape that is grazed by sheep has a low degree of roughness. On

Figure 6.23 Variation of wind speed, $u(z)$, with vertical height, z. Within the height of local obstacles, wind direction changes erratically and there may be large-scale fluctuations in wind speed. Above this erratic region, $u(z)$ is found to vary logarithmically with vertical height.

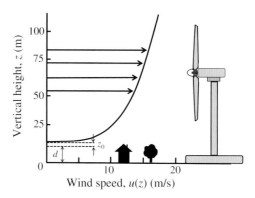

the other hand, long grass and shrubs have a high degree of roughness. In general, the more pronounced the roughness of the terrain, the more the wind will be slowed down. Offshore wind turbines benefit from the fact that the roughness of the water surface is usually low and obstacles are few.

Within the height of local obstacles, wind direction changes erratically and there may be large-scale fluctuations in wind speed. Above this erratic region the wind speed $u(z)$ is found to vary logarithmically with vertical height z, with the form

$$u(z) = V \ln \left(\frac{z - d}{z_0} \right). \tag{6.33}$$

In this expression, V is a characteristic wind speed, d is roughly equal to the height of local obstacles and z_0 is a length that characterizes the roughness of the terrain. For rough pasture z_0 may be ~ 10 mm, while for forest and woodlands it may be ~ 500 mm. Figure 6.23 is a sketch of the variation of $u(z)$ with z, together with typical values of wind speed. Clearly, the taller the turbine, the higher the wind speed. However, because of the logarithmic dependence of $u(z)$ on height, it is considered that there is not much to be gained above a height of about 100 m. Notice from this figure that the wind speed varies across the area swept out by the rotor blades and this causes forces across the turbine that must be taken into account in the design and construction of the turbine.

A *wind farm* consists of many wind turbines and indeed a large wind farm may consist of several hundred individual turbines. The Alta Wind Energy Centre (Mojave Wind Farm) in California, USA, is the largest onshore wind farm outside China, and will see the installation of 600 wind turbines supplying 1.5 GW of power. The London Array is a 175-turbine offshore wind farm located 20 km off the Kent coast in the UK and delivers 630 MW of power. It is the largest offshore wind farm in the world, and the largest wind farm in Europe by megawatt capacity (see Figure 6.24).

Figure 6.24 The London Array offshore wind farm located 20 km off the UK's Kent coast. The London Array has 175 turbines and delivers 630 MW of power. It is the largest wind farm in Europe by megawatt capacity. Courtesy of London Array Limited. http://www.londonarray.com/offshore-2/

A wind turbine produces wind turbulence downstream of the turbine and so individual turbines in a wind farm must be separated by at least five times the rotor diameter, D, to reduce the effects of this turbulence. Figure 6.25 is a schematic diagram of adjacent turbines in a wind farm, separated by five rotor diameters. We see that the area allocated to an individual turbine is proportional to D^2. However, we recall that the power extracted from the wind by a turbine is proportional to the area swept out by the rotor blades, which is also proportional to D^2 (Equation 6.28). Hence we have the important result that the power per unit area of a wind farm is independent of rotor diameter. It does, however, depend on wind speed. For wind speeds that are obtained in practice, the extracted wind power per unit area of wind farm lies within the range of approximately 1–10 W/m^2.

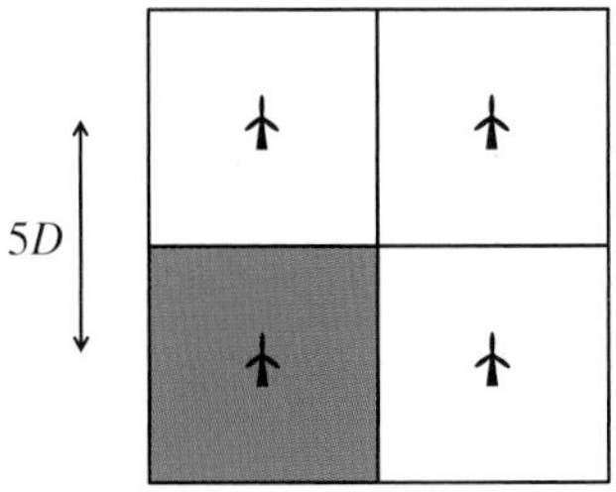

A particularly effective site for a wind turbine is on top of a hill overlooking the surrounding countryside. This gives a wide view of the prevailing wind. In addition the wind speed will increases toward the top of the hill. This effect is illustrated in Figure 6.26. The wind becomes compressed on the hillside facing the prevailing wind direction as it reaches the top of the hill, just as a fluid does in the narrow bore of a Venturi meter. This, in turn, increases the wind speed at the top of the hill.

Figure 6.25 A schematic diagram of adjacent turbines in a wind farm that are separated by $5D$, where D is the rotor diameter.

Figure 6.26 A particularly effective site for a wind turbine is on top of a hill overlooking the surrounding countryside. This gives a wide view of the prevailing wind. In addition, the wind speed will increase towards the top of the hill. The wind becomes compressed on the hillside facing the prevailing wind direction as it reaches the top of the hill. This in turn increases the wind speed.

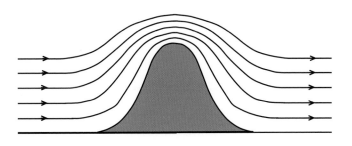

Problems 6

6.1 Each blade of a wind turbine has a length of 35 m. If the wind speed is 15 m/s and the density of the air is 1.1 kg/m^3, calculate the power generated by the turbine if it achieves 75% of the Betz criterion. What power would be generated if: (i) the density of the air is 20% higher; (ii) the blades were 20% longer; and (iii) the wind speed is 20% higher?

6.2 A ball is thrown with uniform velocity v from the centre of a platform that rotates in a counter-clockwise direction at angular velocity Ω. To an observer at the centre of the platform, the ball appears to move towards the right. Show that the ball appears to move to the right with acceleration $2v\Omega$.

6.3

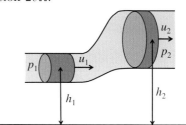

(a) The flow of a liquid through a pipe that has varying bore and height is illustrated in the figure. Use conservation of energy to show that

$$(p_1 - p_2) = \frac{1}{2}\rho\left(u_2^2 - u_1^2\right) + \rho g\left(h_2 - h_1\right).$$

where the symbols have their usual meanings. (b) Water is supplied to a house through an inlet pipe with an internal diameter of 40 mm. A pipe of internal diameter 20 mm delivers water to a bathroom 5.5 m above the inlet pipe at the rate of 4 L/min and at the pressure of 2 atm. Calculate the required pressure at the inlet pipe to the house? Take 1 atm $= 1 \times 10^5$ Pa.

6.4 Suppose the difference between the velocity u_t at the top of an aeroplane wing and the velocity u_b at the bottom of the wing can be written as $(u_t - u_b) = ku_0$, where u_0 is the velocity at which air approaches the wing and k is a constant characteristic of the wing shape and the angle of attack. Show that the lift force on the aeroplane is given by $F_L = A\rho k u_0^2$, to a good approximation when $k \ll 1$, where A is the total area of the wings and ρ is the air density. If the mass of the aeroplane is 3000 kg, $A = 40$ m^2, $\rho = 1.2$ kg/m^3 and $k = 0.1$, what is the value of u_0 when the lift force balances the force of gravity?

6.5 The power coefficient C_P is given by $C_P = 4a(1 - a)^2$. Show that the maximum value of C_P is 0.59 when $a = 1/3$.

6.6 Show that the force per unit area $F_A\ (= F_T/A_T)$ acting on a turbine is given by

$$F_A = \frac{1}{2}\rho u_0^2\left[4a\left(1 - a\right)\right],$$

where the symbols have their usual meanings. At what value of a is F_A a maximum? Calculate the maximum value of F_A for an incident wind speed of 20 m/s. In this case, what is the value of the final wind speed u_f? Compare the maximum value of F_A with the value of F_A when $a = 1/3$. Take the density of air to be 1.2 kg/m^3.

6.7 (a) A wind turbine is maintained at a tip-speed ratio of 6 for all wind speeds. At what wind speed would the blade tip speed be equal to the speed of sound? (b) An offshore wind turbine has a rotor diameter of 100 m. At what frequency (Hz) of the rotor would the tip speed be equal to the speed of sound? Why should operation at a tip speed exceeding the speed of sound be prevented?

6.8 (a) Assuming that the turbines in a wind farm need to be five rotor diameters apart, as in Figure 6.25, calculate the power that can be extracted from the wind per square metre of the wind farm at a wind speed of 8 m/s. Assume that the turbines achieve 80% of the Betz criterion and that the density of air is 1.2 kg/m^3. (b) Using your estimate from part (a), estimate the total area of a wind farm that would generate the same power as a typical nuclear power station.

6.9 The power P in an unobstructed column of air of area A travelling with wind speed u is given by $P = \frac{1}{2}\rho A u^3$. An approximate expression that is often used to determine wind speed u_z at height z is $u_z = u_s(z/h_0)^{0.14}$ m/s, where u_s is the speed at $z = 10$ m and $h_0 = 10$ m. Make a plot of wind power per unit area (P_z/A) in units of (P_s/A) against height z over the range of z from 10 to 250 m, where P_s is the wind power at $z = 10$ m. Comment on the use of turbines with small rotor diameters, say 5 m, above 100 m.

6.10 A conducting disk with radius R rotates about its central axis with constant angular frequency ω in a uniform magnetic field B. Show that the emf between the centre and rim of the disk is given by

$$\varepsilon = \frac{1}{2}\omega B R^2.$$

7

Water power

Earth has a huge amount of water – it covers about three-quarters of the planet. Interestingly, scientists are not sure how and when it got here. One theory is that the water was brought by meteorites or asteroids, both of which contain ice. We derive mechanical power from this water in various ways. This includes hydroelectric power, ocean wave power and tidal power. Water power has been exploited by mankind for thousands of years. For example, the Doomsday Book (AD 1065) records that there were more than 5000 water mills in operation in England at that time. Their main use was to grind corn into flour. Much later, water mills powered the first period of the Industrial Revolution. And although water power later gave way to steam power, it still found use for small-scale operations during the 18th and 19th centuries. In contemporary times, water power has again become significant as it provides a clean, renewable and potentially cheap energy source. In particular, hydroelectric power is now well established. Indeed, in some countries, like Norway and Brazil, hydroelectric power is the dominant source of electricity. Ocean waves and tides are relative newcomers as commercial energy sources and as yet do not provide substantial contributions to world energy demands. This is despite the fact that many hundreds of patents have been taken out on ways of harnessing wave and tidal power.

Hydroelectric power, wave power and tidal power form the three main sections of this chapter. In each case we will describe the main physical principles involved and how the energy is harnessed. This will involve a discussion of the nature of wave motion and the origin of the tides.

Physics of Energy Sources, First Edition. George C. King.
© 2018 John Wiley & Sons, Ltd. Published 2018 by John Wiley & Sons, Ltd.

7.1 Hydroelectric power

Hydroelectric power is by far the most established and widely used renewable energy source for electricity generation and accounts for ~20% of the world's electricity supply. By hydroelectric power, we mean power that is obtained from water that falls through a vertical distance from a reservoir to drive a turbine. This is in contrast to, say, water mills, which draw power from a flowing river. Hydroelectric power has important advantages. Once constructed, hydroelectric plants produce minimal pollution with considerably lower output levels of greenhouse gases than fossil fuel plants. They also have the potential to produce energy at low cost because once it is built, a hydroelectric plant will run for many years with just routine maintenance. The potential energy of the stored water in a reservoir also provides a natural store of energy that can be used to even out variations in seasonal rainfall and varying electricity demand. (In an analogous way, the large ~5000 μF electrical capacitor in the power supply of a domestic audio amplifier evens out the variation in the voltage supplied by the diodes that rectify the mains voltage, and the varying current demands of the amplifier. Not surprisingly, it is called a reservoir capacitor.) The stored energy in the reservoir of a hydroelectric plant is readily available as a turbine/electrical generator combination can be switched on within minutes to meet additional demand. Consequently, hydroelectric power can be used to supply both *base load* and *peak demand* on a national grid supply. This ability of a hydroelectric plant to store energy is taken a step further in *pumped storage*, which we describe in Section 8.4.1. Briefly, water is pumped back into the reservoir from a reservoir that is at a lower height when electricity demand is low, say during the night or at the weekend. Then, when demand increases at peak times, water is released back into the lower reservoir through a turbine.

The main disadvantages of hydroelectric power lie in its impact on the local environment. This includes social impact if there is a displacement of people from the reservoir site, loss of potentially productive land and disruption to surrounding aquatic ecosystems, both upstream and downstream of the plant. In addition, hydroelectric plants have relatively large constructional costs, although once established the cost of hydroelectricity is relatively low, as noted earlier.

7.1.1   The hydroelectric plant and its principles of operation

A schematic diagram of a hydroelectric plant is shown in Figure 7.1. There is a reservoir that collects rainwater from a large catchment area. Often the reservoir is formed from a naturally occurring

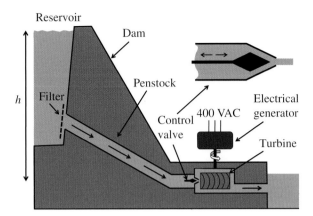

Figure 7.1 A schematic diagram of a hydroelectric plant. The water reservoir is connected to a turbine by a large pipe called a penstock. The filter keeps debris from entering the turbine, while the valve controls the water flow and hence the rotational speed of the turbine and the electrical generator. The vertical distance h between the surface of the water in the reservoir and the turbine is called the head. The turbine drives a generator that produces electrical power at an output voltage of typically 400 VAC.

geographical feature to which a dam is added to confine the water. A large pipe, called the *penstock*, delivers the water to a turbine. The vertical distance between the surface of the water in the reservoir and the turbine is called the *head*. A filter at the input of the penstock stops any debris from entering the turbine. There is also a valve at the output of the penstock that is used to control the flow of water delivered to the turbine and hence its rotational speed. The turbine drives a generator that produces electrical power at an output voltage of typically 400 VAC. This is fed to a step-up transformer (not shown), which converts the voltage to a higher value that is suitable for long-distance transmission. The ability to transport electricity over long distances is very useful in this case as hydroelectric plants are usually remote from urban centres. Although the electricity from a hydroelectric plant is usually delivered to a national grid in this way, some plants are used exclusively for a nearby manufacturing process. A good example is the electrolytic production of aluminium, which requires a lot of electrical power. For large hydroelectric plants like the Itaipu Dam, which is located on the border of Brazil and Paraguay, the output power can be large, greater than 10 GW. (The Itaipu Dam was elected as one of the seven modern wonders of the world by the American Society of Civil Engineers.) But there are also smaller plants producing ~10 MW and even micro plants that deliver ~5–100 kW, which are suitable for small communities.

It is important to note that there is no intermediate thermal process involved in hydroelectric power. The falling water drives a turbine directly. This is in contrast, for example, to a nuclear power plant where there is an intermediate thermal step in which the nuclear energy is first used to produce steam which then drives a turbine. Hence for a hydroelectric plant there is no fundamental thermodynamic limit to the conversion of water power into mechanical power. We can make the same comment about wind turbines, which again drive an electrical generator directly.

The principle of hydroelectric power is straightforward. Water falls a certain vertical distance and its potential energy is transformed into kinetic energy. This kinetic energy is then converted into mechanical energy by the turbine. Suppose that a volume per second, Q, of water falls a vertical distance h. Then the mass of water falling per second is ρQ, where ρ is the density. Then the rate of potential energy lost by the falling water, i.e. the power P, is given by

$$P = \rho Q g h, \tag{7.1}$$

where g is the acceleration due to gravity. This is the fundamental equation for hydroelectric power. Clearly, the larger the flow rate Q and the larger the vertical distance h, the greater the power that is obtained. This result assumes that there are no losses due to friction. In a hydroelectric plant, however, there is friction between the flowing water and the internal surface of the penstock. Thus the actual power that can be obtained from the falling water is less than that given by Equation (7.1), as we describe in the following section.

7.1.2   Flow of a viscous fluid in a pipe

Viscosity is the internal friction in a fluid. All real fluids have viscosity. Some, like water, have relatively low viscosity, while other fluids, such as maple syrup, have relatively high viscosity. We saw in our discussion of wind turbines that air is also a fluid and it, too, has viscosity. The viscosity of air, however, is about 50 times less than that of water. The frictional forces due to viscosity oppose the motion of one portion of the fluid relative to another. These frictional forces are significant and, indeed, the flow of a viscous fluid in a pipe is one of the most important problems in *fluid dynamics*. They arise, for example, in the circulation of blood in the human body.

The effect of viscosity is to reduce the fluid pressure as the fluid travels through a pipe. For example, it is common experience that the water pressure at the outlet of a hosepipe is lower than the water pressure at the tap; the longer the pipe the greater the loss in pressure. Figure 7.2 illustrates the flow of a viscous fluid in a pipe. It shows the velocity profile of the fluid, where the horizontal arrows indicate the magnitude of the fluid velocity at various values of radial distance r. Due to its viscosity, the velocity of the fluid in immediate contact with the walls of the pipe is zero, while the velocity is greatest at the centre of the pipe. The motion of the fluid is rather like a set of concentric tubes sliding relative to each other – the tube at the centre of the pipe moves fastest and the outermost tube is at rest. The viscous forces oppose the sliding of the tubes. The result is that the velocity profile of the fluid is parabolic in shape, as shown in Figure 7.2.

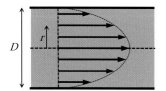

Figure 7.2 The figure represents the flow of a viscous fluid in a pipe of internal diameter D. The horizontal arrows represent the magnitude of the velocity of the fluid at various values of radial distance r. The velocity of the fluid is zero at the walls of the pipe due to frictional forces and reaches a maximum at the centre of the pipe. The motion of the fluid is rather like a set of concentric tubes sliding relative to each other; the tube at the centre of the pipe moves fastest and the outermost tube is at rest. The viscous forces oppose the sliding of the tubes. The result is that the velocity profile of the fluid is parabolic in shape.

Away from the ends of a pipe, the pressure drop per unit length is constant. Thus, neglecting the end effects, we may say that the pressure drop, Δp, along the pipe is linearly proportional to its length L, i.e. $\Delta p \propto L$. It is useful to *scale* the pipe length in terms of the pipe diameter D. We can then write

$$\Delta p \propto \left(\frac{L}{D}\right). \qquad (7.2)$$

The friction increases with fluid velocity u, just as it does for a sphere dropping through a viscous fluid, as described by Stokes' law. Hence, Δp also increases with fluid velocity. From Bernoulli's equation,

$$\frac{p}{\rho} + \frac{1}{2}u^2 = \text{constant}, \qquad (6.10)$$

we see that the quantity $\frac{1}{2}\rho u^2$ has the same dimensions as p. All of these characteristics can be expressed in a single equation:

$$\Delta p \propto \left(\frac{L}{D}\right) \frac{1}{2}\rho u^2. \qquad (7.3)$$

This is not a derivation of this important equation. However, it does illustrate the significant physical quantities involved and it also hints at the use of *dimensional analysis* in its derivation.

Dimensional analysis is widely employed in fluid mechanics, where the complexity of the situation means that it is not possible to solve the problem from first principles. Indeed, the technique is used in many fields of engineering for the same reason. A familiar example of the use of dimensional analysis is to obtain an expression for the period of a simple pendulum. If we say that the period T depends on the length l of the pendulum, the mass m of the pendulum and the acceleration due to gravity g, we have

$$T \propto l^\alpha m^\beta g^\gamma.$$

From the dimensions of these physical quantities we have

$$[T] \equiv [L]^\alpha [M]^\beta [L]^\gamma [T]^{-2\gamma}.$$

Then equating coefficients of the exponents we find:

$$\alpha = 1/2, \beta = 0 \text{ and } \gamma = -1/2.$$

Thus we obtain $T \propto \sqrt{l/g}$, as expected. Of course the dimensional analysis does not give the constant of proportionality.

We now return to Equation (7.3) and recall the general expression $p = \rho g h$, where p is the pressure due to a column of fluid of height h. Thus we can express the drop in pressure, Δp, due to friction in terms of a quantity called the *head loss*, h_f:

$$h_f = \frac{\Delta p}{\rho g}, \qquad (7.4)$$

giving

$$h_f = f_D \left(\frac{L}{D} \right) \frac{\rho u^2}{2g}. \tag{7.5}$$

The constant of proportionality, f_D, is a dimensionless constant that incorporates the fluid viscosity. It is called the *Darcy friction factor*, named after French engineer Henry Darcy and this equation is called the *Darcy–Weisbach equation*. The value of f_D for a given pipe can be obtained from empirical or theoretical relationships or from published tables.

We interpret the head loss h_f as the loss in vertical height due to frictional forces. Thus instead of using the actual head, i.e. vertical height h in Equation (7.1), we use instead the *available head*, h_a, where

$$h_a = h - h_f. \tag{7.6}$$

The value of h_f depends linearly on the length of the penstock, which connects the water reservoir to the turbine in a hydroelectric plant. Hence, the penstock should be as short as possible. In practice, the value of h_f can be kept below $0.1h$ by careful design of the hydroelectric plant.

7.1.3 Hydroelectric turbines

There are two types of hydroelectric turbine. These are the *reaction turbine* and the *impulse turbine*. A reaction turbine is totally immersed in flowing water and is powered from the water pressure drop across the turbine, in a similar manner to the way a wind turbine is powered by the air that flows through it. On the other hand, an impulse turbine is not enclosed in the water. Instead, a jet of water hits the turbine and the power is derived from the rate of loss of momentum of the water. One particular type of impulse turbine in use is called the *Pelton impulse turbine*.

The Pelton impulse turbine

A schematic diagram of a Pelton impulse turbine is shown in Figure 7.3. The turbine has a series of cups attached to a revolving wheel. The potential energy of the water in the reservoir is converted into the kinetic energy of a water jet that is directed at the cups. The jet of water strikes the cups, is deflected and the water suffers a change in momentum – hence the name impulse turbine. The resulting tangential force applied to the wheel causes it to rotate.

Figure 7.3 shows the water jet of density ρ and volume flow rate Q hitting the cup in the *laboratory frame*, i.e. in the frame of the (stationary) hydroelectric plant. The cup moves to the right with

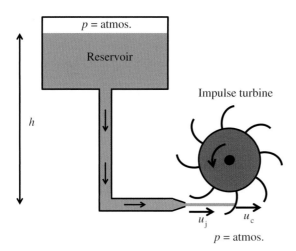

tangential velocity u_c, while the constant velocity of the jet is u_j. In the cup's frame of reference, the cup sees the water jet coming towards it with velocity $(u_j - u_c)$. The shape of the cup is designed so that the water jet is deflected through almost $180°$. Hence in the cup's frame of reference we can say that the water suffers a total change in momentum of $2(u_j - u_c)$. The change in momentum per unit time and hence the force F experienced by the cup is

$$F = 2\rho Q (u_j - u_c). \tag{7.7}$$

Hence the power transferred to the cup is

$$P = F u_c = 2\rho Q (u_j - u_c) u_c. \tag{7.8}$$

By differentiating this equation with respect to u_c, for constant u_j, we see that there is maximum power, P_{max}, when

$$\frac{u_c}{u_j} = 0.5.$$

Then substituting for u_c in Equation (7.8) we obtain for the maximum power:

$$P_{max} = \frac{1}{2}\rho Q u_j^2. \tag{7.9}$$

This is the same result we would obtain if we were to direct the water jet at a fixed wall and its velocity reduced to zero at the wall (see Equation 6.16). Thus, in this ideal case we have extracted all the power of the water jet and the turbine is 100% efficient. The potential efficiency of impulse turbines can be so high because the water is deflected away from the cups. This is in contrast to a wind turbine where the wind has to flow through the rotor and so its velocity cannot reduce to zero. In practice, large commercial impulse turbines have high efficiencies of $\sim$90%.

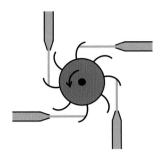

Figure 7.4 The power delivered to an impulse turbine can be increased by the use of additional water jets. Of course, the total flow of water from the jets must be less than the flow of water out of the water reservoir.

The pressures at the top of the reservoir and at the water jet are the same and equal to atmospheric pressure. And the water velocity at the top of the reservoir can be assumed to be zero as the reservoir has a large surface area. Hence from Bernoulli's equation,

$$\frac{p}{\rho} + \frac{1}{2}u^2 + gh = \text{constant}, \tag{6.11}$$

we readily obtain

$$u_j^2 = 2gh, \tag{7.10}$$

where h is the difference in vertical height between the top of the reservoir and the water jet, i.e. the head of water. This result assumes that the water has no viscosity. To take viscosity into account, we replace the head of water h by the available head, $h_a = h - h_f$ (Equation 7.6), to obtain

$$u_j^2 = 2gh_a. \tag{7.11}$$

If the nozzle has area a_j, the flow rate Q through the nozzle is equal to $a_j u_j$. Then, substituting for Q and the jet velocity u_j from Equation (7.11) into Equation (7.9), we see that the maximum obtainable power is given by

$$P_{\text{max}} = \frac{1}{2}\rho Q u_j^2 = \frac{1}{2}\rho a_j u_j^3 = \frac{1}{2}\rho a_j (2gh_a)^{3/2}. \tag{7.12}$$

The 3/2 exponent of h_a emphasises the need to maximise the available head of water.

One way to increase output power is to increase the number of jets striking the turbine cups, as illustrated by Figure 7.4. For n jets, the maximum power is

$$P_{\text{max}} = \frac{1}{2}n\rho a_j (2gh_a)^{3/2}. \tag{7.13}$$

Of course, the total flow, Q, of water through the turbine must be less than the net flow of water out of the reservoir.

The overall efficiency of a hydroelectric plant arises from the combination of the turbine efficiency, the efficiency of the electrical generator and the head loss in the penstock. The efficiencies of the turbine and generator are both typically about 90%, while the available head may be 90% of the actual head. Multiplying these factors together, we obtain an overall efficiency of 73%, which is a typical value for a large hydroelectric plant.

Worked example

The intake of the penstock of a hydroelectric plant is 50 m below the surface of the water reservoir, and the penstock has an available head of 200 m. A Pelton impulse turbine is driven by four water jets at the base of the penstock. If an electrical output power of 1.0 MW is required, what is the required flow of water and what is the radius of the jet nozzles? If the penstock has an internal diameter of 100 cm, what is the water velocity in the penstock? Assume that the efficiency of the turbine/electrical generator combination is 80%.

Solution

Using $u_j^2 = 2gh$, we obtain $u_j = \sqrt{2 \times 9.8 \times 250} = 70$ m/s.

From $P_{max} = \frac{1}{2}\rho Q u_j^2$, with $P_{max} = (1.0 \times 10^6/0.80)$ W, we have

$$Q = \frac{2 \times (1.0 \times 10^6/0.8)}{1.0 \times 10^3 \times 70^2} = 0.51 \text{ m}^3/\text{s}.$$

Then from $P_{max} = \frac{1}{2}\rho a_j u_j^3$, we obtain

$$a_j = \frac{2 \times (1.0 \times 10^6/0.8)}{1.0 \times 10^3 \times 70^3} = 7.9 \times 10^{-3} \text{ m}^2,$$

which gives a radius for each nozzle of 4.8 cm. We can find the water velocity in the penstock from the equation of continuity $a_i u_i = a_j u_j$, where a_i is the area of the penstock. Hence,

$$u_i = \frac{7.9 \times 10^{-3} \times 70}{\pi(0.50)^2} = 0.70 \text{ m/s}$$

7.2 Wave power

Anybody swimming in the sea feels the power of the waves, especially if they are drawn too far away from the coastline. Indeed, ocean waves have the potential to deliver huge amounts of energy; estimates suggest that they could provide a third of the world's energy needs. And they have significant advantages as an energy source. They provide a renewable source of energy, they occur throughout the world's oceans, and waves are, to a degree, more predictable than some other renewable energy sources. Consequently, there have been many schemes to harvest the power of water waves and a number of these schemes have been tried. However, wave power has yet to contribute a significant amount of energy to world demand. The main reason is that an ocean can be a hostile environment. Extraordinarily large waves can occur on occasions and any equipment must be able to withstand the pounding of such waves. Moreover, salt water is corrosive to the mechanical components of the equipment.

Water waves are a familiar phenomenon and have properties that are common to all waves. Indeed, when we visualise a wave we probably think of an ocean wave or a wave breaking on a beach because of their visibility and relatively low speed. Despite their familiarity, however, water waves are quite complicated, more complicated than say a wave on a plucked guitar string. Nevertheless, we can get an understanding of water waves from some basic physics and using reasonable assumptions and simplifications. And we can obtain a good estimate of the power that is delivered by an ocean wave. We begin by describing the general characteristics of wave motion, using transverse travelling waves as an example, as these are the easier to understand. We then extend our discussion to ocean waves.

7.2.1 Wave motion

When we observe a wave it is clear that something, which we may call a disturbance, travels or propagates from one region of a medium to another. This disturbance travels at a definite velocity that is usually determined by the properties of the medium. However, the medium does not travel with the wave. For example, if we tap one end of a solid metal rod, a sound wave propagates along the rod but the rod itself does not move. Rather, the particles of the rod move about their equilibrium positions to which they are bound. Similarly, when we throw a stone into the middle of a pond, the water that is disturbed by the stone does not travel to the edge of the pond and pile up there. Instead, waves propagate because adjacent particles in the medium interact with each other and pass on their energy to their neighbour which, in turn, passes the energy on to its neighbour. We may think of a Mexican wave travelling around a stadium. A spectator jumps up and down as they see their neighbour on one side do the same and this action is propagated around the stadium, but, of course, the spectators do not move away from their seats.

Waves may be either *standing waves* or *travelling waves*. The wave on a plucked guitar string is an example of a standing wave. On the other hand, solar radiation consists of electromagnetic travelling waves. As ocean waves are also travelling waves, we will focus our attention on this type of wave. (We will see an example of standing waves in water in our discussion of tidal motion in Section 7.3.2.) Waves are also characterised as being *transverse* or *longitudinal*. The wave on a guitar string is a transverse wave, where the displacement of the string is perpendicular to the wave's direction of propagation. By contrast, a sound wave is a longitudinal wave, where the disturbance travels as successive compressions and rarefactions of the air. One of the complications of water waves is that they are a combination of both transverse and longitudinal motions; particles in a water wave trace out ellipses about their equilibrium positions, as we shall see. However, both transverse and longitudinal waves are solutions of the *wave equation*, which is one of the fundamental equations in physics.

Imagine that we have a function $y = f(x)$, such as that shown in Figure 7.5. The function has a maximum at $x = x_0$. If we change the

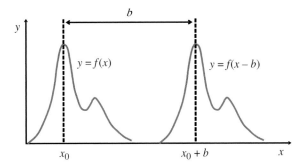

Figure 7.5 Plots of the functions $y = f(x)$ and $y = f(x - b)$. The shapes of the two functions are the same, but $y = f(x - b)$ is displaced by distance b along the positive x-axis with respect to $y = f(x)$.

variable x to, say, $(x - b)$, we obtain the function $y = f(x - b)$, which is also plotted in Figure 7.5. We see that the *shape* of this function is the same as before. We have simply displaced this shape a distance b to the right, so that now its maximum occurs at $x = (x_0 + b)$. Suppose that we now change the variable x to $(x - vt)$ where t is time and v is a constant:

$$y(x, t) = f(x - vt). \tag{7.14}$$

The value of vt increases linearly with time. Consequently, Equation (7.14) describes a *waveform* that moves in the positive x direction at a constant rate. Clearly, the product vt must have the dimensions of length and therefore v has the dimensions of length divided by time, i.e. v is a velocity. This is illustrated in Figure 7.6, which shows the waveform described by function $y = f(x - vt)$ at three successive instants of time, separated by time interval δt. The rate at which it moves is its velocity v.

We can generalise the above by saying that when a wave is travelling in the positive x direction, the dependence of the shape of the wave on x and t must be of the general form $f(x - vt)$, where f is

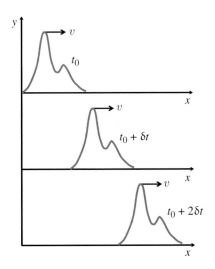

Figure 7.6 The function $y = f(x - vt)$ at three successive instants of time, separated by time interval δt. The rate at which the function moves to the right is the velocity v.

some function of $(x - vt)$. We can obtain the shape of the wave by taking a snapshot of it at a particular instant of time; for example at $t = 0$ when $y = f(x)$. It follows that a wave travelling in the negative x direction must be of the form $y = g(x + vt)$, where g is some function of $(x + vt)$. In general, then, a transverse wave has the form

$$y = f(x - vt) + g(x + vt). \tag{7.15}$$

The one-dimensional wave equation

Equation (7.15) is the general solution of the *one-dimensional wave equation*. To see this, we start with the function $f(x - vt)$ and change variables to $u = (x - vt)$ to obtain the function $f(u)$, which is a function of u only. Then

$$\frac{\partial f}{\partial x} = \frac{\mathrm{d}f}{\mathrm{d}u} \cdot \frac{\partial u}{\partial x}$$

and

$$\frac{\partial^2 f}{\partial x^2} = \frac{\partial}{\partial x}\left(\frac{\mathrm{d}f}{\mathrm{d}u} \cdot \frac{\partial u}{\partial x}\right) = \frac{\mathrm{d}^2 f}{\mathrm{d}u^2}\left(\frac{\partial u}{\partial x}\right)^2 + \frac{\mathrm{d}f}{\mathrm{d}u}\left(\frac{\partial^2 u}{\partial x^2}\right).$$

As $\partial u/\partial x = 1$ and $\partial^2 u/\partial x^2 = 0$, we have

$$\frac{\partial^2 f}{\partial x^2} = \frac{\mathrm{d}^2 f}{\mathrm{d}u^2}. \tag{7.16}$$

Similarly,

$$\frac{\partial^2 f}{\partial t^2} = v^2\frac{\mathrm{d}^2 f}{\mathrm{d}u^2}. \tag{7.17}$$

Combining Equations (7.16) and (7.17) we obtain

$$\frac{\partial^2 f}{\partial t^2} = v^2\frac{\partial^2 f}{\partial x^2}. \tag{7.18}$$

Similarly, we can readily show that $g(x + vt)$ satisfies the equation

$$\frac{\partial^2 g}{\partial t^2} = v^2\frac{\partial^2 g}{\partial x^2}. \tag{7.19}$$

It does not matter that the sign of the velocity has changed between $f(x - vt)$ and $g(x + vt)$ since only the square of the velocity occurs in Equations (7.17) and (7.19). Thus

$$\frac{\partial^2 (f + g)}{\partial t^2} = v^2\frac{\partial^2 (f + g)}{\partial x^2}$$

and hence we can write

$$\frac{\partial^2 y}{\partial t^2} = v^2\frac{\partial^2 y}{\partial x^2}. \tag{7.20}$$

This is a fundamental result. Equation (7.20) is the one-dimensional wave equation with the general solution

$$y(x,t) = f(x - vt) + g(x + vt). \qquad (7.21)$$

The wave equation (7.20) and its general solution (7.21) apply to all waves that travel in one dimension. For example, they describe sound waves in a long tube, voltage waves on a transmission line and temperature fluctuations along a metal rod. Consequently, we write the wave equation more generally as

$$\frac{\partial^2 \psi(x,t)}{\partial t^2} = v^2 \frac{\partial^2 \psi(x,t)}{\partial x^2} \qquad (7.22)$$

and its general solution as

$$\psi(x,t) = f(x - vt) + g(x + vt), \qquad (7.23)$$

where $\psi(x,t)$ represents the relevant physical quantity.

Sinusoidal travelling waves

Sinusoidal waves are important because they occur in many applications of physics and engineering and, indeed, they are used to model ocean waves. In addition, they are important because more complicated wave shapes can usually be decomposed into a combination of sinusoidal waves as described by *Fourier analysis*. So if we understand sinusoidal waves, we can understand these more complicated waves.

A travelling sinusoidal wave is illustrated in Figure 7.7. The dotted parts of the curves indicate that the wave extends a large distance in both directions. A sinusoidal wave is a repeating pattern. The length of one complete pattern is the distance between two successive maxima (crests), or between any two corresponding points. This repeat distance is the *wavelength* λ of the wave. The sinusoidal wave propagates along the x-axis and the displacement is along the y-axis, at right angles to the propagation direction. We represent this travelling sinusoidal wave by

$$y(x,t) = A \sin \frac{2\pi}{\lambda}(x - vt) \qquad (7.24)$$

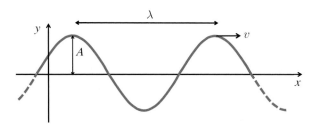

Figure 7.7 The figure illustrates the travelling sinusoidal wave $y = A \sin 2\pi[(x - vt)/\lambda]$, where λ and A are the wavelength and amplitude of the wave, respectively. The wave propagates along the x-axis and the displacement is along the y-axis, at right angles to the propagation direction. The wave travels at velocity v in the positive x-direction. The dotted parts of the curves indicate that the wave extends to large distance in both directions.

where A is the *amplitude* of the wave. The wave travels at velocity v in the positive x direction. The number of times per unit time that a wave crest passes a fixed point is the *frequency*, ν, of the wave. ν is equal to the wave velocity v divided by the wavelength λ. Hence we obtain

$$\nu\lambda = v. \tag{7.25}$$

We see that the important parameters of the wave, wavelength, frequency and velocity are related by this simple equation. The time T that a wave crest takes to travel a distance λ is equal to λ/v, i.e. the reciprocal of the frequency. Hence,

$$\nu = \frac{1}{T}, \tag{7.26}$$

where T is the *period* of the wave.

It is sometimes convenient to use alternative formulations for a travelling sinusoidal wave. Thus,

$$y(x,t) = A\sin\frac{2\pi}{\lambda}(x - vt) \equiv A\sin(kx - \omega t), \tag{7.27}$$

where $k = 2\pi/\lambda$ and $\omega = 2\pi/\lambda v = 2\pi\nu$, and hence

$$v = \frac{\omega}{k}. \tag{7.28}$$

ω is called the *angular frequency* of the wave and k is called the *wavenumber*.

If we consider the displacement of the wave at any particular point, say at $x = 0$ for the sake of simplicity, Equation (7.27) reduces to $y = A\sin(-\omega t)$. We recall that $\sin(-\theta) = -\sin\theta$. Hence, $y = -A\sin\omega t$, which represents simple harmonic motion with angular frequency ω. Hence that point, and indeed all points along the wave, undergo simple harmonic motion about their equilibrium positions at frequency ω. Each point on the wave completes one period of oscillation in time period $T = 2\pi/\omega$.

We have used sine functions to represent waves, but we can equally well use cosine functions such as

$$y(x,t) = A\cos(kx - \omega t), \tag{7.29}$$

since the cosine function is simply the sine function with a phase difference of $\pi/2$.

Transport of energy by a wave

Waves transport energy. This is apparent in sound waves that excite our ear drums and in the heat from sunlight. In mechanical waves, such as sound waves and ocean waves, the energy resides in the kinetic energy and potential energies of the particles in the medium.

The particles have kinetic energy due to their oscillating motion. They have potential energy because there is a restoring force that pulls them back to their equilibrium positions when they are displaced. As a wave propagates, it carries these energies along with it.

Worked example

The wave velocity, ν, of sinusoidal waves on a taut string is equal to $\sqrt{T/\mu}$, where T is the tension in the string and μ is the mass per unit length. Show that the total energy contained in a length λ of the string is given by

$$E_{\text{total}} = \frac{1}{2}\mu\omega^2 A^2\lambda,$$

where ω, A and λ are the angular frequency, amplitude and wavelength of the wave, respectively. At what rate is energy is transported by the wave?

Solution

We imagine the string to be divided into short segments of length δx and mass $\mu\delta x$. A segment between x_0 and $x_0 + \delta x$ is shown in Figure 7.8. As the wave travels along the string, these segments oscillate in the transverse direction and so will have kinetic energy δK given by

$$\delta K = \frac{1}{2}\mu\delta x\left(\frac{\partial y}{\partial t}\right)^2. \tag{7.30}$$

In addition, the segments will be slightly stretched when they are not at their equilibrium positions. As the string is under tension, the segments will therefore have potential energy δU. The potential energy is equal to the extension multiplied by the tension in the string. We can deduce the stretched length, δs, of the segment by applying

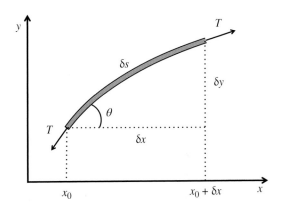

Figure 7.8 Segment of a taut string between x_0 and $x_0 + \delta x$ that carries a wave. The directions of the tension T acting on each end of the segment are indicated.

Pythagoras' theorem, $\delta s^2 \approx \delta x^2 + \delta y^2$, where $\delta y \approx \delta x \, (\partial y/\partial x)$ (see Figure 7.8). Thus the extension of the stretched segment is given by

$$\left[\delta x^2 + \left(\delta x \frac{\partial y}{\partial x}\right)^2\right]^{1/2} - \delta x = \delta x \left\{\left[1 + \left(\frac{\partial y}{\partial x}\right)^2\right]^{1/2} - 1\right\} \approx \frac{1}{2}\delta x \left(\frac{\partial y}{\partial x}\right)^2,$$

since the angle θ of the segment, given by $\partial y/\partial x$, is small. To a good approximation, the potential energy is therefore given by

$$\delta U = \frac{1}{2} T \delta x \left(\frac{\partial y}{\partial x}\right)^2. \tag{7.31}$$

For a sinusoidal wave we can write $y = A\sin(kx - \omega t)$ and we consider a length of the string equal to one wavelength λ. Figure 7.9(a) is a snapshot of the string between $x = x_0$ and $x = x_0 + \lambda$ at a particular instant of time and shows the variation of the instantaneous displacement y with distance x. Figure 7.9(b) shows the variation of the instantaneous velocity $\partial y/\partial t = -\omega A\cos(kx - \omega t)$ with x. From Equation (7.30) the kinetic energy of a segment δx of the string at position x and time t is given by

$$\delta K = \frac{1}{2}\mu \delta x \omega^2 A^2 \cos^2(kx - \omega t). \tag{7.32}$$

The resultant variation of the kinetic energy K with distance x is shown in Figure 7.9(c). The total kinetic energy contained in a wavelength λ is given by

$$K_{\text{total}} = \frac{1}{2}\mu\omega^2 A^2 \int_0^\lambda \cos^2(kx - \omega t)\mathrm{d}x. \tag{7.33}$$

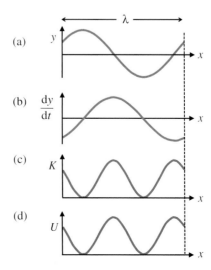

Figure 7.9 (a) Snapshot of a portion of a taut string carrying a travelling sinusoidal wave, over one complete wavelength λ. (b) Variation of the instantaneous velocity dy/dt of the wave. (c) Variation of the instantaneous kinetic energy, K. (d) Variation of the instantaneous potential energy, U.

At any given instant of time, t has a fixed value and so t is a constant in the integration of this equation. Hence,

$$\int_0^\lambda \cos^2(kx - \omega t) = \frac{\lambda}{2},$$

giving

$$K_{\text{total}} = \frac{1}{4}\mu\omega^2 A^2 \lambda. \tag{7.34}$$

Similarly, we find from Equation (7.31) that the instantaneous potential energy, δU, of a string segment at position x and time t is given by

$$\delta U = \frac{1}{2}T\delta x \left(\frac{\partial y}{\partial t}\right)^2 = \frac{1}{2}v^2\mu\delta x k^2 A^2 \cos^2(kx - \omega t),$$
$$= \frac{1}{2}\mu\delta x \omega^2 A^2 \cos^2(kx - \omega t), \tag{7.35}$$

where we have used $v = \sqrt{T/\mu}$ and $v = \omega/k$ (Equation 7.28). The variation of the instantaneous potential energy U with x is shown in Figure 7.9(d). The total potential energy is obtained by integrating Equation (7.35) over the complete wavelength. The result is

$$U_{\text{total}} = \frac{1}{4}\mu\omega^2 A^2 \lambda. \tag{7.36}$$

We see that the total kinetic energy and the total potential energy contained in a wavelength of the string are equal. This is an example of the *virial theorem* which relates the kinetic energy of a system to its potential energy. In this particular case, these two energies are equal. The total energy in a wavelength is then given by

$$E_{\text{total}} = \frac{1}{2}\mu\omega^2 A^2 \lambda. \tag{7.37}$$

The total energy varies as the square of the amplitude of the wave, which is a common feature of waves.

The time it takes for one wavelength of the wave to pass a given point is equal to λ/v. Hence the rate at which energy passes this given point, i.e. the power P, is

$$P = \frac{E_{\text{total}}}{\lambda/v} = \frac{1}{2}\mu\omega^2 A^2 v. \tag{7.38}$$

The power transmitted by the wave again depends on the square of the amplitude. Figure 7.10 illustrates (a) the displacement y and (b) the energy distribution of a sinusoidal wave travelling to the right with velocity v. This figure serves to show that the energy of contained in the wave is transported at the velocity v of the wave.

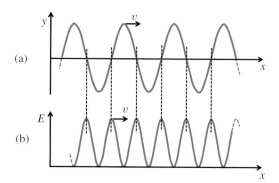

Figure 7.10 Part of a sinusoidal wave travelling on a taut string at velocity v towards the right. (a) Displacement of the wave. (b) Energy distribution in the wave. The energy is transported with the wave at velocity v.

From Equation (7.37) we find that the energy per unit length of the string is given by

$$\frac{E_{\text{total}}}{\lambda} = \frac{1}{2}\mu\omega^2 A^2, \tag{7.39}$$

and is called the *energy density*. Then from Equation (7.38) we have:

transmitted power = energy density × velocity of wave.

Dispersion of waves, phase and group velocities and wave groups

In some media, the velocity v of a transmitted wave does not depend on its wavelength, i.e. $v = \nu\lambda = \omega/k = $ constant, where the symbols have their usual meanings. Sound waves in air are an example of this. Such media are said to be *non-dispersive*. In some other media, the velocity of a wave does depend on wavelength. Perhaps the most familiar example of this is light waves in glass. Because different wavelengths travel at different velocities in glass, white light is dispersed by a glass prism into the colours of the rainbow. Such media are said to be *dispersive*. In the case of ocean waves, it is found that the velocity of waves on shallow water is independent of wavelength. However, for waves on deep water, the velocity does depend on wavelength.

In our discussion of sinusoidal waves so far, we considered the case of a *monochromatic* wave, i.e. one with a single, well defined wavelength and frequency. In most practical situations, however, we usually deal with a *group* of waves that have different wavelengths. In a storm, for example, waves of different wavelength are generated. On deep water these waves travel at different velocities; at some distance from the storm it is found that water waves of long wavelength reach an observer before waves of shorter wavelength do. And experienced sailors can use this to deduce how far the storm is away. When we deal with a group of waves containing different wavelengths, two different velocities arise. These are the *phase velocity* v_{p} and the *group velocity* v_g as we will shortly describe. The distinction between the two velocities is important because when we have a group of waves, *the energy is carried at the group velocity.* In our previous discussion of sinusoidal waves, $y(x, t) = A \sin \frac{2\pi}{\lambda}(x - vt)$, there was a single wavelength λ. In that case, v is the phase velocity and the concept of group velocity did not arise.

We illustrate the difference between phase and group velocities by considering the superposition of two sinusoidal waves, ψ_1 and of ψ_2, having slightly different frequencies that travel in a dispersive medium. The frequencies of the waves are ω_1 and ω_2, respectively, with corresponding wavenumbers k_1 and k_2:

$$\psi_1 = A \cos (k_1 x - \omega_1 t) \,, \psi_2 = A \cos (k_2 x - \omega_2 t) \,,$$

where A is the amplitude of both waves. When we sum these two waves together we obtain

$$\begin{aligned} \Psi = \psi_1 + \psi_2 = 2A \cos &\left[\frac{(k_2 - k_1)}{2} x - \frac{(\omega_2 - \omega_1)}{2} t \right] \\ \times \cos &\left[\frac{(k_2 + k_1)}{2} x - \frac{(\omega_2 + \omega_1)}{2} t \right] . \end{aligned} \tag{7.40}$$

As the medium in which the two waves travel is dispersive, they travel at different velocities given by $v_1 = \omega_1/k_1$, and $v_2 = \omega_2/k_2$, respectively. We let

$$k_0 = \frac{(k_2 + k_1)}{2}, \qquad \omega_0 = \frac{(\omega_2 + \omega_1)}{2}, \tag{7.41}$$

where k_0 and ω_0 are the mean values of the wavenumbers and frequencies, respectively. As the differences between ω_1 and ω_2 and between k_1 and k_2 are small, we write

$$\frac{(k_2 - k_1)}{2} = \Delta k, \qquad \frac{(\omega_2 - \omega_1)}{2} = \Delta \omega. \tag{7.42}$$

We can then write Equation (7.40) as

$$\Psi(x, t) = 2A(x, t) \cos [k_0 x - \omega_0 t] \,, \tag{7.43}$$

where

$$A(x, t) = 2A \cos [x \Delta k - t \Delta \omega] . \tag{7.44}$$

Equation (7.43) represents a wave $\Psi(x, t)$ that has frequency ω_0, wavenumber k_0 and velocity v_p given by

$$v_p = \frac{\omega_0}{k_0}, \tag{7.45}$$

where v_p is the *phase velocity.* This wave looks like a sinusoidal wave but one whose amplitude $A(x, t)$ is *modulated* according to Equation (7.44). This modulation forms an *envelope* that contains the sinusoidal-like wave.

Figure 7.11 The propagation of the modulated wave $\Psi(x,t)$ in a dispersive medium. The wave is shown at several successive instants of time, separated by time interval δt. The wave is shown as the solid line and is contained within the envelope of the modulation, which is represented by the dashed black lines. The vertical arrows indicate a particular crest of the wave, which travels at the phase velocity v_p. The black dots indicate a particular maximum of the envelope, which travels at group velocity v_g. In this example, $v_\mathrm{p} > v_\mathrm{g}$ and so the wave crest moves forward through the envelope as the wave propagates. This can be seen from the changing relative positions of the arrows and the bold dots.

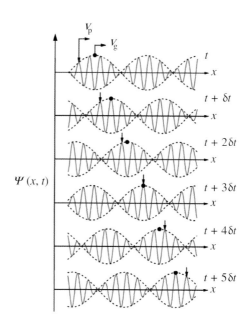

Propagation of the modulated wave is illustrated in Figure 7.11. This shows plots of $\Psi(x,t)$ at successive instants of time that are separated by time interval δt. The wave is shown as a solid line, while the envelope of the wave is represented by the dashed lines. As can be seen, the envelope travels forward with the wave but it does so at a different velocity. This can be discerned from the relative movement of a particular maximum of the wave, marked by the vertical arrows, with respect to a particular maximum of the envelope, which is marked by the black dot. While the crest of the wave travels at the phase velocity $v_\mathrm{p} = \omega_0/k_0$, the envelope travels at the *group velocity* v_g.

We can obtain an expression for the group velocity in the following way. Consider for a moment the maximum in the envelope marked with the black dot in the top waveform shown in Figure 7.11, at time t. At this point in the envelope, the function $A(x,t)$ must have its maximum value and this occurs when the cosine term $\cos\left[x\Delta k - t\Delta\omega\right] = 1$, with $\left[x\Delta k - t\Delta\omega\right] = 0$. Indeed, as the envelope moves along with the wave, it will always be the case that $x\Delta k - t\Delta\omega = 0$ at the maximum in the envelope; the values of x and t change but they always obey this equation. By differentiating this equation with respect to t and remembering that Δk and $\Delta\omega$ have fixed values, we obtain the group velocity v_g at which the maximum in the envelope, and indeed the whole envelope, travels.

$$\frac{\mathrm{d}}{\mathrm{d}t}\left[x\Delta k - t\Delta\omega = 0\right] \Rightarrow \frac{\mathrm{d}x}{\mathrm{d}t} = \frac{\Delta\omega}{\Delta k}.$$

Hence,

$$v_\mathrm{g} = \frac{\mathrm{d}x}{\mathrm{d}t} = \frac{\Delta\omega}{\Delta k} = \frac{\omega_2 - \omega_1}{k_2 - k_1}. \tag{7.46}$$

As ω is a function of wavenumber k in a dispersive medium, we write Equation (7.46) as

$$v_g = \frac{\omega_2(k) - \omega(k)}{k_2 - k_1}.$$ (7.47)

Using Taylor's theorem, we have

$$\omega(k_0 \pm \Delta k) = \omega(k_0) \pm (\Delta k) \left(\frac{d\omega}{dk}\right)_{k=k_0}$$
$$+ \text{ terms proportional to } (\Delta k)^2, (\Delta k)^3, \ldots,$$ (7.48)

where $\Delta k = (k_2 - k_1)/2$ (Equation 7.42). When Δk is small compared with k_0, we need only retain linear terms in Equation (7.48), which we can write as

$$\omega(k_0 \pm \Delta k) = \omega(k_0) \pm (\Delta k) \left(\frac{d\omega}{dk}\right)_{k=k_0}.$$ (7.49)

Substituting for k_0 and Δk from Equations (7.41) and (7.42), respectively, we obtain

$$\omega(k_2) - \omega(k_1) = (k_2 - k_1)\left(\frac{d\omega}{dk}\right)_{k=k_0},$$

and hence Equation (7.47) for the group velocity becomes

$$v_g = \left(\frac{d\omega}{dk}\right)_{k=k_0}.$$ (7.50)

We see that the group velocity is equal to the derivative of ω with respect to k, evaluated at the wavenumber k_0.

We have obtained expressions for the phase and group velocities using the example of the sum or superposition of just two single-frequency sinusoidal waves. These expressions, however, apply to any group of waves so long as their frequency range is narrow compared with their mean frequency. Thus, for the general case, we define the phase velocity, v_p as

$$v_p = \frac{\omega}{k}$$ (7.51)

and the group velocity, v_g, as

$$v_g = \frac{d\omega}{dk}.$$ (7.52)

Wave groups

We can imagine summing a large number of sinusoidal waves that cover a narrow range of angular frequencies centred about a mean frequency ω_0. In that case we would obtain a *wave group* or *wave packet*, which looks like that shown in Figure 7.12. Over a particular, narrow spatial range, all the individual sinusoidal waves are in phase

Figure 7.12 The figure shows a wave group that results from summing together a large number of sinusoidal waves that cover a narrow range of frequencies centred about a mean frequency. The figure shows the wave group at three successive instants of time, separated by time interval δt for the case where the phase velocity v_{p} is larger than the group velocity v_{g}. A crest of the wave, indicated by the arrows, travels a distance $2v_{\mathrm{p}} \times \delta t$, while the envelope travels a distance $2v_{\mathrm{g}} \times \delta t$. We can imagine crests of the wave appearing at the back (left-hand) side of the envelope, passing through the envelope and disappearing at the front of the envelope. As the amplitude is only non-zero within the spatial extent of the group, it follows that the wave energy is transported at the group velocity v_{g}.

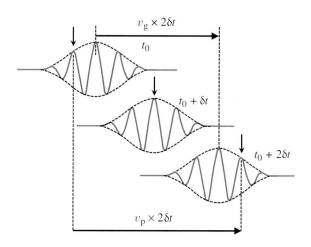

and interfere constructively. Outside this narrow range, however, the waves rapidly go out of phase and interfere destructively, reducing the wave amplitude to zero. As the individual sinusoidal waves travel along the propagation direction, they do so at slightly different velocities in a dispersive medium. The result is that the narrow spatial region over which they interfere travels at a different velocity to the individual waves. Again we can think of the resultant waveform as a sinusoidal-like wave of frequency ω_0 and corresponding wavenumber k_0 whose amplitude is heavily modulated. As before, the wave travels at the phase velocity $v_{\mathrm{p}} = \omega_0/k_0$, and the envelope containing the wave travels at the group velocity $v_{\mathrm{g}} = \mathrm{d}\omega/\mathrm{d}k$.

Figure 7.12 shows the wave group at three instants of time, separated by time interval δt, for the case where the phase velocity v_{p} is larger than the group velocity v_{g}. A crest of the wave, indicated by the arrows, travels a distance $v_{\mathrm{p}} \times 2\delta t$, while the envelope travels a distance $v_{\mathrm{g}} \times 2\delta t$; the wave crest moves through the envelope. We have seen that the energy contained in a wave is proportional to the square of its amplitude (see Equation 7.38). Looking at the wave group in Figure 7.12 we see that the amplitude is only non-zero within the spatial extent of the group as defined by its envelope. It follows that the energy is transported at the same velocity as for the envelope of the wave, i.e. at the group velocity.

Worked example

For waves on a liquid, where the wavelength is small compared with the depth of the liquid, the angular frequency ω and wavenumber k are related by the dispersion relation

$$\omega^2 = gk + \frac{Sk^3}{\rho}$$

where g is the acceleration due to gravity, and ρ and S are the density and surface tension of the liquid, respectively. (a) Deduce the ratio of the group and phase velocities for: (i) the limit of short wavelength, and (ii) the limit of long wavelength. (b) At what wavelength is the group velocity equal to the phase velocity for water? (The density and surface tension of water are 1.0×10^3 kg/m^3 and 7.2×10^{-2} N/m, respectively, and the acceleration due to gravity is 9.8 m/s^2.)

Solution

(a) Since

$$\omega = \left(gk + \frac{Sk^3}{\rho} \right)^{1/2},$$

$$v_{\mathrm{p}} = \frac{\omega}{k} = \left(\frac{g}{k} + \frac{Sk}{\rho} \right)^{1/2}.$$

(i) In the limit of short wavelength, $\lambda \to 0$ and $k \to \infty$, and

$$v_{\mathrm{p}} = \left(\frac{Sk}{\rho} \right)^{1/2} = \frac{\omega}{k}, \text{ giving } \omega = \left(\frac{Sk^3}{\rho} \right)^{1/2}.$$

Hence,

$$v_{\mathrm{g}} = \frac{\mathrm{d}\omega}{\mathrm{d}k} = \frac{3}{2} \left(\frac{Sk}{\rho} \right)^{1/2} = \frac{3}{2} v_{\mathrm{p}}.$$

(ii) In the limit of long wavelength, $k \to 0$, and

$$v_{\mathrm{p}} = \left(\frac{g}{k} \right)^{1/2}, \text{ giving } \omega = (gk)^{1/2}.$$

Hence,

$$v_{\mathrm{g}} = \frac{1}{2} \left(\frac{g}{k} \right)^{1/2} = \frac{1}{2} v_{\mathrm{p}}.$$

(b)

$$v_{\mathrm{g}} = \frac{\mathrm{d}\omega}{\mathrm{d}k} = \frac{1}{2} \left(gk + \frac{Sk^3}{\rho} \right)^{-1/2} \left(g + \frac{3Sk^2}{\rho} \right).$$

Putting $v_{\mathrm{g}} = v_{\mathrm{p}}$ from part (a) and simplifying, we obtain

$$k = \left(\frac{g\rho}{S} \right)^{1/2},$$

giving

$$\lambda = \frac{2\pi}{k} = 2\pi \left(\frac{7.2 \times 10^{-2}}{9.8 \times 1.0 \times 10^3} \right)^{1/2} = 1.7 \times 10^{-2} \text{ m.}$$

For wavelengths much greater than 17 mm, the wave motion is dominated by gravity. For wavelengths much less than this, it is dominated by surface tension. The ocean waves that are useful as a source of energy have wavelengths that are much greater than 17 mm. Consequently, we can usually neglect surface tension in our discussion of them.

7.2.2 Water waves

Physical characteristics of water waves

Water waves are generated by wind passing over the surface of the sea. On a calm sea, the wind has little grip on the water. As it moves over the water surface, however, the wind starts to form eddies and small ripples. Once the surface becomes uneven in this way, the wind has an increasing grip on it. It becomes easier for the wind to transfer its energy to the water and the ripples turn into small waves. As long as the wind velocity is greater than the velocity of the waves, there is a transfer of energy from the wind to the waves and the height of the waves steadily increases. There is, however, a limit to this energy transfer and when a wave has absorbed the maximum possible energy from the wind, it is said to be *fully developed*. The eventual height of a wave is determined by wind velocity, the duration of time the wind has been blowing, and the area of contact between the wind and the water, which is called the *fetch*. The depth and topography of the seafloor, which can focus or disperse the energy of the waves, also affect the wave height. And the greater the wave height, the greater is the energy that is contained in the wave.

One important characteristic of water waves that we have already noted is that they exhibit both transverse and longitudinal motion. Imagine that we watch a small ball that floats on the surface of deep water. What we see as a wave passes by is that the ball bobs up and down on the water. What we also see is that the ball moves forwards and backwards along the propagation direction of the wave. This is illustrated by Figure 7.13, which shows the position of the ball at nine equally spaced instants of time. The ball never leaves the surface of the water, which is shown in blue. Initially it is on the crest of the wave. As the wave moves forward, the ball does too. However, as the ball reaches the wave trough, it starts to move backwards and eventually ends up at its initial position, which coincides with the position of the following wave crest. The ball moves in a uniform circular motion in a clockwise direction, and there is no net forward movement of the ball as the wave propagates. This orbital motion of the ball is in contrast to the situation for waves on a taut string, where the motion of each particle of the string is confined to the

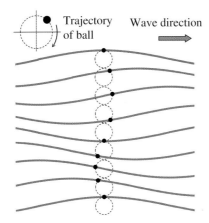

Figure 7.13 The figure illustrates the motion of a small ball floating on a water surface at nine equally spaced instants of time as a wave passes by. As the wave moves forward, so too does the ball. However, as the ball reaches the wave trough, it starts to move backwards and eventually ends up at its initial position, which coincides with the position of the following wave crest. The ball moves in a uniform circular motion in a clockwise direction as shown in the inset of the figure and there is no net forward movement of the ball as the wave propagates. Notice that the ball never leaves the surface of the water, which is depicted as the blue lines.

transverse direction. The period of the orbital motion of the ball, $T = 2\pi/\omega$, is equal to the period of the wave, $T = \lambda/v$, and hence the angular frequency, ω, of the circular motion is given by

$$\omega = \frac{2\pi v}{\lambda}, \tag{7.53}$$

where v is the wave velocity and λ is the wavelength. Ocean waves typically have wavelengths within the range of tens to hundreds of metres and periods of typically 5–20 s.

We have a similar situation for the water particles on the surface of deep water when a wave passes by. This is illustrated in Figure 7.14, which shows the motion of a number of water particles that are located on the water surface; the surface is shown in blue. The figure shows the positions of these particles at three equally spaced instants of time. We see that as a wave passes by, the water particles on the surface move clockwise in circular orbits. Again there is no net movement of the water particles in the forward direction. There is a *phase difference* between the respective motions of the water particles and as a water particle falls when a wave crest passes it by, so the next particle rises to take its place on the crest.

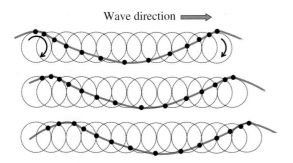

Figure 7.14 Motion of a number of water particles that are located on the water surface, which is shown in blue. The figure shows the particles at three equally spaced instants of time. All the particles move clockwise in circular orbits, but there is a phase difference between their respective motions. As a water particle falls when a wave crest passes it by, so the next particle rises to take its place on the crest.

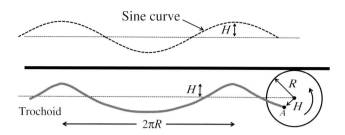

Figure 7.15 The circulating motion of particles in a water wave results in a wave profile that is described as a trochoid. A trochoidal curve can be generated as shown in this figure. Here, the circle of radius R rolls anti-clockwise beneath the black horizontal line. The point A, at a distance H from the centre of the circle, traces out a trochoid as indicated by the blue curve. As can be seen, the shape of the trochoid resembles a sinusoidal wave of wavelength $2\pi R$ and amplitude H.

The circulating motion of the water particles results in a wave profile that is called a *trochoid*. A trochoid is a curve that can be generated as illustrated in Figure 7.15. This shows a circle of radius R that rolls anti-clockwise along the heavy horizontal line, keeping in contact with the line. The point A that is at a distance H from the centre of the circle traces out a trochoid as shown by the blue curve. The distance between adjacent maxima is equal to $2\pi R$ and the amplitude of the curve is equal to H. As can be seen, the shape of the trochoid resembles a sinusoidal wave, and a sinusoidal wave of wavelength $2\pi R$ and amplitude H is shown for comparison. Water waves are often modelled as sinusoidal waves, which we will do, because these are easy to handle mathematically.

The water particles below the surface of deep water also move in uniform circular motion, as illustrated in Figure 7.16. However, the orbital radius r decreases exponentially with vertical distance z below the surface and is given by

$$r = He^{-kz} \tag{7.54}$$

where H is the amplitude of the circular wave motion at the surface and k is the wavenumber ($= 2\pi/\lambda$). Thus, by the time $z \sim \lambda/2$, the radius of the orbit has reduced almost to zero. One aspect of this rapid decrease in r is that at depths below about $\lambda/2$, the water is not appreciably affected by the motion of the waves. So, a diver below this depth in deep water would not experience any wave motion. And buoyancy tanks used to support say an offshore oil rig in deep water

Figure 7.16 In a deep-water wave, the water particles that lie below the water surface also move in a uniform circular motion, but the orbital radius decreases exponentially with vertical distance below the surface.

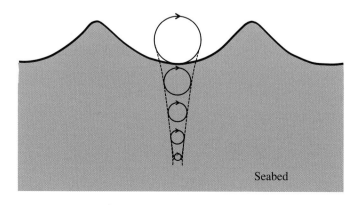

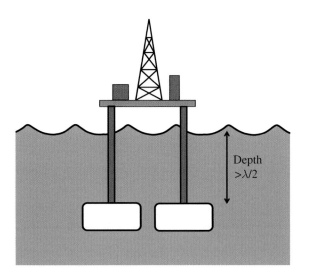

Figure 7.17 As the radius of orbiting water particles in deep water falls of exponentially, the water at depths below about $\lambda/2$ is not affected by the motion of the waves. Hence, buoyancy tanks used to support, say, an offshore oil rig in deep water are not appreciably affected by waves if their depth is greater than about half the wavelength of the waves.

(see Figure 7.17) are not appreciably affected by waves if their depth is greater than about half the wavelength of the waves.

For the case where the water is *shallow*, i.e. where the water depth $d < \lambda/2$, the water particles are affected by the presence of the seabed. Then the water particles move in elliptical orbits as shown in Figure 7.18. As the depth of a water particle increases, its vertical component of velocity decreases but its horizontal component remains the same. This explains why the ellipses become flatter towards the seabed. In this case, a diver close to the shore in shallow water, would feel the horizontal motion of the waves.

Velocity of a water wave

We have seen that the motion of the particles in a water wave is strongly influenced by the water depth d. The velocity of a water wave is also strongly influenced by water depth. We will consider two extremes: shallow-water waves and deep-water waves. Waves on shallow water, i.e. $d < \lambda/2$, are the easiest to analyse and we deal with this case first. In our discussions of both types of water waves, we make the following assumptions:

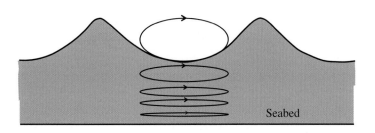

Figure 7.18 In shallow water, the water particles in a wave are affected by the presence of the sea bed. This results in the particles moving in elliptical orbits. As the depth of a water particle increases, its vertical component of velocity decreases but its horizontal component remains the same, and so the ellipses become flatter towards the sea bed.

- the water is incompressible;

- the water has no viscosity;

- any variation in the height of the water as the wave passes by is small compared with the depth of the water and small compared with the wavelength of the wave;

- the seabed is level.

Water waves on shallow water, $d < \lambda/2$ We describe shallow-water waves in terms of a one-dimensional model in which the waves propagate along the x-axis and the *wavefronts* are linear, i.e. the wave crests and troughs are linear and perpendicular to the propagation direction. Thus the height, h, of the water varies only along the x-axis. In Figure 7.19 we show the situation where the water is still and the water surface is flat. We imagine the water to be divided into thin slices, each contained between two vertical sides. These slices have width δx along the x-axis, length l along the wavefront, and height h_0, equal to the depth of the still water. The width δx of the slices shown in Figure 7.19 is exaggerated for the sake of clarity. In this model, there can be no flow of water into or out of the slices, i.e. the volume of the slices remains constant. When the surface of the water is perturbed by a wave, the slices change shape. In particular, the height, h, of a slice changes to follow the profile of the wave. And as the volume of the slice does not change, the slices will expand and contract as the wave passes by. However, the model assumes that the sides of the slice remain vertical.

Figure 7.20 illustrates the situation where a wave propagates along the water surface. It shows how an unperturbed slice of water is perturbed to follow the wave profile. We are assuming that any variation in the height of the water is small compared with the depth of the water and the wavelength of the wave. Hence, we neglect any variation in the height of the perturbed slice across its width. Then, if the two sides of the slice are displaced by ξ_1 and ξ_2, respectively, we have

$$h \left(\delta x + \xi_2 - \xi_1 \right) l = \text{constant}, \tag{7.55}$$

Figure 7.19 In considering wave motion, we imagine shallow water to be composed of thin slices of water, each contained between two vertical sides. These slices have width δx along the x-axis, length l along the wavefront, and height h_0, equal to the depth of the still water. The width δx of the slices is exaggerated for the sake of clarity. When the surface of the water is perturbed by a wave, the slices change shape to follow the profile of the wave. However, the volume of each slice remains constant.

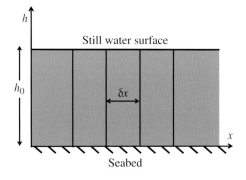

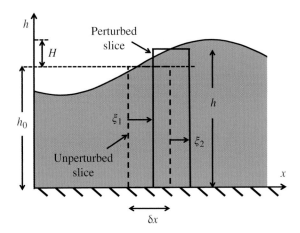

Figure 7.20 The figure illustrates a wave propagating along the surface of shallow water. It shows how a slice of the water is perturbed to follow the wave profile. The height of the slice increases to h and the width changes to $\delta x + (\xi_2 - \xi_1)$. Any variation in the height of the water is small compared with the depth of the water and the wavelength of the wave. Hence, we neglect any variation in the height of the perturbed slice across its width.

as the volume of the slice is constant. Differentiating this equation with respect to time t, we obtain

$$(\delta x + \xi_2 - \xi_1)\, l\frac{\partial h}{\partial t} + hl\left(\frac{\partial \xi_2}{\partial t} - \frac{\partial \xi_1}{\partial t}\right) = 0. \qquad (7.56)$$

We take the distances ξ_1 and ξ_2 to be small in comparison to δx and neglect them. Moreover, as we are assuming that any variation in the height of the water is small compared with the depth h_0 of the unperturbed water, we can replace h by h_0 in the second term of Equation (7.56), which then becomes

$$\frac{\partial h}{\partial t} = -h_0\left(\frac{\partial \xi_2/\partial t - \partial \xi_1/\partial t}{\delta x}\right). \qquad (7.57)$$

The derivatives $\partial \xi_1/\partial t$ and $\partial \xi_2/\partial t$ are just the velocities of the two sides of the slice, v_{x1} and v_{x2} respectively:

$$\frac{\partial h}{\partial t} = -h_0\left(\frac{v_{x2} - v_{x1}}{\delta x}\right).$$

In the limit $\delta x \to 0$, we obtain

$$\frac{\partial h}{\partial t} = -h_0\frac{\partial v_x}{\partial x}. \qquad (7.58)$$

We thus have a relationship between the *temporal* derivative of the height h of the slice and the *spatial* derivative of its horizontal velocity v_x.

We can obtain another relationship between these derivatives by considering the difference in hydrostatic pressure resulting from the variation in height of the water surface. Considering Figure 7.21, we see that the difference in pressure p across the slice is given by

$$(p_2 - p_1) = (h_2 - h_1)\,\rho g, \qquad (7.59)$$

where ρ is the density of water and g is the acceleration due to gravity. This pressure difference results in a net force on the slice

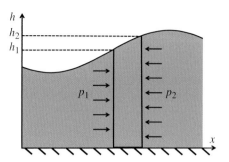

Figure 7.21 This figure illustrates the difference in hydrostatic pressure across a perturbed slice of water resulting from the variation in height of the water surface as a wave passes by.

equal to $(p_1 - p_2) h_0 l$, acting in the positive x direction. Taking the volume of the slice to be $\delta x h_0 l$, we have, from Newton's second law:

$$\rho \delta x h_0 l \frac{\partial v_x}{\partial t} = (p_1 - p_2) h_0 l = (h_1 - h_2) \rho g h_0 l. \qquad (7.60)$$

This simplifies and rearranges to

$$\frac{(h_2 - h_1)}{\delta x} g = -\frac{\partial v_x}{\partial t}. \qquad (7.61)$$

Then in the limit $\delta x \to 0$, we obtain

$$\frac{\partial h}{\partial x} g = -\frac{\partial v_x}{\partial t}. \qquad (7.62)$$

Differentiating this equation with respect to x, gives

$$\frac{\partial^2 h}{\partial x^2} g = -\frac{\partial^2 v_x}{\partial x \partial t}, \qquad (7.63)$$

and differentiating Equation (7.58) with respect to t, we obtain

$$\frac{\partial^2 h}{\partial t^2} = -h_0 \frac{\partial^2 v_x}{\partial t \partial x}. \qquad (7.64)$$

Since

$$\frac{\partial^2 v_x}{\partial t \partial x} = \frac{\partial^2 v_x}{\partial x \partial t},$$

we have

$$\frac{\partial^2 h}{\partial t^2} = g h_0 \frac{\partial h}{\partial x^2}. \qquad (7.65)$$

This is a statement of the wave equation (Equation 7.22), with phase velocity v_{p} given by

$$v_{\mathrm{p}} = \frac{\omega}{k} = \sqrt{g h_0}. \qquad (7.66)$$

We see that the phase velocity of shallow-water waves depends upon the depth of the water. For example, if the water depth is 10 m, $v_{\mathrm{p}} \approx 10 \, \mathrm{m/s}$.

We emphasise that the motion of shallow-water waves occurs predominantly in the horizontal direction and that vertical motion is relatively small by comparison. This is consistent with our picture of water particles following rather flat, elliptical orbits in shallow water (see Figure 7.18).

Water waves on deep water, $d > \lambda/2$ Water waves on deep water are more difficult to analyse, but we can use a dimensional argument to deduce an expression for the wave velocity. For this case, we suppose that the phase velocity depends on wavelength λ and on the acceleration due to gravity g, i.e.

$$v \propto \lambda^{\alpha} g^{\beta}.$$

Dimensionally, this gives

$$\left[\frac{L}{T}\right] \equiv L^{\alpha}\left[\frac{L}{T^2}\right]^{\beta}.$$

Equating coefficients for length and time, we readily obtain

$$v_{\mathrm{p}} \propto \sqrt{g\lambda}. \tag{7.67}$$

In this case, we see that the phase velocity has a strong dependence on wavelength. Further analysis shows that the constant of proportionality is $1/\sqrt{2\pi}$. Hence,

$$v_{\mathrm{p}} = \sqrt{\frac{g\lambda}{2\pi}}. \tag{7.68}$$

For example, if the wavelength is 250 m, $v_{\mathrm{p}} \approx 20\,\mathrm{m/s}$.

Energy of a water wave

In Section 7.2.1 we determined the energy in a wave on a taut string by considering the potential and kinetic energies of a segment of the string as it oscillates about its equilibrium position in simple harmonic motion. For a wave on a taut string, the tension in the string provides the restoring force on the displaced segments. For water waves, it is gravity or surface tension that provides the restoring force, although, as we have already seen, the wavelengths of the ocean waves that we are interested in are too long for them to be appreciably affected by surface tension. Again, we will deal with the two separate cases of waves on shallow water and waves on deep water.

Waves on shallow water   As before, we consider a thin slice of water that, in the absence of a wave, has width δx along the propagation direction, length l along the wavefront, and height h_0. As a wave passes by, the slice is perturbed so that its height increases from h_0

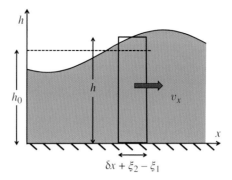

Figure 7.22 A perturbed slice of shallow water gains potential energy because its centre of mass rises from $h_0/2$ to $h/2$. It also has kinetic energy because it has horizontal velocity v_x.

to h, its width becomes $(\delta x + \xi_2 - \xi_1)$ and it has horizontal velocity v_x, as illustrated by Figure 7.22. Again, we ignore any variation in wave height across the width of the perturbed slice.

The mass of the unperturbed slice is $\rho \delta x l h_0$ and its centre of mass is at height $h_0/2$ above the seabed. Thus its potential energy with respect to the seabed is

$$\rho \delta x l h_0 \times g \times \frac{h_0}{2} = \frac{1}{2}\rho g \delta x l h_0^2,$$

where ρ is the water density. Similarly, the potential energy of the perturbed slice is

$$\frac{1}{2}\rho g \left(\delta x + \xi_2 - \xi_1\right) l h^2.$$

Again, the distances ξ_1 and ξ_2 are small in comparison to δx and we neglect them. The perturbed slice has thus gained an amount of potential energy δU given by

$$\delta U = \frac{1}{2}\rho g \delta x l (h - h_0)^2. \tag{7.69}$$

Note that if $(h - h_0)$ is a negative quantity, i.e. $h < h_0$, the potential energy must also increase. As the total volume of the water in the ocean is constant, any reduction in the height of the slice must be accompanied by a rise in the water level elsewhere and an equivalent increase in potential energy.

In order to find the total potential energy, U_{total}, that is contained within a wavelength of the wave, we integrate over all the slices in a wavelength λ. In the limit $\delta x \to 0$, this is given by the integral

$$\frac{1}{2}\rho g l \int_0^{\lambda} (h - h_0)^2 \mathrm{d}x.$$

If we model the water wave as a sinusoidal travelling wave that propagates along the water surface, we have

$$h(x,t) = H \sin(kx - \omega t) + h_0, \tag{7.70}$$

where H is the wave amplitude. Then

$$U_{\text{total}} = \frac{1}{2}\rho g l H^2 \int_0^\lambda \sin^2(kx - \omega t)\mathrm{d}x.$$

The integral has the value $\lambda/2$ and hence the total potential energy in wavelength λ over a length l of the wavefront is given by

$$U_{\text{total}} = \frac{1}{4}\rho g l H^2 \lambda, \qquad (7.71)$$

and hence the potential energy *per unit area of the seabed* is

$$\frac{U_{\text{total}}}{l\lambda} = \frac{1}{4}\rho g H^2. \qquad (7.72)$$

The slice also has kinetic energy δK as it is moving along the x-axis with velocity v_x, where

$$\delta K = \frac{1}{2}\rho \delta x l h v_x^2 = \frac{1}{2}\rho \delta x l h_0 v_x^2, \qquad (7.73)$$

to a good approximation. It is useful to obtain an expression for δK in terms of the height h of the slice so that we can compare its kinetic and potential energies. We have already obtained the following relationship for waves on shallow water:

$$\frac{\partial h}{\partial x}g = -\frac{\partial v_x}{\partial t}. \qquad (7.62)$$

The function $h(x,t)$ describing the water wave must be of the general form: $h(x,t) = h(x - v_{\text{p}}t)$, where v_{p} is the phase velocity. Then:

$$\frac{\partial h}{\partial x} = -\frac{1}{v_{\text{p}}}\frac{\partial h}{\partial t}. \qquad (7.74)$$

Substituting for $\partial h/\partial x$ from Equation (7.74) into Equation (7.62), we obtain

$$\frac{\partial v_x}{\partial t} = \frac{g}{v_{\text{p}}}\frac{\partial h}{\partial t}. \qquad (7.75)$$

Integrating this equation with respect to t, we have

$$v_x = \frac{g}{v_{\text{p}}}h + \text{constant}. \qquad (7.76)$$

For still water, when $h = h_0$, $v_x = 0$. Hence we can write

$$v_x = \frac{g}{v_{\text{p}}}(h - h_0). \qquad (7.77)$$

Substituting for v_x from this equation into Equation (7.73), we obtain the following expression for the kinetic energy δK of the slice:

$$\delta K = \frac{1}{2}\rho \delta x l h_0 \left(\frac{g}{v_{\text{p}}}\right)^2 (h - h_0)^2. \qquad (7.78)$$

We recall that the phase velocity v_p of a shallow-water wave is given by

$$v_\mathrm{p} = \sqrt{gh_0}. \tag{7.66}$$

Hence, substituting for v_p in Equation (7.78) we obtain

$$\delta K = \frac{1}{2}\rho\delta xlg(h - h_0)^2. \tag{7.79}$$

Comparing Equations (7.79) and (7.69), we see that the potential and kinetic energies of the slice are equal, just as the potential and kinetic energies of a segment of a taut string are equal when a wave propagates. Thus we can immediately say that the kinetic energy per unit area of seabed is

$$\frac{K_\mathrm{total}}{l\lambda} = \frac{1}{4}\rho gH^2. \tag{7.80}$$

The total energy per unit area of seabed E_total, i.e. the energy density, is then

$$\frac{E_\mathrm{total}}{\lambda l} = \frac{1}{2}\rho gH^2. \tag{7.81}$$

We emphasise again that the predominant motion of shallow-water waves is in the horizontal direction and that the kinetic energy resides in this horizontal motion.

Waves on deep water   In the case of waves on deep water, the water particles move in circular motion. And this must be taken into account in determining the kinetic energy of a wave. We can do this in the following way. Figure 7.23 shows an elemental volume of water in the wave that is at a vertical depth z below the still-water surface. The width, length and height of the elemental volume are dx, dl and dz, respectively; dl is measured along the wavefront. Figure 7.23 is a snapshot of the element at the instant of time when its instantaneous velocity v is pointing upwards.

Figure 7.23 The water particles in a wave on deep water follow circular paths with velocity $v = r\omega$, where r is the radius and ω is the angular frequency. The figure shows an elemental volume of water in the wave at the instant of time when its instantaneous velocity v is pointing upwards.

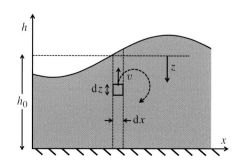

The mass of the element is $\rho \mathrm{d}l \mathrm{d}z \mathrm{d}x$, where ρ is the water density. Its kinetic energy is given by

$$\delta K = \frac{1}{2} \left(\rho \mathrm{d}l \mathrm{d}z \mathrm{d}x\right) v^2 = \frac{1}{2} \left(\rho \mathrm{d}l \mathrm{d}z \mathrm{d}x\right) (r\omega)^2 \qquad (7.82)$$

as it is moving in a circular motion with velocity $v = r\omega$, where r is the radius and ω is the angular frequency. Substituting for $r = H\mathrm{e}^{-kz}$, from Equation (7.54), we obtain

$$\delta K = \frac{1}{2} \left(\rho \mathrm{d}l \mathrm{d}z \mathrm{d}x\right) \left(H\mathrm{e}^{-kz}\omega\right)^2. \qquad (7.83)$$

To obtain the total kinetic K energy in a vertical slice of water of width $\mathrm{d}x$ we integrate this equation from $z = 0$ to $z = \infty$. (We can use the limit $z = \infty$ because we are dealing with deep-water waves where z is large and where the term e^{-2kz} decreases very rapidly with z. Then

$$K = \frac{1}{2} \rho \mathrm{d}l \mathrm{d}x H^2 \omega^2 \int_0^\infty \mathrm{e}^{-2kz} \mathrm{d}z = \frac{1}{2} \rho \mathrm{d}l \mathrm{d}x H^2 \omega^2 \times \frac{1}{2k}$$
$$= \frac{1}{4k} \rho \mathrm{d}l \mathrm{d}x H^2 \omega^2. \qquad (7.84)$$

To find the total kinetic energy in a wavelength λ of the wave along a length l of the wavefront, we integrate this equation over $\mathrm{d}x$ from $x = 0$ to λ, and over $\mathrm{d}l$ from $l = 0$ to l:

$$K_{\text{total}} = \frac{1}{4k} \rho H^2 \omega^2 \int_0^\lambda \mathrm{d}x \int_0^l \mathrm{d}l. \qquad (7.85)$$

The result is

$$K_{\text{total}} = \frac{1}{4k} \rho H^2 \omega^2 \lambda l. \qquad (7.86)$$

We have $k = 2\pi/\lambda$ and $\omega = 2\pi v_{\mathrm{p}}/\lambda$ (see Equation 7.28). The phase velocity of waves on deep water is given by $v_{\mathrm{p}} = \sqrt{g\lambda/2\pi}$ (Equation 7.68). Thus we obtain $\omega^2 = 2\pi g/\lambda$ and, substituting for k and ω^2 in Equation (7.86), we have

$$K_{\text{total}} = \frac{1}{4} \rho g H^2 \lambda l, \qquad (7.87)$$

and the kinetic energy per unit area of the seabed is

$$\frac{K_{\text{total}}}{l\lambda} = \frac{1}{4} \rho g H^2. \qquad (7.88)$$

This is the same result that we obtained for the kinetic energy density of a shallow-water wave.

The potential energy per unit area of seabed for a deep-water wave is obtained in exactly the same way as that for a wave on shallow water with the same result:

$$\frac{U}{l\lambda} = \frac{1}{4}\rho g H^2. \tag{7.72}$$

Thus the total energy E_{total} per unit area, i.e. the energy density in a deep-water wave is equal to $\frac{1}{2}\rho g H^2$, just as it is for shallow-water waves.

Power of a water wave

From Equation (7.81) the wave energy per unit wavefront, i.e. per metre, is

$$\frac{E_{\text{total}}}{l} = \frac{1}{2}\rho g \lambda H^2.$$

This result applies to both shallow- and deep-water waves. The time it takes for one wavelength of the wave to pass a given point is equal to the wavelength divided by the velocity of the wave. When we are dealing with a group of waves, as we usually are in ocean waves, the wave energy is transported at the group velocity v_{g} (see Section 7.2.1). Hence, the rate at which energy per unit wavefront passes a given point, i.e. the wave power P per unit wavefront is

$$\frac{E_{\text{total}}/l}{\lambda/v_{\text{g}}} = \frac{1}{2}\rho g H^2 v_{\text{g}}. \tag{7.89}$$

For waves on shallow water, the phase velocity $v_{\text{p}} = \omega/k = \sqrt{gh_0}$ (Equation 7.66). Then,

$$v_{\text{g}} = \frac{\mathrm{d}\omega}{\mathrm{d}k} = \sqrt{gh_0} = v_{\text{p}}, \tag{7.90}$$

and we see that the group velocity is equal to the phase velocity for shallow-water waves. Hence the wave power per unit length of wavefront is given by

$$P = \frac{1}{2}\rho H^2 \sqrt{\frac{h_0}{g^3}}. \tag{7.91}$$

Water waves on deep water, on the other hand, are dispersive with phase velocity $v_{\text{p}} = \sqrt{g\lambda/2\pi}$ (Equation 7.68), which gives

$$\frac{\omega^2}{k^2} = \frac{g}{k}. \tag{7.92}$$

Hence,

$$v_{\text{g}} = \frac{\mathrm{d}\omega}{\mathrm{d}k} = \frac{g}{2\omega} = \frac{1}{2}\frac{g}{\sqrt{gk}} = \frac{1}{2}\sqrt{\frac{g\lambda}{2\pi}} = \frac{1}{2}v_{\text{p}},$$

and in this case, the group velocity is equal to half the phase velocity. Hence the power P delivered by a deep water wave is given by

$$P = \frac{1}{2}\rho g H^2 v_{\mathrm{g}} = \frac{1}{2}\rho g H^2 \left(\frac{1}{2}\sqrt{\frac{g\lambda}{2\pi}}\right). \tag{7.93}$$

Worked example

Calculate the power in a deep-water wave of wavelength 100 m and amplitude 1.5 m, which are realistic values for Atlantic waves. Take the density of seawater to be 1030 kg/m^3.

Solution

The group velocity, v_{g}, of the wave

$$= \frac{1}{2}\sqrt{\frac{g\lambda}{2\pi}} = \frac{1}{2}\sqrt{\frac{9.8 \times 100}{2\pi}} = 6.2 \text{ m/s}.$$

$$P = \frac{1}{2}\rho g H^2 v_{\mathrm{g}} = \frac{1}{2} \times 1030 \times 9.8 \times 1.5^2 \times 6.2 = 70 \text{ kW/m}.$$

Suppose that we have a device that harvests this wave power over a wavefront of length 100 m and converts it into electricity with an efficiency of 50%. This would produce an electrical power of 3.5 MW. This is about the same as is produced by a commercial wind turbine, operating at its rated output.

We have used sinusoidal waves of amplitude H to model water waves. In this case the wave height, trough to crest, is equal to twice the amplitude, i.e. equal to $2H$. In reality, there is a distribution of wave heights. In oceanography this is taken into account by using a representative height H_{s}, which is called the *significant wave height*. It is defined as the mean height of the highest third of the waves and is a parameter that can be measured empirically. In terms of this parameter, the expression for the wave power per unit length of wavefront for deep-water waves is

$$P = \frac{1}{32}\rho g H_{\mathrm{s}}^2 \left(\frac{1}{2}\sqrt{\frac{g\lambda}{2\pi}}\right). \tag{7.94}$$

7.2.3   Wave energy converters

A machine that harnesses wave energy is known as a *wave energy converter*. A large number of different methods of wave energy conversion has been proposed and some typical methods are described below. No main design has been adopted, unlike the situation for wind power where most wind turbines have adopted the same basic design. Indeed, wave power technology is a few decades behind wind power technology. However, because of the issue of climate change, there is growing interest in wave power, which has significant advantages: it is a renewable energy source, it has the potential to deliver huge amounts of energy, and it does not produce any pollutant or greenhouse gases in operation. However, until one or two main designs are established, it seems unlikely that large-scale manufacture of wave energy convertors will begin, and at present there are no commercial-scale wave energy conversion systems in operation around the world.

The challenges for wave energy converters arise from the fact that they are usually located out to sea. There they must be able to withstand storm conditions with energies hundreds of times larger than they were designed to capture, while seawater is corrosive to mechanical components. Furthermore, the electrical power generated by the converter must be transported to land, as is also the case for offshore wind turbines. In addition, waves at a particular location vary in height, velocity and direction and so it is a challenge to find an optimum design for all sea conditions. It may be that two main designs will emerge, one for deep water and one for shallow water found closer to shore.

Wave power can be expected to be greatest in those regions where strong winds blow consistently in a particular direction. We saw in Section 6.2 that increased wind activity is found between latitudes of 30° and 60° and so the best sites for wave power are found in these regions. Important research centres for wave power are also located in these regions, and Scotland and Oregon on the north west coast of North America have important such centres.

Overtopping devices

An overtopping device is perhaps the simplest wave energy conversion system conceptually. The principle of operation is illustrated in Figure 7.24. Waves break over a sea wall, whose height is above the mean sea level, and the water is collected in a reservoir. The collected water is returned to the sea via a turbine, which drives an electrical generator in a similar way to a conventional hydroelectric plant. In order to maximise the amount of collected water, a tapered channel funnels the incoming waves into the reservoir. Friction between the waves and the sides of the channel is small and so there is insignificant dissipation of wave energy. As the width of the channel decreases, the height of the waves increases and this enables more of the waves to reach above the sea wall. Such a system, called the tapered channel (*Tapchan*) method was successfully demonstrated in Norway in 1985, and delivered a nominal electrical power of 350 kW. As there

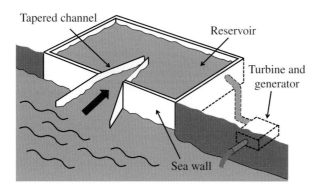

Figure 7.24 In an overtopping energy converter, waves break over the sea wall whose height is above the mean sea level and the water is collected in a reservoir. The collected water is returned to the sea via a turbine, which drives an electrical generator. In order to maximise the amount of collected water, a tapered channel funnels the incoming waves into the reservoir.

are few moving parts, maintenance costs are relatively low and in addition, the reservoir provides in-built energy storage. The main disadvantage of this system is the large capital building cost.

Oscillating water column systems

The principle of this system is illustrated in Figure 7.25 for the case of a shore-based application; analogous systems can be located offshore. The system has an enclosed air chamber that is securely mounted on the land side. The chamber has an open bottom that is submerged in the sea, while the other end is connected to a turbine. The water level in the chamber rises and falls in synchronism with the incoming waves and sets up an oscillating column of air. As a result, the air in the chamber is forced forwards and backwards through the turbine and this causes the turbine to rotate. The turbine is a *Wells turbine*, which has the important feature that it is driven in the same direction by both forward and backward air flow. The turbine is connected to a conventional generator that produces the electricity.

The channel that connects the air chamber to the turbine has a reducing cross-sectional area. This increases the air speed and gives better matching of the relatively slow motion of the water to the fast rotation of the turbine without mechanical gearing. The shape

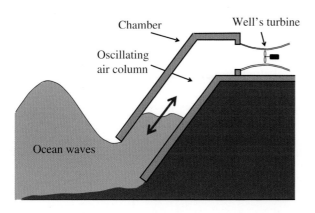

Figure 7.25 Operating principle of an oscillating water column energy converter. The water level in the chamber rises and falls in synchronism with the incoming waves and sets up an oscillating column of air. As a result, the air in the chamber is forced forwards and backwards through a Well's turbine and this causes the turbine to rotate. The turbine is connected to a conventional generator that produces the electricity.

Figure 7.26 The Edinburgh duck, also known as Salter's Duck. It floats in the water and as waves pass beneath it, the duck rocks back and forth. This rocking motion absorbs the mechanical energy in the wave. The absorbed energy is converted into electrical energy by a combination of hydraulic rams and an electrical generator that are contained within the duck.

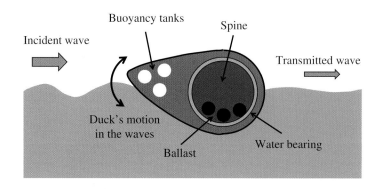

and size of the air chamber determine its frequency response to the incoming waves, i.e. resonance frequency. These parameters can be tuned to best match the frequency of the waves so that maximum power is extracted from the waves. Important advantages of such an onshore device are that the turbine is remote from the corrosive action of the seawater and there is no need for underwater electrical components or cables. An example of this type of energy converter is *Limpet*, which is installed on the Scottish Isle of Islay. It was the first commercial wave energy converter to be connected to the UK national grid and it has provided much of Islay's electricity needs since it became operational in the year 2000.

The Edinburgh Duck

The Edinburgh Duck, also known as Salter's Duck, was one of the earliest wave energy converter designs, following the oil crisis of the 1970s. The device is shown schematically in Figure 7.26. It floats in the water and as waves pass beneath it, the duck rocks back and forth – hence its name. This rocking motion absorbs the mechanical energy in the wave. The shape of the duck is critical to maximise the amount of energy absorbed and was designed by Stephen Salter at Edinburgh University. Theoretically this design has an energy conversion efficiency of 90%, although inevitably in real sea conditions the efficiency would be lower than this. The absorbed energy is converted into electrical energy by a combination of *hydraulic rams* and an electrical generator that are contained within the duck.

The original duck never went into commercial operation, partly because of the fall in oil prices in the 1980s. However, development of the original duck and duck-like devices has continued, spurred on by the increasing interest in renewable energy sources. It is envisaged that a long string of, say, 24 duck-like devices could be linked together through a central spine and moored parallel to the incoming wavefronts. Clearly, a wave energy converter captures energy from the waves. As a result the waves will be of lower height in the region

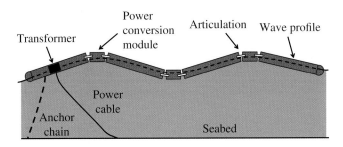

Transformer • Power conversion module • Articulation • Wave profile • Power cable • Anchor chain • Seabed

Figure 7.27 Illustration of the Pelamis energy converter, which is a semi-submerged, articulated structure. It consists of long cylindrical sections that are linked to shorter sections by hinged joints. The short sections contain the machinery that generates the electricity. Pelamis is tethered to the seabed by an anchor chain and undulates up and down with the motion of the waves, conforming to the local shape of the wave. The undulating motion of the device is converted into electrical energy by a combination of hydraulic rams and an electrical generator.

behind the wave power device, and this may have environmental effects.

Pelamis energy converter

Perhaps the most developed wave energy converter is the *Pelamis* system. It is a semi-submerged, articulated structure as illustrated schematically in Figure 7.27. This snake-like structure is 150 m long and consists of long cylindrical sections that are linked to shorter sections by hinged joints. The short sections contain the machinery that generates the electricity. Pelamis is tethered to the seabed by an anchor chain and undulates up and down with the motion of the waves, conforming to the local shape of the wave. It is partially free to rotate and can align itself with the direction of the waves for maximum absorption of energy.

As the Pelamis undulates with the waves, hydraulic pumps in the power-conversion sections drive high-pressure fluid through *hydraulic motors*, as illustrated in Figure 7.28; the action of an individual motor is illustrated schematically in Figure 7.29. The high-pressure fluid passes through the input port and travels between the gear teeth and the motor housing, before leaving via the output port. This flow of fluid causes the gears to rotate. The shaft of one of the gears is connected to an electrical generator and its output is connected to a transformer that delivers three-phase electricity at a voltage of 15 kV. This is fed down an electrical cable to the seabed and another high voltage cable transfers the electricity to shore. For a wave height of 5 m, the Pelamis can produce ~750 kW of electricity. A Pelamis has been successfully tested at the European Marine Energy Centre in Orkney, Scotland, and another has been installed off the coast of

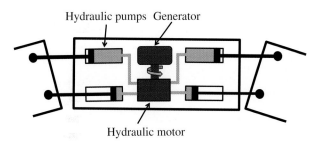

Hydraulic pumps Generator • Hydraulic motor

Figure 7.28 Schematic diagram of a hydraulic motor. As the sections of the Pelamis undulate with the action of the waves, high-pressure fluid is forced through pumps that are connected to a hydraulic motor. These hydraulic components are located in the short sections of the Pelamis.

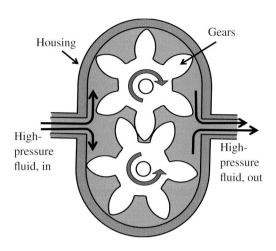

Figure 7.29 Action of a hydraulic motor. High-pressure fluid passes through the input port and travels between the gear teeth and the motor housing, before leaving via the output port. This flow of fluid causes the gears to rotate. The shaft of one of the gears is connected to an electrical generator.

Portugal. A photograph of a Pelamis wave energy converter at sea is shown in Figure 7.30.

7.3 Tidal power

The daily rising and falling of the sea level is a familiar sight. Close to a coastline, the difference in height between low tide and high tide, called the *range*, can be considerable, ~ several metres or more. Clearly this involves huge changes in the potential energy of the

Figure 7.30 The Pelamis wave energy converter at sea. Courtesy of Umanji Solar. http://buildipedia.com/aec -pros/public-infrastructure/ pelamis-wave-energy- converter-renewable-energy- from-ocean-waves

water. If the water at high tide is trapped behind a dam or barrier, this potential energy can be harnessed in an analogous way to a hydroelectric plant. Tides may also result in the flow of water through narrow channels that connect large areas of water. Such channels occur, for example, between the mainland and a nearby island. These tidal *currents* may reach speeds of ~5 m/s and provide an alternative way to harvest tidal power. This power can be harnessed in a similar manner to the way wind turbines harness wind power although, of course, the water turbines are located under water.

We remark straight away that tidal motions are complicated and so are difficult to predict. This is not least because of the topography of a particular location, for example the shape of the seabed close to a coastline or the occurrence of estuaries or bays. However, detailed observations have been made of tidal motion over many decades at many locations because knowledge of local tides is essential for the navigation of shipping. And it is these long-term observations that are the basis upon which tide motion can now be successfully predicted. Thus any variations in output power from a tidal power plant can be foreseen and taken into account when delivering power to a national grid. Tidal power has other important advantages. It is a renewable energy source and it has the potential to deliver enormous amounts of energy. It is estimated that the sites of greatest potential throughout the world could provide ~120 GW of power and that, in principle, 25% of the UK's energy needs could be supplied by tidal power. The main drawback of tidal power has to do with the very high costs involved in building the power plants. Moreover, the sites for these plants have very specific requirements with regard to their topography, and the availability of suitable sites is limited. An additional issue is the large ecological impact that a tidal power plant has on the local environment. For these reasons, only a few tidal power plants have been constructed so far. However, further plants are being proposed including one to be sited in the Bristol Channel, UK. In this section we present simplified models to describe the origin and behaviour of the tides. We also describe ways in which tidal power can be harnessed.

7.3.1   Origin of the tides

It is well known that the tides are caused by gravitational forces that the Moon and Sun exert on the Earth and its oceans. Indeed the link between the tides and the motion of the Moon was recognised in ancient times. But why does the Moon play the dominant role even though the Sun's gravity is almost 200 times greater? And why are there two tides each day? It was Newton who first successfully described the origin of the tides in his *Principia* and it was a great

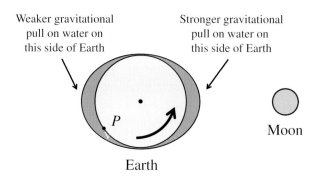

Figure 7.31 The Earth and the Moon mutually attract each other as described by the universal law of gravity. The gravitational pull on the water on the side facing the Moon is greater than the gravitational pull on the Earth because it is closer to the Moon. As the water is fluid, it flows to produce a bulge of water on the side facing the Moon. The gravitational pull on the water on the far side of the Earth is less than the gravitational pull on the Earth and so water flows to produce a bulge of water on that side too. Then as the Earth revolves about its axis, a given geographical location on Earth, say point P, experiences two high tides and two low tides each day.

victory for his universal law of gravity to provide the answers to these questions. Since then, many scientists have contributed to a more detailed understanding of tidal motions, including George Airy, Pierre-Simon Laplace, George Darwin (son of Charles Darwin) and Lord Kelvin.

The gravitation force between two spheres is the same as if the mass of the each sphere were concentrated at its centre. However, the gravitational force exerted by sphere A upon sphere B is not constant over the surface of sphere B. So, the occurrence of the tides is not simply the gravitational attraction of the Moon and the Sun; *the tides result from the way in which the gravitational attraction of the Moon and the Sun vary over the surface of the Earth.* All particles on Earth are attracted to the Moon and Sun by gravity, but some are attracted more and some are attracted less, depending on their location.

Figure 7.31 shows schematically the Earth–Moon system, as it is the Moon that dominates tidal motion. The Earth and the Moon mutually attract each other as described by the universal law of gravity. The gravitational pull on the water on the near side to the Moon is greater than the gravitational pull on the Earth because it is closer to the Moon. Because the water is fluid, it flows to produce a bulge of water on the side facing the Moon. The gravitational pull on the water on the far side of the Earth is less than the gravitational pull on the Earth and so water flows to produce a bulge of water on that side too. It is not that the water is pushed away from the Moon, but that the gravitational pull is less. Then as the Earth revolves about its axis, a given geographical location on Earth, say point P in Figure 7.31, experiences two high tides and two low tides each day.

To be more quantitative about the variation in the gravitational pull of the Moon, we consider its gravitational attraction on two identical particles of mass m that are on opposite sides of Earth, as illustrated in Figure 7.32. We are only interested in the gravitational attraction of the Moon on these two particles and so we ignore the gravitational attraction that the Earth exerts on them. Mass m is

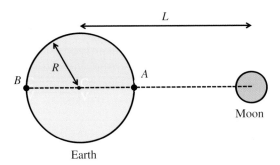

Figure 7.32 A particle on the side of the Earth closest to the Moon experiences a larger gravitational force of attraction from the Moon than an identical particle on the far-side. The difference in these gravitational forces depends on the cube of the distance L.

small, say 1 kg, and so the gravitational force of attraction between the two particles can also be ignored. The Moon's gravitational force of attraction F_A on the particle at point A is given by

$$F_A = G\frac{mM_{\text{Moon}}}{(L - R)^2}, \tag{7.95}$$

where G is the gravitational constant, M_{Moon} is the mass of the Moon, R is the radius of the Earth and L is the distance between the centres of the Earth and the Moon. The gravitational force of attraction F_B of the Moon on the particle at point B is given by

$$F_B = G\frac{mM_{\text{Moon}}}{(L + R)^2}. \tag{7.96}$$

The difference between the two is

$$F_A - F_B = G\frac{mM_{\text{Moon}}}{(L - R)^2} - G\frac{mM_{\text{Moon}}}{(L + R)^2}, \tag{7.97}$$

which we can write as

$$F_A - F_B = \frac{GmM_{\text{Moon}}}{L^2}\left[\left(1 - \frac{R}{L}\right)^{-2} - \left(1 + \frac{R}{L}\right)^{-2}\right].$$

As $R \ll L$, this reduces to

$$F_A - F_B \approx \frac{GmM_{\text{Moon}}}{L^2}\left[\left(1 + \frac{2R}{L}\right) - \left(1 - \frac{2R}{L}\right)\right] = 4GRm\left(\frac{M_{\text{Moon}}}{L^3}\right). \tag{7.98}$$

We see that the difference in the Moon's gravitational attraction on the two particles depends on the *cube* of the distance L. This inverse-cubic dependence on distance rather than the more familiar inverse-square dependence arises because we are dealing with a *difference* between two forces. The inverse-cubic dependence explains why the Moon has a greater influence on the tides than the Sun. Although the Sun is much more massive than the Moon, the *variation* in its gravitational field across the Earth is smaller.

Worked example

Determine the relative influence of the Moon and the Sun on the Earth's tides. The masses of the Sun and Moon are 2.0×10^{30} and 7.3×10^{22} kg, respectively, the Sun–Earth distance is 1.5×10^8 km and the Moon–Earth distance is 3.8×10^5 km.

Solution

Inspection of Equation (7.98) shows that the gravitational influence of the Moon or Sun scales as M/L^3. Hence, the relative influences are given by the ratio

$$\frac{M_{\text{Moon}} \times L_{\text{Sun}}^3}{M_{\text{Sun}} \times L_{\text{Moon}}^3} = \frac{7.3 \times 10^{22} \times \left(1.5 \times 10^8\right)^3}{2.0 \times 10^{30} \times \left(3.8 \times 10^5\right)^3} = 2.2.$$

The Sun does influence the Earth's tides, but we see that the influence of the Moon is just over twice as strong. Of course, the Sun's gravitational force on an object on the Earth is very much greater than that of the Moon. Gravitational forces scale as M/L^2. Thus the strengths of the gravitational force of the Moon and Sun are in the ratio

$$\frac{M_{\text{Moon}} \times L_{\text{Sun}}^2}{M_{\text{Sun}} \times L_{\text{Moon}}^2} = \frac{7.3 \times 10^{22} \times \left(1.5 \times 10^8\right)^2}{2.0 \times 10^{30} \times \left(3.8 \times 10^5\right)^2} = 5.7 \times 10^{-3}.$$

We see that the gravitational strength of the Moon is a factor of 5.7×10^{-3} less than that of the Sun.

Tidal force

We have seen that the gravitational force of the Moon is not constant over the surface of the Earth, the nearest side is attracted more strongly than the furthest side. Such differences in gravitational attraction give rise to the so-called *tidal force*. The tidal force is not of itself a force but rather it is a *differential* force, being the difference between the gravitational force acting at different locations. It is the tidal force that is responsible for the tides. For our discussion of the tides, we define the tidal force acting on a particle at a given location as *the vector difference between the gravitational force acting on the particle and the gravitational force acting on an identical particle located at the Earth's centre.*

We can explain this result as follows. For a particle located at the centre of the Earth, the Moon's gravitational force of attraction is exactly balanced by the centrifugal force the particle experiences. Hence there is no net force tending to pull or push the particle away

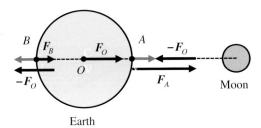

Earth

Figure 7.33 The figure shows the positions of three identical particles, one at the Earth's centre and the others at points A and B on opposite sides of the Earth. The Moon's gravitational forces, F_0, F_A and F_B, that act on these particles are represented by black arrows. The tidal force acting on a particle at a given point is equal to the vector difference between the gravitational force exerted on the particle and the gravitational force, F_0, exerted on an identical particle at the Earth's centre. The resulting tidal forces at points A and B are indicated by the blue arrows and point away from Earth's centre.

from this location. *All* identical particles on or in the Earth experience the same centrifugal force; the same in magnitude and the same in direction. This direction lies parallel to the line connecting the Earth and Moon centres. However, the Moon's gravitational force of attraction varies over the Earth. Consequently, for particles at locations other than the Earth's centre, the Moon's gravitational pull is not balanced by the centrifugal force with the result that the particle will experience a net force that pushes or pulls it. The net force is the vector addition of the centrifugal force and Moon's gravitational force of attraction. Those particles that can move will do so and consequently the waters of the Earth flow to produce the tidal bulges. Since the centrifugal force and the gravitational force acting on a particle at the Earth's centre are equal and opposite, it follows that the net force, i.e. the tidal force, experienced by an identical particle anywhere else is equal to the vector difference between the gravitational force it experiences and the gravitational force experienced by the particle at the Earth's centre. We now use this result to deduce the tidal force at different locations on Earth and we will see how the waters of the Earth move under the influence of the tidal force to produce the two tidal bulges we see in Figure 7.31.

Figure 7.33 shows the positions of three identical particles, one at the Earth's centre and the others at points A and B on opposite sides of the Earth. The gravitational forces, F_O, F_A and F_B, that act on these particles are represented by black arrows. These arrows point towards the centre of the Moon, of course. The arrow at point A is longer than the one at point B, indicating the relative magnitudes of the two forces. The particle at the centre of the Earth experiences an attractive force of magnitude F_O, which is essentially the mean of the magnitudes of the forces at points A and B, a result we can readily obtain using the previous analysis. The tidal force acting on the particle at A is the vector difference between F_O and F_A, as indicated by the figure; this tidal force is shown as a blue arrow on Figure 7.33. We see that tidal force at A points away from the Earth's centre. The same procedure gives the tidal force at B, which is also represented by a blue arrow. At that point, the tidal force again points away from the Earth's centre.

We can extend this procedure to obtain the tidal force at an arbitrary point P on the Earth's surface. For this we adopt the

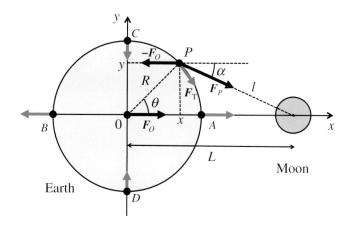

Figure 7.34 The figure shows the construction to obtain the tidal force acting on a particle at an arbitrary point P on the Earth's surface. $\boldsymbol{F}_P$ is the gravitational force acting on the particle, while $\boldsymbol{F}_O$ is the gravitational force acting on a particle at the Earth's centre. The blue arrow represents the resultant tidal force $\boldsymbol{F}_T$.

representation shown in Figure 7.34. In this representation, point P has coordinates (x, y), with $x = R\cos\theta$ and $y = R\sin\theta$ with respect to the Earth's centre. Again we compare the Moon's force of attraction on identical particles at P and at the Earth's centre. The magnitude F_O of the gravitation force on a particle at the centre of the Earth is given by

$$F_O = G\frac{mM_{\mathrm{Moon}}}{L^2}, \tag{7.99}$$

where m is the mass of the particle. The magnitude F_P of the gravitation force, on the particle at P is given by

$$F_P = G\frac{mM_{\mathrm{Moon}}}{l^2}, \tag{7.100}$$

where l is the distance of P to the Moon's centre. The component of this force in the x-direction is

$$G\frac{mM_{\mathrm{Moon}}}{l^2}\cos\alpha, \tag{7.101}$$

where the angle α is defined in Figure 7.34. Hence the tidal force $\boldsymbol{F}_T$ at P has a component $\boldsymbol{F}_{T,x}$ in the x-direction that is given by

$$\begin{aligned}
F_{T,x} &= \left(G\frac{mM_{\mathrm{Moon}}}{l^2}\cos\alpha\right) - \left(G\frac{mM_{\mathrm{Moon}}}{L^2}\right) \\
&= GmM_{\mathrm{Moon}}\left(\frac{\cos\alpha}{l^2} - \frac{1}{L^2}\right).
\end{aligned} \tag{7.102}$$

We have

$$\cos\alpha = \frac{(L - x)}{l} = \frac{L}{l} - \frac{x}{l} \approx 1,$$

since $L \approx l$ and $x \ll l$. Hence,

$$F_{T,x} \approx GmM_{\mathrm{Moon}}\left\{\frac{1}{l^2} - \frac{1}{L^2}\right\}. \tag{7.103}$$

$$l^2 = (L - x)^2 + y^2 = (L - x)^2 + R^2 - x^2.$$

This rearranges to give:

$$l^2 = L^2 \left[1 + \frac{R^2}{L^2} - \frac{2x}{L}\right] \approx L^2 \left[1 - \frac{2x}{L}\right], \qquad (7.104)$$

as $R^2/L^2 \ll 2x/L$. Then, because $x \ll L$, we can write

$$\frac{1}{l^2} \approx \frac{(1 + 2x/L)}{L^2}.$$

Substituting for l^2 in Equation (7.103), we obtain

$$F_{T,x} \approx \frac{2GmM_{\text{Moon}}}{L^3} x = \frac{2GmM_{\text{Moon}}}{L^3} R \cos \theta. \qquad (7.105)$$

The y-component of the gravitation force $\boldsymbol{F}_P$ on the particle at P is

$$- G\frac{mM_{\text{Moon}}}{l^2} \sin \alpha. \qquad (7.106)$$

The gravitational force $\boldsymbol{F}_O$ acting on the particle at the Earth's centre has no component in the y-direction and so the y-component of the tidal force at P is simply given by

$$F_{T,y} = -G\frac{mM_{\text{Moon}}}{l^2} \sin \alpha. \qquad (7.107)$$

As $\sin\alpha$ is very small, $\sin \alpha \approx \tan \alpha \approx y/(L - x) \approx y/L$. Hence, taking $l \approx L$, we obtain

$$F_{T,y} = -G\frac{mM_{\text{Moon}}}{L^3} y = -G\frac{mM_{\text{Moon}}}{L^3} R \sin \theta. \qquad (7.108)$$

Hence the x- and y-components of the tidal force at point P are given by

$$F_{T,x} = \frac{2GmM_{\text{Moon}}}{L^3} R \cos \theta; \qquad F_{T,y} = -G\frac{mM_{\text{Moon}}}{L^3} R \sin \theta,$$

within the approximations we have made.

At point A on Figure 7.34, $\theta = 0$. Hence,

$$F_{T,x} = \frac{2GmM_{\text{Moon}}}{L^3} R, \quad \text{and} \quad F_{T,y} = 0.$$

Thus the magnitude of the tidal force is

$$\frac{2GmM_{\text{Moon}} R}{L^3}$$

and the force points away from the Earth's centre, as represented by the blue arrow at that point. At point B, with $\theta = \pi$, the magnitude of the tidal force is also

$$\frac{2GmM_{\text{Moon}} R}{L^3}$$

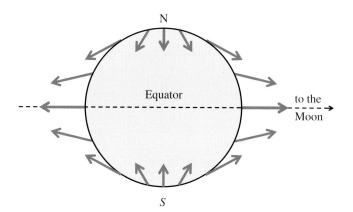

Figure 7.35 The blue arrows represent the direction and relative strength of the tidal force at various places on the Earth's surface.

and again the force points away from the Earth's centre, as $\cos\left(\pi\right) = -1$. At point C, $\theta = \pi/2$. Hence, $F_{\mathrm{T},x} = 0$ and

$$F_{\mathrm{T},y} = -G\frac{mM_{\mathrm{Moon}}}{L^3}R.$$

We see that the tidal force has a magnitude that is half its value at A and B and is directed inwards. At point D, $\theta = 3\pi/2$. The tidal force has the same magnitude as at C, but points in the opposite direction, again directed inwards.

Using Equations (7.105) and (7.108), we can obtain the strength and direction of the tidal force for all positions on the surface of the Earth. When we do this we obtain the diagram shown in Figure 7.35, where the blue arrows indicate the relative magnitude and direction of the tidal force at various places on the Earth's surface.

We model the Earth as a sphere of solid matter that is covered by a relatively thin layer of water; the average depth of water in the oceans is ∼4000 m, which is less than 1% of the Earth's radius. The tidal force has relatively little effect on the solid Earth, but the water can readily flow under the influence of the force. The tidal force, illustrated by the blue arrows in Figure 7.35, has a component parallel to the Earth's surface and a component normal to the surface. It is the parallel component that produces the tidal bulges. This component 'pushes' the water towards either side of the Earth, forming the two water bulges seen in Figure 7.31. The perpendicular component of the tidal force has little effect on the water. We can see this as follows. From Equations (7.105) and (7.108), the strength of the tidal force F_{T} on a particle of mass m is

$$\sim\frac{GmM_{\mathrm{Moon}}}{L^3}R,$$

while the strength of the gravitational force F_g on an identical particle at the Earth's surface is

$$\sim \frac{GmM_{\text{Earth}}}{R^2}.$$

Hence, their relative strengths are given by

$$\frac{F_T}{F_g} \sim \frac{GmM_{\text{Moon}}}{GmM_{\text{Earth}}L^3}R^3 \sim \frac{M_{\text{Moon}}}{M_{\text{Earth}}}\left(\frac{R}{L}\right)^3.$$

Putting in appropriate values:

$$\frac{F_T}{F_g} \sim \frac{7.34 \times 10^{22}}{5.98 \times 10^{24}}\left(\frac{6.37 \times 10^6}{3.84 \times 10^8}\right)^3 \sim 6 \times 10^{-8}.$$

We see that the tidal force is far too weak compared with the Earth's gravitational attraction for it to 'raise' the level of the oceans. On the other hand, the Earth's gravitation force does not have a component parallel to the Earth's surface and so the parallel component of the tidal force is effective in creating the water bulges.

Worked example

Deduce the height, h, between high tide and low tide in the middle of an ocean.

Solution

The height h is the difference in water levels between points C and A in Figure 7.34. We can deduce this height by considering the work done by the tidal force in moving a particle of water from C to A. The work dW is given by

$$dW = F_{T,x}dx + F_{T,y}dy,$$

where $F_{T,x}$ and $F_{T,y}$ are the x- and y-components of the tidal force. Hence, substituting for $F_{T,x}$ and $F_{T,y}$ from Equations (7.105) and (7.108), we obtain

$$dW = G\frac{mM_{\text{Moon}}}{L^3}\left(2xdx - ydy\right),$$

where m is the mass of the water particle. x varies from 0 to R and y varies from R to 0. Hence total work done by the tidal force is given by

$$W = G\frac{mM_{\text{Moon}}}{L^3}\left[\int_0^R 2xdx - \int_R^0 ydy\right],$$

giving

$$W = \frac{3GmM_{\text{Moon}}}{2L^3}R^2.$$

The particle gains gravitational potential energy equal to mgh. Setting this equal to W the work done, we obtain

$$h = \frac{3GM_{\text{Moon}}}{2gL^3}R^2.$$

Taking appropriate values of the quantities in this equation, gives

$$h = \frac{3 \times 6.67 \times 10^{-11} \times 7.34 \times 10^{22} \times (6.37 \times 10^6)^2}{2 \times 9.8 \times (3.84 \times 10^8)^3} = 0.54 \text{ m}.$$

This value of 0.54 m is surprisingly small, compared with tidal ranges that are observed at a coastline. In Section 7.3.2 we will see how tidal range can be enhanced by the topography of a coastline.

The reader may expect tidal forces to distort the shape of the solid Earth: to stretch the planet along the line connecting the Earth and Moon. They do, but because of the Earth's rigidity, the extent of such distortions is small compared with the tidal range of an ocean. The reader may also expect that the rotation of the Earth about its own axis would cause it to bulge around the equator. It does. In fact the Earth is described as an oblate spheroid, where the distance from Earth's centre to sea level is about 20 km greater at the equator, the polar radius being approximately 0.3% shorter than the equatorial radius. However this is a constant effect that is independent of the motion of the Moon or Sun. It simply forms the underlying shape of the Earth's surface upon which the oceans move.

Apart from being responsible for the Earth's tides, tidal forces also arise in many astronomical situations. Their effects are evident where the gravitational force of attraction due to a large body varies across the spatial extent of another body, causing that body to become distorted in shape or even to pull apart. A particular example is a moon of Jupiter called Io. The mass of Jupiter is more than 300 times the mass of Earth and it exerts a correspondingly large gravitational pull. Io is the moon of Jupiter that lies closest to the planet. Jupiter, in conjunction with its other moons, exerts huge tidal forces on Io, causing it to be continuously distorted, rather like a squash ball being rapidly squeezed and released. These deformations generate friction and enough heat to give Io a molten interior. As a result, molten lava spews out of numerous volcanoes on its surface, making Io the most volcanically active world in the solar system. This volcanic activity was revealed by the high-resolution images of Io received from the *Voyager* spacecraft in 1979. The gravitational

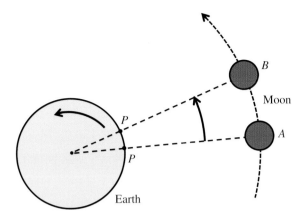

Figure 7.36 This figure shows the view of the Earth–Moon system from Polaris, the north star. From this direction the Earth rotates about its axis in an anti-clockwise direction, while the Moon orbits the Earth also in an anti-clockwise direction. Initially, point P on earth is directly below the Moon, which is at position A. After one full revolution of the Earth about its axis, the Moon has moved from A to B. So the Earth must then rotate a further amount until point P is once again directly 'under' the Moon. It takes the Earth about 50 minutes to catch up with the Moon.

pull of Jupiter also led to the break-up of comet Shoemaker-Levy 9. In 1992, the comet passed close to Jupiter and the tidal force exerted by the planet on the comet caused it to pull apart into more than 20 fragments. Then in 1994, these fragments spectacularly collided with Jupiter, providing the first direct observation of an extraterrestrial collision of solar-system objects.

Period of lunar tides

We have seen that at most geographical locations there are two high tides each day. This suggests that high tides are separated by 12 hr. In fact they are separated by 12 hr 25 min. This is because, as well as the Earth rotating about its own axis, the Moon moves in its orbit around the Earth. This is illustrated schematically in Figure 7.36, which is the view of the Earth–Moon system from Polaris, the north star. From this direction the Earth rotates about its own axis in an anti-clockwise direction, while the Moon orbits the Earth also in an anti-clockwise direction. The figure shows two positions of the Moon, A and B. Initially point P on Earth is directly below the Moon, which is at position A. After one full revolution of the Earth about its axis, which takes 24 hr, the Moon has moved from A to B. Then it takes the Earth about 50 minutes to catch up with the Moon so that point P is once again directly 'under' the Moon. As there are two high tides per day, it follows that they are separated by approximately 12 hr 25 min.

7.3.2   Variation and enhancement of tidal range

At the beginning of our discussion of the tides, we emphasised that tidal motions are complicated. Not surprisingly the tidal range at a particular location is invariably very different from that predicted for the middle of an ocean. Indeed, the tidal range at a coastal location may be an order of magnitude or more larger, say ~5m. There are various additional factors that may affect tidal range. These include:

(i) the Moon is not usually in the equatorial plane of the Earth;

(ii) the Sun's gravitational attraction, which we have so far neglected, also influences the tides;

(iii) the tidal bulges do not align with the Earth–Moon axis;

(iv) the topography of the coastal location – this is usually the most important factor of all.

We deal with these factors in turn. So far it has been convenient to assume that the Moon is situated directly above the Earth's equator. However, this is usually not the case. Firstly the orbit of the Moon is inclined at an angle of 5° with respect to the Earth's equatorial plane – the plane of the Earth's orbit about the Sun. And more importantly, the axis of the Earth's daily rotation is inclined at an angle of 23.5° with respect to its equatorial plane. In Section 4.3.2 we considered how the inclination of the Earth's axis leads to the occurrence of the seasons (see Figure 4.15). We defined the Sun's angle of declination as the angle between the Earth's equatorial plane and a line drawn from the centre of the Earth to the centre of the Sun. We saw how this angle varied from +23.5° at the summer solstice to −23.5° at the winter solstice. Similarly, the Moon has an angle of declination δ with respect to the Earth's equatorial plane, which changes continuously as it orbits the Earth. If, for the sake of simplicity, we neglect the relatively small angle that the Moon's orbit makes with the Earth's orbit, the angle δ varies continuously from +23.5° to −23.5° during the ~28 day cycle of the Moon's orbit.

Figure 7.37 shows the Earth–Moon system when δ has a positive value and when the Moon is directly above the equator, with $\delta = 0$. The figure also compares the orientations of the tidal bulges for these two cases; the tidal bulges have been greatly exaggerated for the sake

Figure 7.37 This figure shows the Earth–Moon system when the moon's angle of declination δ has a positive value and when the Moon is directly above the equator, with $\delta = 0$. The figure also compares the orientations of the tidal bulges for these two cases; the tidal bulges have been greatly exaggerated for the sake of clarity. As δ increases, the tidal bulges move to follow the Moon and no longer align with the Earth's equator. From the figure we can see that the two high tides at latitude P will have different heights. Furthermore, at high latitudes such as at Q, there will be only one tide rather than two.

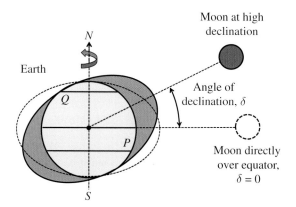

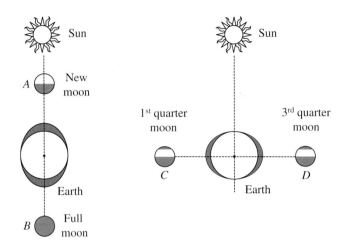

Figure 7.38 When the Moon is in line with the Sun, at positions A and B, the gravitational pull of the Sun enhances the gravitational pull of the Moon. At these times, the tidal range is greatest and the tides are called *spring tides*. When the Moon is on a line that is perpendicular to the direction of the Sun, at positions C and D, the gravitational pull of the Sun and of the Moon act in different directions. The effect of this is to reduce the tidal range and such tides are called *neap tides*. The figure shows the corresponding phases of the Moon.

of clarity. As δ increases, the tidal bulges move to follow the Moon and are no longer aligned with the Earth's equator. Then, as we can see from the figure, the two high tides at latitude P will have different heights. Furthermore, at high latitudes such as at Q, there will be only one tide rather than two. An analogous situation occurs for negative values of δ.

The gravitational influence of the Sun on the tides is less than half as strong as that of the Moon, as we have seen, but the Sun still plays a significant role in determining tidal range. We can see this pictorially in Figure 7.38, which is a view of the Sun–Earth–Moon system from Polaris. When the Moon is line with the Sun, at positions A or B, the gravitational pull of the Sun enhances the gravitational pull of the Moon. At these times the tidal range is greatest and the tides are called *spring tides*. When the Moon is on a line that is perpendicular to the direction of the Sun, at positions C or D, the gravitational pull of the Sun and of the Moon act in different directions. The effect of this is to reduce the tidal range and such tides are called *neap tides*. Figure 7.38 also shows the corresponding phases of the Moon.

The model of the tides that we described in Section 7.3.1 is called the *equilibrium tide* model. In this model we have tidal bulges either side of the Earth, which are fairly static since the Moon takes a month to revolve about the Earth, and the Earth rotates within these tidal bulges. From a different viewpoint, we can think of ocean tides as a dynamic system. From this viewpoint, the Moon produces a periodic force that drives the oceans to and fro between the continents. We may recall the familiar example of a forced oscillator of a mass on a spring that is driven by a periodic force. In general there is a phase difference between the displacement of the mass and the driving force. So it is with the tides. In reality the tidal bulges are not aligned with the Earth–Moon axis as shown in Figure 7.39.

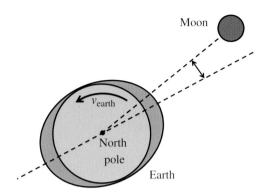

Figure 7.39 In reality, the tidal bulges are not aligned with the Earth–Moon axis. There is a phase lag between the peak of the bulge and the 'overhead' position of the Moon because the tidal bulge cannot move quickly enough to keep up with the motion of the Moon.

There is a phase lag between the peak of the bulge and the 'overhead' position of the Moon.

We can get some physical insight into the dynamic behaviour of the tides by considering the tidal bulge to behave like a water wave as it moves across the oceans and indeed it is called a *tidal wave*. (This is the correct meaning of the term tidal wave; Tsunamis are generated by abrupt movements in the Earth's surface layer and are not tidal waves, although they are often described in this way.) For the tidal bulge to remain directly below the Moon, the bulge would have to travel at the same speed as the surface of the Earth that rotates beneath it. However, the tidal bulge cannot move fast enough to achieve this. Taking the Earth to rotate once on its axis in 24 hr and the radius of the Earth to be 6400 km we find the linear velocity of a point on the Earth's equator to be

$$\frac{6400 \times 2\pi}{24} \approx 1670 \text{ km/hr or about } 465 \text{ m/s.}$$

We can assume here that the Moon is stationary since it takes about 28 days to orbit the Earth. In Section 7.2.2, we found that the group velocity of a shallow water wave is given by

$$v_{\text{g}} = \sqrt{gh_0}, \tag{7.90}$$

where g is the acceleration due to gravity and h_0 is the depth of the water. By shallow we mean that the wavelength λ of the wave is long compared with the water depth h_0. The average depth of the world's oceans at about 4000 m certainly cannot be called shallow, but the wavelength of the tidal bulge is much larger than this; in principle its wavelength is about half the circumference of the Earth. Thus the requirement $\lambda \gg h_0$ is obtained. Thus, according to Equation (7.90), the velocity of the tidal bulge is $\sim\sqrt{40\,000}$ m/s, i.e. about 200 m/s ($\sim$700 km/hr). This is much less than the velocity of the Earth's surface and the tidal wave cannot keep up with the Earth's rotation. The picture we thus obtain is as shown in Figure 7.39; there is a time lag between the peak of the bulge and the Moon's position. We also note that as a tidal wave propagates across an ocean, it experiences a Coriolis force, just as the global winds do. The tidal waves are similarly deflected and indeed tidal waves are observed to follow circular patterns on a global scale.

In reality, the topography of a particular coastline is usually the dominant factor in determining tidal range. Close to coastlines, the water depth usually decreases significantly. When the tidal wave runs into shallow waters, its velocity is reduced (see Equation 7.90). Then the energy of the wave accumulates in a smaller volume and the rise and fall of the tide are amplified. In addition, estuaries and bays can have a particularly large effect on tidal range. They funnel the incoming tide and thus greatly amplify tidal range. As we saw in Section 7.2.3, such an effect is exploited in an overtopping power plant.

There may also be *resonant enhancement* of the tidal range in a bay or an estuary. When we push a playground swing, we time the pushes so that they match the natural frequency of the swing. At this resonance frequency a small effort causes a large amplitude of swing. The rising and falling tide at the entrance to a bay acts as a driving force that causes the body of water within the bay to rise and fall. The body of water also has a resonant frequency. If the frequency of the driving tide is close to this resonant frequency, a standing wave is set up in the bay and this amplifies the tidal range. If we swing a bucket of water to and fro, the water also sloshes to and fro and, in particular, the water rises abruptly at the sides of the bucket. Similarly, in the resonant enhancement of the tides in a bay, the maximum tidal range occurs at the *closed* end of the bay.

A simple model of a bay is shown in Figure 7.40, where the bay has length L. The tidal motion at the entrance to the bay sets up a standing wave of water with wavelength λ. The first *resonant mode* in the bay occurs for $L = \lambda/4$, as illustrated in Figure 7.40, where the blue dotted curve represents the elevation of the water in the bay. The period T of the resulting standing wave is given by

$$T = \frac{\lambda}{v},\qquad(7.109)$$

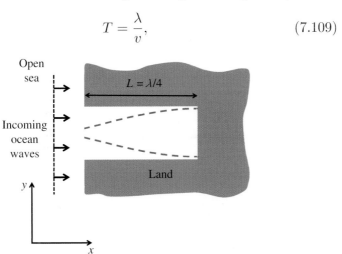

Figure 7.40 The figure shows a simple model of a bay. The rising and falling tide at the entrance to the bay acts as periodic driving force that causes the body of water within the bay to rise and fall. The body of water also has a resonant frequency. If the frequency of the driving tide ($\sim$12.5 hr) is close to this resonant frequency, a standing wave of water is set up in the bay and this amplifies the tidal range. The first resonant mode in the bay occurs for $L = \lambda/4$, where λ is the wavelength of the standing wave. The blue dashed curves represents the elevation of the water in the bay at resonance. Maximum elevation is obtained at the *closed* end of the bay.

where v is the wave velocity. As the water in a bay is usually shallow, we can take the wave velocity to be given by $v = \sqrt{gh_0}$ (Equation 7.90). Hence,

$$T = \frac{\lambda}{\sqrt{gh_0}} = \frac{4L}{\sqrt{gh_0}}. \tag{7.110}$$

Resonance enhancement occurs when the period of the tide in the open sea matches the period, T, of this resonant mode. Of course, real estuaries and bays do not have the uniform dimensions of our simple model. However, the Bay of Fundy, New Brunswick, Canada, has a shape that is not too dissimilar to that shown in Figure 7.40. The length of the Bay of Fundy is $\sim$300 km and the average depth is $\sim$70 m. Using these values of L and h_0, respectively, we find that the value of the period T for the bay is

$$= \frac{4 \times 3 \times 10^5}{\sqrt{9.8 \times 70}} = 4.8 \times 10^4 \text{ s} = 12.7 \text{ hr.}$$

This period is close to the period $\sim$12.5 hr at which the Moon drives the tides – the time between high tides in the North Atlantic. Consequently, the Bay of Fundy has a tidal range that exceeds 15 m at some times in the year, the highest tidal range in the world. This value can be compared with the typical tidal range in open ocean, which is $\sim$0.5 m.

Worked example

For the one-dimensional model of a bay, shown in Figure 7.40, an appropriate solution for a standing wave in the bay is given by

$$y(x, t) = A \cos(kx - \omega t) + B \cos(kx + \omega t), \qquad 0 < x < L$$

where the wavenumber $k = \omega/v$, v is the wave velocity and ω is the angular frequency, and the x–y axes are indicated on the figure. (This solution represents a travelling wave of amplitude A travelling in one direction and a travelling wave of amplitude B travelling in the opposite direction.) Show that the period T of the first normal mode of the water in the bay is given by $T = \frac{4L}{\sqrt{gh_0}}$.

Solution

The standing wave is a solution of the one-dimensional wave equation

$$\frac{\partial^2 y}{\partial t^2} = v^2 \frac{\partial^2 y}{\partial x^2}, \tag{7.21}$$

where $y(x, t)$ is the elevation of the water in the bay. The boundary conditions for this case are: (i) the elevation at the mouth of the bay must follow the tidal motion, i.e. $y(0, t) = y_0 \cos \omega t$; and (ii) there is no flux of water passing through the closed end of the bay, i.e. $\frac{\partial y}{\partial x}(L, t) = 0$. By applying these boundary conditions to the given solution for $y(x, t)$, we find

$$A = B = \frac{y_0}{\cos(kL)}.$$

Then substituting for A and B, we find

$$y(x, t) = \frac{2y_0}{\cos(kL)} \cos\left[k(x - L)\right] \cos \omega t.$$

Resonance enhancement occurs when the elevation of the tides at the closed end of the bay is maximized with respect to the elevation at the entrance to the bay. This ratio is given by

$$\frac{y(L, t)}{y(0, t)} = \frac{1}{\cos(kL)}.$$

This ratio approaches infinity if $kL = \left(n + \frac{1}{2}\right)\pi$, where $n = 0, 1, 2, \ldots$. (This simple model does not take account of damping in the system, which stops this ratio going to infinity in real bays. In the case of the Bay of Fundy, the ratio can be $\sim$10.) Hence, the natural modes for the bay have frequencies given by

$$\omega_n = \left(n + \frac{1}{2}\right)\pi\frac{\sqrt{gh}}{L}.$$

Hence, the period of a normal mode is given by

$$T_n = \frac{2\pi}{\omega_n} = \frac{2L}{\left(n + \frac{1}{2}\right)\sqrt{gh}}.$$

The first mode with $n = 0$ gives $T_0 = \frac{4L}{\sqrt{gh}}$.

An example of the variation of tidal range over the period of a month is shown in Figure 7.41. It was recorded at Bridgeport, Connecticut, USA. The sea level rises and falls in a sinusoidal-like fashion with a period of $\sim$12.5 hr, due to the Earth's rotation. The tidal range varies over the course of the month due to the relative motions of the Moon and Sun, giving rise to two spring tides and two neap tides per month. In this example, the tidal range of the spring tides is about twice as large as that of the neap tides, which is quite typical.

7.3.3   Harnessing tidal power

There are two main ways in which tidal power is harnessed. One exploits the potential energy of water at high tide and is called

Figure 7.41 An example of the variation of tidal range over the period of a month, which was recorded at Bridgeport, Connecticut, USA. The sea level rises and falls in a sinusoidal-like fashion with a period of ~12.5 hr, due to the Earth's rotation. The tidal range varies over the course of the month due to the relative motions of the Moon and Sun, giving rise to two spring tides and two neap tides per month.

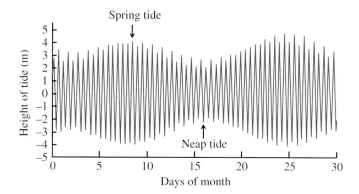

Figure 7.41 An example of the variation of tidal range over the period of a month, which was recorded at Bridgeport, Connecticut, USA. The sea level rises and falls in a sinusoidal-like fashion with a period of ~12.5 hr, due to the Earth's rotation. The tidal range varies over the course of the month due to the relative motions of the Moon and Sun, giving rise to two spring tides and two neap tides per month.

tidal range power. The other exploits the kinetic energy of tidal water that flows through narrow channels and is called *tidal current* power.

Tidal range power

The principle of operation of a tidal range power plant is similar to that of a hydroelectric plant. A barrier is built across the mouth of an estuary or bay, thus forming a reservoir behind the barrier. As the sea level rises with the high tide, water is allowed to pass into the reservoir through a sluice gate. When high tide is reached the sluice gate is closed, trapping water at the high-tide level in the reservoir. Then when the sea level drops at low tide, the trapped water is passed through a turbine, which is connected to an electrical generator. In contrast to a conventional hydroelectric plant, the head of water is low ~5 m and the water flow rate is high, and the turbines must be designed to take this into account.

If the surface area of the *tide pool* containing the collected water is A and the tidal range is D then the potential energy U of the stored water is given by

$$U = AD\rho \times \frac{Dg}{2} = \frac{1}{2}AD^2\rho g, \tag{7.111}$$

where ρ is the density of seawater and g is the acceleration due to gravity. The factor of $\frac{1}{2}$ arises because the centre of mass of the body of water falls through distance $\frac{1}{2}D$. The quadratic dependence on D emphasises the importance of having as large a value of D as possible. The average power $\bar{P}_{av}$ is then given by

$$\bar{P}_{av} = \frac{AD^2\rho g}{2\tau}, \tag{7.112}$$

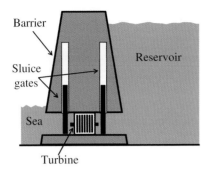

Figure 7.42 A schematic diagram of a tidal range power plant. By suitable manipulation of the sluice gates, the turbine is driven by water that flows into the reservoir as well as by water that flows out as the tide rises and falls.

where τ is the tidal period. As we have seen, tidal range, and hence the available power, varies throughout the month and year and an approximation that is used for the average power is

$$\bar{P}_{av} = \frac{A\rho g}{2\tau}\left(\bar{D}_{av}\right)^2, \tag{7.113}$$

where $\bar{D}_{av}$ is the mean range of all tides. Taking $\bar{D} = 4$ m, $A = 10$ km^2, $\rho = 1.03 \times 10^3$ kg/m^3, $g = 9.8$ m/s^2 and $\tau = 12$ hr 25 min $= 4.47 \times 10^4$ s, we obtain

$$\bar{P}_{av} = \frac{10.0 \times 10^6 \times 1.03 \times 10^3 \times 9.8 \times 4.0^2}{2 \times 4.47 \times 10^4} = 18 \text{ MW}.$$

As usual, the maximum possible power cannot be achieved in practice and the overall efficiency of the system for producing electricity may be ~80% of this value.

The output tidal range power can be doubled if the plant uses reversible turbines that are driven by water that flows *into* the reservoir as well as by water that flows out. The arrangement for this is illustrated in Figure 7.42, where the barrier contains two sluice gates. The cycle of events is as follows. Let us say that the water level in the reservoir is initially at the high-tide level and that the tide is low on the sea side. The sluice gates are opened and water from the reservoir flows through the turbine. Then when the water level in the reservoir has fallen to low-tide level, the sluice gates are closed. After low tide has been reached, the sea level steadily rises to high-tide level, high above the water level in the reservoir. When high tide is reached, the sluice gates are opened and water flows through the turbines from the sea to the reservoir. The reservoir steadily fills up and when high-tide level is reached, the sluice gates are closed. The cycle then repeats.

A notable example of a tidal range plant is the Rance Tidal Power Station that is located on the estuary of the Rance River in Brittany, France, where the average tidal range is a massive 8 m, the highest in France. It was opened in 1966 as the world's first tidal range power station and was for 45 years the largest such power station, until the

Figure 7.43 A photograph of the Rance Tidal Power Station which is located on the estuary of the Rance River in Brittany, France, where the average tidal range is a massive 8 m. The Rance tidal basin has a total area of 22.5 km^2, and the power station has 24 turbines which produce a peak output power of 240 MW and an average power of 62 MW. Source: http://theearthproject.com/wp-content/uploads/2016/01/La-Rance-Tidal-Power-Plant.jpg

South Korean Sihwa Lake Tidal Power Station was opened in 2011. The Rance tidal basin has a total area of 22.5 km^2. Its power station has 24 turbines and they produce a peak output power of 240 MW and an average power of 62 MW. A photograph of the Rance Tidal Power Station is shown in Figure 7.43.

Tidal current power

As previously noted, tides can also result in the flow of water through narrow channels that connect large areas of water, for example a channel between the mainland and a nearby island. Moreover, these tidal currents may reach speeds of ∼5 m/s. Energy can be harvested from the flowing water just as energy is harvested from the wind; the water is passed through turbines that drive electrical generators. The potential power, P, from a water turbine is described by the same equation as for a wind turbine (see Equation 6.15):

$$P = \frac{1}{2}\rho A v^3, \tag{7.114}$$

where A is the area of the turbine, ρ is the density of seawater and v is the velocity of the water. The important difference between water and wind turbines is that the density of water is about 800 times greater than that of air. This higher density compensates for the fact that water velocities are generally smaller than wind velocities. And a water turbine is immersed in seawater, which is a harsh environment. Taking a turbine diameter of 15 m, a water velocity of 3.0 m/s and a seawater density of 1.03×10^3 kg/m^3, we obtain

$$P = \frac{1.03 \times 10^3 \times \pi \times 7.5^2 \times 3^3}{2} = 2.5 \text{ MW},$$

which is of the same order as the power from a wind turbine.

An important example of a tidal current power plant is called SeaGen, which was installed in Strangford Lough in Northern Ireland in 2008. One of the principal features of this lough is a restricted channel that connects it to the Irish Sea and through which seawater flows in and out of the lough. The channel is $\sim$8 km long, with a minimum width of $\sim$1 km and a depth varying between 30 and 60 m. This constriction results in very strong currents that peak in excess of 4.5 m/s on spring tides and it is estimated that $\sim 3.5 \times 10^8$ m^3 of water enters and leaves the lough at each tide. Interestingly, it is recorded that Strangford Lough was exploited as a tidal energy source as far back as 532 AD, when a *tidal mill* was used as part of a monastic settlement.

A schematic diagram of the SeaGen plant is shown in Figure 7.44. It consists of twin turbines mounted on a crossbeam. The rotor diameter of each turbine is 16 m and each turbine drives a generator. The turbines have a patented feature that allows the rotor blades to be pitched through more than 180°. This allows the turbines to operate for water flowing into or out of the lough. The crossbeam is attached to a collar that slides over a central column and this allows

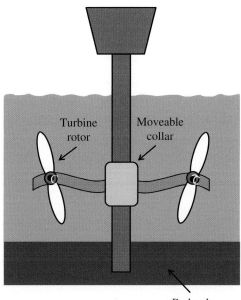

Figure 7.44 A schematic diagram of the SeaGen tidal current plant at Strangford Lough in Northern Ireland. The rotor diameter of each turbine is 16 m and each turbine drives a generator. The turbines have a feature that allows them to operate for water flowing into or out of the lough. The crossbeam is attached to a collar which slides over a central column. This allows the turbines to be lifted above the water line for routine maintenance.

the turbines to be lifted above the water line for routine maintenance. At full power, SeaGen delivers a maximum power of 1.2 MW and delivers up to 20 MWh/day. Currently, the SeaGen plant is the only commercial tidal current plant in the world, although similar plants are planned for other locations.

Problems 7

7.1 A transverse wave travelling along a taut string is described by the function $y = 25\cos(0.15x + 75t)$, where x and y are in millimetres and t is in seconds. (a) Find (i) the amplitude, (ii) the wavelength, (iii) the frequency (Hz), and (iv) the velocity of the wave. In what direction is the wave travelling? (b) What is the maximum transverse velocity and acceleration of an element of the string?

7.2 One end of a long taut string is moved up and down in simple harmonic motion with an amplitude of 0.10 m and a frequency of 10 Hz. At time $t = 0$, the end of the string (at $x = 0$) has its maximum upward displacement. The resultant wave travels down the string in the positive x-direction with velocity 40 m/s. Obtain an equation describing the wave.

7.3 You move one end of a long rope in simple harmonic motion with amplitude 0.15 m and frequency 4.0 Hz, while maintaining a tension of 50 N in the rope. The other end of the rope is connected to a distant tree. The rope has a mass of 0.5 kg/m. (a) How much energy is contained in a length of the rope equal to one wavelength of the transmitted wave? (b) How much power are you expending? (c) What will be the required power if: (i) the frequency is doubled; (ii) the amplitude is halved?

7.4 Show that $y = A\sin(\omega t - kx) + B\sin(\omega t + kx)$ is a solution of the one-dimensional wave equation.

7.5 The Glendoe hydroelectric plant in Scotland has a catchment area for rainfall of 75 km^2, and the annual rainfall is approximately 2 m. The plant has a head of 600 m, the highest in the UK. Make an estimate of the average power that this plant can provide.

7.6 The flow of liquid through a pipe depends on: (i) the coefficient of viscosity η; (ii) the radius a of the pipe; and (iii) the pressure gradient g set up along the pipe. The pressure gradient is p/l, where p is the pressure difference between the ends of the pipe and l is its length. Use the method of dimensions to find an expression for the flow rate in terms of η, a, and g, and then in terms of η, a, p and l. Note that the units of η are Ns/m^2.

7.7 Compare the difference in the gravitational pull of the Earth at the nearest and furthest sides of the Moon with the difference in the gravitational pull of the Moon at the nearest and furthest sides of the Earth.

7.8 Show that the latitude on Earth at which the tidal force due to the Moon is parallel to the Earth's surface is 54.7°, i.e. the latitude at which the vertical component of the tidal force is zero.

7.9   (a) Calculate the distance of the centre of mass of the Earth–Moon system from the Earth's centre as a fraction of the Earth's radius. (b) Calculate the distance of the centre of mass of the Sun–Earth system from the Sun's centre as a fraction of the Sun's radius. The masses of the Earth, Moon and Sun are 5.97×10^{24} kg, 7.35×10^{22} kg and 1.99×10^{30} kg, respectively; the distance between the Earth and Moon is 3.84×10^5 km and that between the Sun and Earth is 1.50×10^8 km; the radii of the Earth and Sun are 6.38×10^3 km and 6.96×10^5 km, respectively.

7.10 For waves on deep water, the angular frequency ω and wavenumber k are related by the dispersion relation

$$\omega^2 = gk + \frac{Sk^3}{\rho}$$

where g is the acceleration due to gravity, and ρ and S are the density and surface tension of the liquid, respectively. Determine the minimum value of the phase velocity for water waves on deep water. The density and surface tension of water are 1.0×10^3 kg/m^3 and 7.2×10^{-2} N/m. Assume $g = 9.8$ m/s^2.

7.11 For a canal 4.0 m deep, calculate the phase velocity and group velocity for water waves of wavelength (i) 10 mm, (ii) 1.0 m and (iii) 50 m.

7.12 The British Isles has about 1000 km of Atlantic coastline. Calculate the average wave power that this could provide, assuming the amplitude, H, of the waves is 1.0 m and the group velocity v_g is 8.0 m/s. Assuming 50% of this could be converted into electrical power, approximately how much electrical power would be provided per person in the British Isles?

7.13 A tidal power plant utilising the incoming and outgoing tides has a tide pool of area 22.5 km. Taking the average tidal range to be 8 m, calculate the maximum possible output power of the plant. Calculate the power per square metre of the tide pool. How does this energy density compare with the same quantity for a wind farm? How large must the tide pool be for the generation of 2.5 MW? Take the density of seawater to be 1030 kg/m^3.

7.14 Suppose that we build a water farm of underwater turbines in the Irish Sea. What power per square metre of the sea should we obtain? Assume that the underwater turbines must be separated in the same way that wind turbines are separated on land. Take the average speed of the water to be 1.0 m/s and assume an overall efficiency of 50%.

8

Energy storage

Energy has to be provided when and where it is needed. Some of the energy sources that we have described, such as nuclear power stations and hydroelectric plants, can provide a continuous supply of energy and their power output is controllable. On the other hand, some energy sources such as wind turbines and solar cells produce energy intermittently – turbines need the wind and solar cells need sunlight. Consequently, these sources may not generate enough energy when it is needed or, alternatively, they may generate excess energy. Energy storage systems allow the excess energy to be stored and used at a later time. And with increasing use of intermittent sources, energy storage becomes increasingly important. The other side of the coin is that demand for energy varies substantially throughout the seasons and throughout the day and there are sharp peaks in power consumption. Indeed, one method for estimating television-viewing figures is to measure the peaks in power consumption during advertisement breaks when viewers go to switch on the kettle. Power stations, however, should ideally be operated at a fairly constant output level, close to where they operate most efficiently. Energy storage allows energy providers to balance supply and demand, in order to provide the extra power required during peak times. It is often necessary to bring this extra power on stream rapidly and most energy storage systems have the ability to do this; some can be switched on in a timescale of seconds. In contrast, thermal power stations are less able to respond to rapid changes. Changes in demand can cause frequency and voltage instability in the *national grid*. The ability of stored energy systems to respond to sudden changes in demand also helps to stabilise frequency and voltage.

Physics of Energy Sources, First Edition. George C. King.
© 2018 John Wiley & Sons, Ltd. Published 2018 by John Wiley & Sons, Ltd.

In this chapter we describe various types of energy storage together with their main characteristics.

8.1 Types of energy storage

Most forms of energy can be used for the purpose of energy storage, including chemical, thermal, mechanical and electrical energies. And sometimes energy storage involves the conversion of energy from one kind to another. Table 8.1 lists various forms of energy together with examples of how these are used to store energy in practical applications. This table also gives typical values of energy density and values of *round-trip efficiency*. The units of energy density are MJ/kg, unless otherwise stated – a useful conversion factor is $3.6 \, \text{MJ} = 1 \, \text{kWh}$

Important parameters of any energy storage system include the following:

- Capacity – here we include the amount of energy that is stored and the rate at which this energy can be delivered, i.e. the power output.

- Energy density – the amount of energy stored per unit mass or per unit volume.

- Round-trip efficiency – the amount of energy released compared with the amount of energy that was put in.

- The time it takes to switch on the power output.

Table 8.1 Various forms of energy and examples of how this energy is stored. The table also gives typical values of energy density and round-trip efficiency for various storage systems

Form of energy	Energy storage	Energy density (MJ/kg)	Round-trip efficiency (%)
Chemical energy	Fossil fuels; e.g. coal, oil.	35	–
	hydrogen gas storage	140	~60
Thermal energy	Hot water storage	0.13	> 50
	Ice storage	0.33	~80
Potential energy	Pumped hydroelectric	0.001	~75
	Compressed air storage	0.33	~50
Kinetic energy	Flywheels	0.04–0.4	~80
Electrical energy	Supercapacitors	~0.02	~95
	Superconducting magnets	0.001[a]	95
	Lead-acid battery	0.13	~80
	Fuel cells	~2[b]	–

[a] The units of energy density in this case are MJ/L.
[b] Includes the mass of the gas storage tank.

8.2   Chemical energy storage

8.2.1   Biological energy storage

By chemical energy we mean the potential energy that is associated with the chemical bonds of the molecules of a fuel such as wood or coal. In the case of biological systems, the chemical bonds are formed in the process of photosynthesis and hence are a store of solar energy. A way to release this energy is by combustion, e.g. by burning the wood. In fact, fossil fuels provide a natural store of energy. Moreover, their energy density is high. For example, burning 1 L of oil produces ~35 MJ of energy. (Similarly, we can say that naturally occurring radioactive nuclei such as ^{235}U and ^{238}U are stores of nuclear energy. And as we saw in Chapter 3, the energy density of these radioactive nuclei is extremely high, $\sim 10^{5\text{-}6}$ greater than the energy density of fossil fuels.)

8.2.2   Hydrogen energy storage

A way to store chemical energy is to produce and store hydrogen gas. Then, when needed, the hydrogen gas is used as a fuel. The hydrogen gas can be burnt to produce thermal energy that drives a conventional steam turbine. Alternatively, as described in Section 8.5.4, the hydrogen gas can be reacted with oxygen in a *fuel cell*. (The term *hydrogen economy* refers to the concept of using hydrogen gas as a *low-carbon* energy source replacing, for example, petrol as a transport fuel or natural gas as a heating fuel.) Hydrogen energy storage is attractive because, whether hydrogen is burnt in air or reacted with oxygen in a fuel cell, the only by-product is water and so there is no pollution. Moreover, hydrogen gas has a high energy density of 140 MJ/kg. But, of course, hydrogen is a flammable gas and its safe storage and transport must be taken into account.

Hydrogen gas can be produced by the electrolysis of water. This method uses a DC power supply and has the advantage that it works over a wide range of applied voltage. This makes it particularly suitable for energy sources such as wind turbines and solar cells, where the output voltage that is generated may vary. In practice, however, most of the commercial production of hydrogen gas employs the method of *steam reforming*. In this process, high-temperature steam (700–1000°C) is used to produce hydrogen from a methane source, such as natural gas. The methane reacts with steam under a pressure of 3–25 bar in the presence of a nickel catalyst to produce hydrogen and carbon monoxide:

$$CH_4 + H_2O \rightarrow CO + 3H_2.$$

This is an *endothermic reaction*, so that heat must be supplied for the reaction to proceed. Then in a second stage at a temperature of about 360°C, and in the presence of a copper or iron catalyst, additional hydrogen is generated through the reaction

$$CO + H_2O \rightarrow CO_2 + H_2.$$

In this reaction an oxygen atom is removed from a water molecule and the atom oxidises CO to CO_2. This is an *exothermic reaction*, i.e. it releases heat to the surroundings. Producing hydrogen from natural gas results in the emission of the greenhouse gas CO_2, as we can see. However, when compared, for example, with internal combustion engines using petroleum, vehicles using hydrogen gas as fuel produce fewer greenhouse gas emissions overall.

Small amounts of hydrogen can be stored in pressurised vessels at 100–300 bar or liquefied at 20 K. Alternatively, solid *metal hydrides* or *nanotubes* can absorb and hence store hydrogen at very high density. For very large amounts of hydrogen, the gas can be stored underground in salt mines with a volume of up to $\sim 5 \times 10^5 \, \text{m}^3$ and at a pressure of 200 bar. This corresponds to an energy storage capacity of $\sim$360 GJ.

8.3 Thermal energy storage

In this storage method, thermal energy is stored by heating (or cooling) a storage medium. We have already encountered the storage of thermal energy in the case of thermal solar power systems (Section 4.5.3). In that system, excess power generation is used to heat salt to high temperature and the molten salt is sent to a storage tank for later use. There we noted that it is easier to store thermal energy than electrical energy and, in general, electrical energy is less easy to store than other forms of energy. Storage of thermal energy is also used in domestic storage heaters, which contain thermally insulated clay bricks. Low-tariff electricity is used to heat the clay bricks during the night. Then during the day the heated bricks steadily release heat to the living space.

Disadvantages of storing heat in a medium are that the temperature of the medium changes as thermal energy is extracted, which complicates delivery of the stored energy, and energy densities are relatively low. For example, if we have water at 50°C and the stored energy is to be delivered at a temperature of 20°C, the initial stored energy density is $4200 \times (50 - 20) = 0.13 \, \text{MJ/kg}$, where we have taken the heat capacity of water to be 4200 J/kg K. The energy density is increased if a phase change in the medium is involved when the energy is stored or released. For example, some air conditioning systems use low-tariff electricity to freeze water during the night. The ice is then used to cool the air during the day. The latent heat of fusion of ice is $330\,\mathrm{kJ/kg}$, i.e. 1 kg of ice is able to absorb $330\,\mathrm{kJ}$ of heat when it melts. Moreover, the ice remains at a temperature of $0°C$ while it is melting and this simplifies delivery of the stored energy.

Worked example

One method of thermal energy storage is to heat the rock beneath a house by solar energy in the summer months and use the stored energy to heat the house in the winter months. It is important that a substantial fraction of the stored energy does not diffuse away from the rock on a timescale of several months. Figure 8.1(a) represents a one-dimensional model of heat diffusion in a rock. It shows a cylindrical block of granite of area A, with the centre of the block at $x = 0$. A quantity Q_0 of thermal energy is deposited at the centre of the rock at time $t = 0$. The thermal energy diffuses away from the centre and the temperature distribution along the rock at $t > 0$ is given by

$$T\left(x,t\right) = \frac{B}{\sqrt{t}} \exp\left(-\frac{x^2}{at}\right),$$

where B and a are constants and $T\left(x,t\right)$ is measured with respect to the temperature of the rock before the deposition of the energy. (a) Starting with the diffusion equation (Equation 4.45)

$$\frac{\partial T}{\partial t} = \frac{\kappa}{\rho C}\frac{\partial^2 T}{\partial x^2},$$

show that

$$a = \frac{4\kappa}{\rho C},$$

where the symbols have their usual meanings. Obtain a value for the width w of the function $T\left(x,t\right)$ at $t = 3\,\mathrm{months}$. For granite, $C = 0.82\,\mathrm{kJ/kg\,K}$, $\rho = 2500\,\mathrm{kg/m^3}$ and $\kappa = 2.1\,\mathrm{W/m\,K}$. (b) By

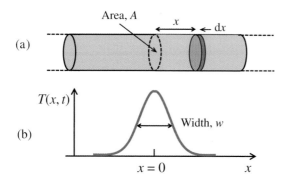

Area, A

(a)

$T(x,t)$

Width, w

(b)

$x = 0$ x

Figure 8.1 (a) A one-dimensional model of heat diffusion in a cylindrical block of granite rock. A quantity of heat is deposited at the centre of the rock, $x = 0$, at time $t = 0$. (b) This heat then diffuses along the rock, setting up a variation in temperature along the rock given by the function $T(x, t)$.

considering the resultant thermal energy in a slice of the cylindrical block, and integrating over the length of the block, assuming it to be infinitely long, show that the constant B is given by

$$B = \frac{Q_0}{A(4\pi\kappa\rho C)^{1/2}}.$$

Solution

(a) $$\frac{\partial T(x,t)}{\partial t} = \frac{Bx^2}{at^{5/2}} \exp\left(-\frac{x^2}{at}\right) - \frac{B}{2t^{3/2}} \exp\left(-\frac{x^2}{at}\right),$$

$$\frac{\partial^2 T(x,t)}{\partial x^2} = \frac{4Bx^2}{a^2 t^{5/2}} \exp\left(-\frac{x^2}{at}\right) - \frac{2B}{at^{3/2}} \exp\left(-\frac{x^2}{at}\right).$$

Substituting for $\partial T/\partial t$ and $\partial^2 T/\partial x^2$ in diffusion equation and equating coefficients of

$$\frac{x^2}{t^{5/2}} \exp\left(-\frac{x^2}{at}\right) \quad \text{or} \quad \frac{1}{t^{3/2}} \exp\left(-\frac{x^2}{at}\right),$$

obtain $a = 4\kappa/\rho C$. The function $T(x,t)$ is a Gaussian function (see Figure 8.1b) whose width we take to be

$$2\sqrt{at} = 4\sqrt{\frac{\kappa t}{\rho C}}$$

$$= 4\sqrt{\frac{2.1 \times 3 \times 30 \times 24 \times 60^2}{2500 \times 0.82 \times 10^3}} = 11 \, \text{m}.$$

This result indicates that energy storage in this way is entirely feasible.

(b) Considering a slice of the block at distance x with thickness dx and area A, the thermal energy contained in the slice is $\rho A C T(x,t)\, dx$, and hence,

$$\frac{\rho AC}{t^{1/2}} \int_{-\infty}^{+\infty} T(x,t)\, dx = Q_0.$$

$$\frac{\rho AC}{t^{1/2}} \int_{-\infty}^{+\infty} T(x,t)\, dx = \rho AC \frac{B}{t^{1/2}} \int_{-\infty}^{+\infty} \exp\left(-\frac{x^2}{at}\right) dx,$$

where

$$a = \frac{4\kappa}{\rho C}.$$

$$\int_{-\infty}^{+\infty} \exp\left(-\frac{x^2}{at}\right) dx = \sqrt{\pi at}.$$

Hence,

$$\rho AC \frac{B}{t^{1/2}} \int_{-\infty}^{+\infty} \exp\left(-\frac{x^2}{at}\right) \mathrm{d}x = \rho AC \frac{B}{t^{1/2}} \sqrt{\frac{4\pi\kappa t}{\rho C}} = Q_0,$$

and

$$B = \frac{Q_0}{A(4\pi\kappa\rho C)^{1/2}}.$$

8.4 Mechanical energy storage

Mechanical energy can be stored as either potential or kinetic energy. A simple example of stored potential energy is the hanging weight in a grandfather clock. The weight provides the energy to drive the clock mechanism for a period a week or so. On a much larger scale, potential energy is stored as water in a pumped hydroelectric plant or as compressed gas in an underground salt mine. For the case of kinetic energy, this can be stored as the rotational energy of a spinning flywheel.

8.4.1 Pumped hydroelectric energy storage

We described hydroelectric power in Section 7.1 and saw how the gravitational potential energy of water is used to drive a turbine. The dam of a reservoir serves to contain a large volume of water and, hence, a hydroelectric plant has in-built energy storage. In a pumped hydroelectric storage plant there are two reservoirs, one at a higher elevation than the other as shown schematically in Figure 8.2. At times of low electricity demand, for example during the night or the weekend, excess electricity is used to pump water into the upper reservoir. Then when there is high demand, water is released back into the lower reservoir through a turbine, thereby generating electricity. In practice, a reversible pump–turbine unit performs

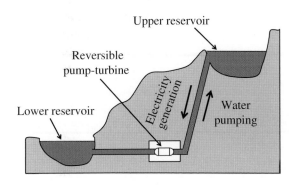

Upper reservoir

Reversible pump-turbine

Lower reservoir

Electricity generation

Water pumping

Figure 8.2 Schematic diagram of a pumped hydroelectric storage plant. During periods of low electricity demand, water is pumped to the upper reservoir. Then during periods of high demand the water in the upper reservoir is used to drive a turbine. Often the pump and turbine are combined into a single unit, as shown.

both functions. The amount of stored energy is given by the equation $E = mgh$, where m is the mass of raised water, h is the distance over which the centre of gravity of the raised mass of water is moved vertically and g is the acceleration due to gravity. The energy density of pumped storage systems is rather low. For example, the energy density is 0.98 kJ/kg for a height difference between reservoirs of 100 m. However, this low value of energy density is offset by the large amount of water that can be moved between the two reservoirs. Indeed, pumped hydroelectric storage has the largest energy capacity of all storage systems and is also the most cost-effective. Consequently, it is the most widely used method of large-scale energy storage and, in particular, is used to supply additional energy to a national grid. It also has the advantage that it can be switched on relatively quickly and can respond to rapid changes in power demand within seconds. Taking into account evaporation losses from the exposed water surface and conversion losses, approximately 70–85% of the electrical energy used to pump the water into the upper reservoir can be regained.

Clearly, mass m and height difference h should be as large as possible and, in addition, the head loss should be as small as possible. This is achieved by having a large body of water located on a hill relatively near, but as high as possible above a second body of water. In some places, this occurs naturally, while in others one or both bodies of water have been man-made. An example of a pumped hydroelectric plant is the Ffestiniog Power Station in North Wales, in the UK. The upper reservoir is Llyn Stwalan and water is discharged into the Tan-y-Grisan reservoir, about 300 m below. There are four generating units that have a combined output of 360 MW of electricity, which is enough to supply the entire power needs of North Wales for several hours. Figure 8.3 shows a photograph of the Llyn Stwlan upper reservoir.

Figure 8.3 Llyn Stwlan, the upper reservoir of the Ffestiniog Pumped Power Station in North Wales. Courtesy of Arpingstone. https://commons.wikimedia .org/wiki/File:Stwlan.dam.jpg

---

Worked example

Calculate the temperature difference between the top and bottom of a waterfall that is 150 m high. The heat capacity of water is 4200 J/kg K and the acceleration due to gravity is 9.8 m/s^2.

Solution

Consider a mass m of water that falls a distance h. The potential energy is converted into kinetic energy that is equal to mgh as the water hits the bottom of the waterfall. Assuming that the kinetic energy is converted completely into thermal energy, the heat gained by the water will be $Cm\Delta T$, where C is the heat capacity of water.

$$\text{Hence } \Delta T = \frac{gh}{C} = \frac{9.8 \times 150}{4200} = 0.35\,\text{K}.$$

The Manchester physicist James Joule spent part of his honeymoon trying to measure the temperature difference between the top and bottom of the Cascade de Sallanches waterfall in France, although this proved to be impractical.

8.4.2   Compressed air energy storage

A compressed gas exerts a force on its surroundings and so has the potential to do work. Indeed, we can compare a compressed gas to a compressed spring and Boyle, famous for his gas law, spoke about the 'springiness' of air. Just as a compressed spring stores energy in the form of potential energy, so a compressed gas is also a store of energy. In Section 4.6 we discussed the expansion and compression of ideal gases and saw that these processes may be isothermal or adiabatic. We considered the example of an ideal gas contained in a cylinder with a frictionless piston and processes that are quasi-static, i.e. where the piston is moved slowly so that the gas is never far from an equilibrium state. In an isothermal compression and expansion of a gas, the gas and its surroundings remain at constant temperature. In the compression stage, the work done on the gas is converted into heat that is passed to the surroundings. The work, W, we have to do to compress the gas from volume V_1 to V_2 is given by

$$W = \int_{V_1}^{V_2} p\mathrm{d}V. \tag{4.60}$$

Using the fundamental gas law $pV = nRkT$ (Equation 4.55), with T constant, we readily obtain

$$W = nRT \ln\left(\frac{V_2}{V_1}\right). \tag{8.1}$$

where n is the number of moles of gas and R is the gas constant. When we then expand the gas to its original volume V_1, again at constant temperature, the gas does an amount of work that is exactly the same as the amount we used to compress the gas; this work comes from heat given up by the surroundings. We thus have a means of storing energy. Note that the internal energy of an ideal gas, which is the kinetic energy of its molecules, remains the same as it depends only on the temperature of the gas and this is constant. This *ideal* isothermal and reversible process is 100% efficient; we get as much energy out as we put in. An ideal adiabatic compression and expansion cycle of a gas is also 100% efficient. As we recall, in an adiabatic process the system is thermally insulated so that no heat is lost to the surroundings. In the compression stage, the work we do on the gas goes into raising the temperature of the gas, i.e. into increasing its internal energy. When we then expand the gas to its original volume the gas returns to its original temperature as it gives up energy to do the work to expand the gas. Of course in real-life, compression–expansion cycles are not ideal isothermal or adiabatic processes and their efficiencies will less than 100%. However, by considering ideal situations, we can determine the maximum amount of energy that a compressed gas can store.

Worked example

(a) One mole of air at 300 K initially at atmospheric pressure is compressed isothermally by a factor of 10. Calculate the amount of energy stored in the compressed air and its energy density. (b) If the air is compressed adiabatically by the same factor, calculate the amount of energy stored in the compressed air and its energy density. Assume that the air behaves as an ideal gas with $\gamma = 1/4$ and $C_V = 20.8\,\mathrm{J/mole}$, and that the processes are reversible.

Solution

(a) From $W = nRT \ln(V_2/V_1)$, work done

$$= 1.0 \times 8.31 \times 300 \times \ln(10) = 5.74 \times 10^3 \text{ J}.$$

This is the amount of energy we can regain from the compressed air. Using $pV = nRT$, we find that the final volume of air

$$= \frac{1.0 \times 8.31 \times 300}{10 \times 1.01 \times 10^5} = 2.47 \times 10^{-3} \text{ m}^3.$$

Hence, energy density of compressed air is $2.3\,\mathrm{MJ/m^3}$.

(b) The final temperature of the air in the adiabatic compression is given by

$$T_2 = T_1(V_1/V_2)^{\gamma-1},$$

giving $T_2 = 754$ K. In an adiabatic compression, the work, W, done on the gas goes into heating the gas, i.e. increasing its internal energy U. We recall from Section 4.6.3 that $C_V = dU/dT$. Hence,

$$W = nC_V (T_2 - T_1) = 1.0 \times 20.8 \times (754 - 300) = 9.43 \times 10^3 \text{ J},$$

and energy density = 3.82 MJ/m^3. The mass of 1 mole of air is 29 g. Taking a value of stored energy of 9.43×10^3 J, we obtain an energy density of 0.33 MJ/kg.

The basic idea of compressed air energy storage is straightforward. When demand for electricity is low, air is compressed and stored at high pressure. Then, when demand is high, the compressed air is used to drive a turbine that, in turn, drives a generator that delivers electricity back to the national grid. Large-scale storage of energy for a national grid requires very large gas containers and very high compression ratios. In practice, compressed air plants have used disused salt mines to store compressed air at a pressure of ∼70 bar.

The main features of a compressed air storage plant are shown schematically in Figure 8.4. Air is compressed and stored in an underground cavern or mine with a large volume ∼ 3×10^5 m^3. This compression happens rapidly so that the process is not isothermal and consequently the gas heats up. The plant machinery cannot function at high temperatures and the heat produced in the compression

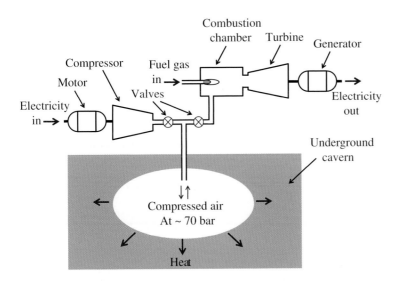

Figure 8.4 Schematic diagram of a compressed air storage plant. During periods of low electricity demand, air is compressed and stored in an underground salt mine or cavern. Heat generated in this compression stage is dissipated in the surrounding rock. During periods of high demand, the compressed air is released to turbines to generate electricity. The compressed air is heated by natural gas before it enters the turbines to compensate for the heat that it lost during the compression stage.

step must be removed. This is done by heat exchangers that transport the heat away or by dissipation of the heat in the surrounding rock, as indicated in Figure 8.4. These cooling processes maintain the stored gas close to ambient temperature. This represents a substantial loss of energy and reduces the round-trip efficiency of the system. When the compressed air is released to the turbines, the gas cools down because, again, the expansion is rapid. The degree of cooling is substantial and would stop the turbines functioning. To prevent this, the compressed air is first heated. This is done in a combustion chamber where a fuel, typically natural gas, burns in the compressed air. The need to provide external energy to heat the gas further reduces the overall efficiency of the storage system. However, the amount of natural gas the system burns is roughly one-third the amount used by conventional combustion turbines, thus producing only about one-third of the pollutants. A typical value of round-trip efficiency is ~50%. Although this figure may seem to be low, compressed air storage is the only system that has a storage capacity that is comparable to pumped hydroelectric storage. Presently there are two large-scale compressed air storage plants. These are the Huntorf compressed air storage plant in Germany, built in 1978, and the McIntosh plant in Alabama, USA, built in 1991. These are capable of delivering electrical powers of 290 and 110 MW, respectively.

A new type of compressed air storage system is being developed with the potential for increased efficiency. In this new type, the heat that is produced in the compression stage is used to heat oil or molten salt that is stored in thermally insulated containers. This stored thermal energy is then used to heat the compressed air before it is passed to the turbines. This system is called *advanced adiabatic compressed air energy* storage, and its principle of operation is illustrated schematically in Figure 8.5. As it minimises the loss of heat

Figure 8.5 Schematic diagram of an advanced adiabatic compressed air energy storage plant. In this system, the heat produced in the compression stage is used to heat oil or molten salt that is stored in thermally insulated containers. This stored thermal energy is then used to heat the compressed air before it enters the turbines.

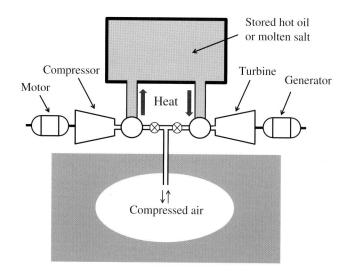

during the compression stage and removes the need for fuel gas to heat the compressed air during the expansion stage, it is anticipated that the overall efficiency of this new kind of storage plant will be $\sim 70\%$. It has the additional advantage that no CO_2 gas is produced.

8.4.3 Flywheel energy storage

A rotating body has rotational kinetic energy and hence can be used as a store of this energy. If a mass m rotates about a fixed axis, it has rotational kinetic energy

$$K = \frac{1}{2}mr^2\omega^2, \tag{8.2}$$

where r is the perpendicular distance from the axis and ω is the angular frequency. If a rigid body consists of a number of individual masses, the total kinetic energy of the body is the sum of the kinetic energies of the individual masses. The kinetic energy of the ith particle with mass m_i, radial distance r_i and angular frequency ω is

$$K_i = \frac{1}{2}m_i r_i^2 \omega^2. \tag{8.3}$$

Summing over all particles, we obtain

$$K = \frac{1}{2}\sum_i \left(m_i r_i^2 \omega^2\right) = \frac{1}{2}\left(\sum_i m_i r_i^2\right)\omega^2. \tag{8.4}$$

The sum in the term on the right is the body's moment of inertia I about the axis of rotation:

$$I = \sum_i m_i r_i^2, \tag{8.5}$$

so that $K = \frac{1}{2}I\omega^2$. To calculate the moment of inertia of a continuous body, such as a solid disc, we imagine dividing the body into elements of mass dm so that all points in a particular element can be taken to have the same perpendicular distance from the axis of rotation. Then the finite sum in Equation (8.5) becomes the integral

$$I = \int r^2 dm. \tag{8.6}$$

Worked example

Show that the moment of inertia of a uniform disc of radius R and mass M is given by $I = \frac{1}{2}MR^2$. Calculate the stored rotational energy, K, in a uniform disc that has a mass of 5.0 kg, a radius of 0.25 m and spins at the frequency of 10 000 rpm.

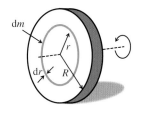

Figure 8.6 A solid uniform disc has a moment of inertia equal to $\frac{1}{2}MR^2$, where M is its mass and R is its radius.

Solution

Figure 8.6 shows the disc. The mass element $\mathrm{d}m$ is a hoop of radius r and thickness $\mathrm{d}r$. The moment of inertia of the mass element is $r^2\mathrm{d}m$. As the disc is uniform, the mass per unit area, ρ_A, is constant and equal to M/A, where $A = \pi R^2$. As the area of each mass element is $\mathrm{d}A = 2\pi r\mathrm{d}r$, the mass of each element is

$$\mathrm{d}m = \rho_A\mathrm{d}A = 2\pi\rho_A r\mathrm{d}r.$$

We thus have

$$I = \int r^2\,\mathrm{d}m = \int_0^R r^2 2\pi\rho_A r\mathrm{d}r = 2\pi\rho_A \int_0^R r^3\,\mathrm{d}r$$

$$= 2\pi\frac{M}{\pi R^2}\frac{R^4}{4} = \frac{1}{2}MR^2.$$

$$K = \frac{1}{2}\left(\frac{1}{2}MR^2\right)\omega^2 = \frac{1}{2}\left(\frac{1}{2}\times 5\times 0.25^2\right)\left(\frac{2\pi\times 10{,}000}{60}\right)^2 = 85.7\,\mathrm{kJ}.$$

K scales linearly with M, but scales as the square of ω; doubling the mass doubles the stored rotational energy, while doubling the angular frequency increases the stored energy by a factor of 4.

Also of practical importance is the energy density of a flywheel. Taking again the flywheel to be a uniform disc, the energy density is given by

$$\frac{K}{M} = \frac{1}{4}R^2\omega^2. \tag{8.7}$$

It follows that to maximise the energy density (and also the stored energy), a flywheel must spin as fast as is practicable. However, a spinning disc develops substantial stresses at high speeds due to the centrifugal force tending to pull it apart and this limits its rotational speed. The maximum stress, σ_{max}, in a rotating disc is proportional to the square of the angular frequency ω, the square of the disc radius R and also the density ρ of the disc material:

$$\sigma_{\mathrm{max}} \propto \rho\omega^2 R^2. \tag{8.8}$$

Substituting for $\omega^2 R^2$ from Equation (8.8) into Equation (8.7) we obtain

$$\frac{K}{M} \propto \frac{\sigma_{\mathrm{max}}}{\rho} = S\frac{\sigma_{\mathrm{max}}}{\rho}, \tag{8.9}$$

where S is a dimensionless constant called the *shape factor*. The value of S depends on the shape of the flywheel and this shape can be designed to maximise the value of S. For the case of a uniform disc, $S = 0.6$. The ratio σ_{max}/ρ is the important limitation to energy density and so materials with high tensile strength and low density are desirable. Composite materials such as carbon-fibre composites have these characteristics and provide much larger energy densities than steel. Thus, although first-generation flywheel energy storage systems used a large steel flywheel, newer systems use carbon-fibre composite flywheels, which are also an order of magnitude lighter. Taking σ_{max} and ρ for steel to be 5.5×10^8 N/m^2 and 7.8×10^3 kg/m^3, respectively, we find that the maximum energy density for a uniform steel disc is 0.042 MJ/kg.

A schematic diagram of a flywheel energy storage system is shown in Figure 8.7. The flywheel is mounted on the same shaft as a combination electric motor/generator. The electric motor accelerates the flywheel to high angular velocity in a matter of minutes, reaching the energy capacity much more quickly than some other methods of energy storage. The generator produces electrical power on demand by decelerating the flywheel. It is important to minimise energy losses due to frictional forces. This is achieved by mounting the flywheel assembly on magnetic bearings. This prevents any contact between the stationary and rotating parts and eliminates friction. In addition, the assembly is contained in an evacuated chamber to eliminate air resistance. With these features, the round-trip efficiency can be as high as 90%. Flywheels for energy storage typically spin at 25 000 to over 50 000 rpm and carbon-fibre flywheels achieve energy storage densities of ~0.25 MJ/kg, while typical storage capacities range from ~10 MJ to 2 GJ. Flywheel storage systems need little or no maintenance and have long lifetimes, lasting several decades or more. They can also undergo a very large number of charge/discharge cycles, ~10^6 cycles. Their main disadvantage is their relatively

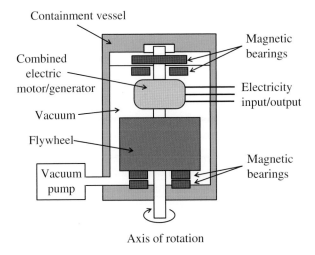

Containment vessel

Magnetic bearings

Combined electric motor/generator

Electricity input/output

Vacuum

Flywheel

Vacuum pump

Magnetic bearings

Axis of rotation

Figure 8.7 Schematic diagram of a flywheel energy storage system. The flywheel is mounted on the same shaft as a combination electric motor/generator. The electric motor accelerates the flywheel to high angular velocity. The generator produces electrical power on demand by decelerating the flywheel. Energy losses due to frictional forces are minimised by mounting the flywheel assembly on magnetic bearings in an evacuated chamber.

high capital cost. They also require strong containment vessels as a safety precaution, and this increases the total mass of the system.

Flywheel energy storage is used in the operation of the tokamak at the Joint European Torus (JET) facility for nuclear fusion (see also Section 3.5.6). Pulses of electrical power are used to produce the plasma in the tokamak. Each pulse requires a total energy of 10 GJ with peak powers exceeding 1000 MW. These pulses are ~20 s long and are followed by much longer periods during which the tokamak cools down. If these pulses of electrical power were drawn directly from the national grid, electric lights in the surrounding area would go out! Thus JET employs two massive flywheels. These can deliver the huge power levels required on the short timescale of the pulses, while energy can be accumulated in the flywheels during the longer, intervening periods. Each of the JET flywheels has a diameter of 9 m and a weight of 800 tonne. Each is spun at 225 rpm so that the flywheel edge rotates at a speed of 382 km/hour, and each flywheel is capable of storing ~4 GJ of energy with a power output of ~400MW. One flywheel supplies the toroidal field coils of the tokamak and the other supplies the poloidal coils. The remaining power required during the pulse, namely part of the toroidal coils' consumption and all the additional heating, is obtained directly from the national grid.

We have described the use of flywheel energy storage in the supply of electricity. However, the use of flywheels is more widespread. Indeed, most of flywheel technology was developed in the motor vehicle and aerospace industries. In their application to vehicles, a flywheel absorbs energy as the vehicle breaks and then supplies the stored energy as required, for example, when the vehicle travels uphill. One aspect of this application is that the rotating flywheel acts like a gyroscope and must be orientated so that it does not interfere with the steering of the vehicle.

8.5   Electrical energy storage

A charged capacitor is a store of electrical energy. The energy is stored in the electric field between the plates of the capacitor. Similarly, an inductor carrying a steady current is a store of energy. In this case, the energy is stored in the magnetic field of the inductor. Both these types of energy storage have been developed in recent times and are now used in many practical applications. One important aspect of electrical energy storage is that electrical energy is put in and electrical energy is taken out. There is no intermediate thermal energy step and so thermodynamic limitations are avoided and consequently electrical storage systems have large round-trip efficiencies, ~95%. Electrical energy can also be stored in a rechargeable battery. As we will see, this involves electrochemical reactions for the storage and subsequent delivery of the electrical energy. Fuel cells, which we will also describe, have some similarity to batteries in that they also use electrochemical reactions to deliver electrical energy.

8.5.1 Capacitors and super-capacitors

The charge Q on each plate of a capacitor and the voltage V applied across the plates are related by

$$Q = CV, \tag{8.10}$$

where C is the capacitance. To find the energy stored, imagine an intermediate stage in the charging process, where the charge on the capacitor is q and the potential difference is v. Then $v = q/C$. At this stage, the work $\mathrm{d}W$ required to transfer an additional element of charge $\mathrm{d}q$ from one plate to the other is

$$\mathrm{d}W = v\mathrm{d}q = \frac{q\mathrm{d}q}{C}. \tag{8.11}$$

The total work W needed to increase the capacitor's charge q from zero to Q is

$$W = \int_0^W \mathrm{d}W = \frac{1}{C} \int_0^Q q\mathrm{d}q = \frac{1}{2}\frac{Q^2}{C} = \frac{1}{2}CV^2. \tag{8.12}$$

If we define the energy of an uncharged capacitor to be zero, then W is the energy of the charged capacitor, where W is in joules, Q is in coulombs (C), C is in farads and V is in volts.

The capacitance of a parallel-plate capacitor is

$$C = \frac{\varepsilon\varepsilon_0 A}{D}, \tag{8.13}$$

where A is the area of the plates, D is their separation, ε_0 is the permittivity of free space and ε is the relative permittivity of the dielectric material that separates the plates. It follows from Equation (8.13) that in order to store substantial amounts of energy, it is necessary to have a large value of area A, a low value of separation D and a high value of relative permittivity ε. This is achieved in a *supercapacitor*. One type of supercapacitor is illustrated schematically in Figure 8.8(a). It has two principal differences from conventional capacitors. The capacitor plates are formed from porous carbon that is electrochemically etched so that its effective surface area is $\sim 10^5$ times larger than a smooth surface. This leads to a correspondingly large value of A. In addition, instead of a solid dielectric

Figure 8.8 Schematic diagram of a supercapacitor. The capacitor plates are formed from porous carbon with an exceptionally high effective surface area. The plates are separated by an electrolyte containing a mix of positive and negative ions. These form layers of electric charge of opposite polarity to the electrode's polarity, and these layers are at an extremely small distance from the electrodes, ~0.5 nm. These two features, together with the high relative permittivity of the electrolyte, lead to extremely high values of capacitance.

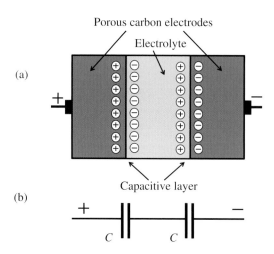

material between the capacitor plates, the plates are immersed in an electrolytic solution. This is typically potassium hydroxide or sulphuric acid in a water solvent. This electrolyte contains a mix of positive and negative ions. When the electrodes are polarised with an applied voltage, ions in the electrolyte migrate to them to form layers of electric charge of opposite polarity to the electrode's polarity, as shown in Figure 8.8(a). These layers are balanced by layers of opposite charge on the electrodes. The charge layers are separated from each other by a monolayer of solvent molecules. This makes the distance between the two charge layers extremely small, ~0.5 nm. Consequently, at each electrode, there is a capacitor with an extremely large value of area A and an extremely small value of separation D. In addition, the relative permittivity of the electrolyte is about 10. Hence, typical values of capacitance for a supercapacitor are three to six orders of magnitude larger than those of conventional electrolytic capacitors, with values measured in farads rather than microfarads. As a capacitor is formed at both electrodes, this structure effectively creates two equivalent capacitors (between each plate and the electrolyte) connected in series, as shown in Figure 8.8(b). If the capacitance at each electrolyte/plate interface is C, the total capacitance of the combination is $C/2$.

Supercapacitors have the advantages that they have no moving parts and are relatively compact. They can be charged and discharged very quickly and this cycle can be performed many hundreds of thousands of times. Moreover, the round-trip efficiency is high, < 95%. One limitation of a supercapacitor is that the voltage difference that can be applied between its plates is limited to about 2 V, as higher voltages cause the electrolyte to decompose. Supercapacitors are used in a wide range of practical applications. For example, they are used in consumer electronics such as in laptop

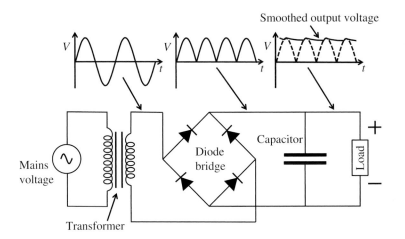

Figure 8.9 Circuit diagram for a DC power supply. The diodes rectify the AC output voltage of the transformer. The capacitor acts as a store of electrical energy that serves to smooth the rectified voltage. The upper part of the figure shows the voltage waveform at various points in the circuit.

computers, LED lighting that is energised by solar cells during the day, and hybrid motor vehicles.

Conventional electrolytic capacitors are widely used in power supplies where AC voltages are converted into DC voltages. Strictly speaking, the capacitor is a smoothing device in this application. Nevertheless it does act to store energy and is often called a reservoir capacitor. A circuit for this application is shown in Figure 8.9. The transformer reduces the AC voltage from mains voltage to the required value. The diode bridge consists of four diodes (see Section 5.3.3) and rectifies the AC voltage. The capacitor acts as a reservoir of charge and smoothes the rectified voltage as shown. The value of the capacitor in a domestic audio amplifier is about 5000 μF charged to a voltage of about 50 V. These values give a stored energy of

$$\frac{1}{2}(5000 \times 10^{-6} \times 50^2) = 6.3\,\text{J}.$$

8.5.2   Superconducting magnetic storage

An expression for the stored energy in a current-carrying inductor can be obtained in an analogous way to the energy stored in a capacitor. The result is that the stored energy (J) in an inductor of self-inductance L (farads) carrying current i (A) is $\frac{1}{2}Li^2$. This energy is stored in the magnetic field generated by the current in the inductor. What makes energy storage in inductors a practical proposition is superconductivity. When the inductor is cooled below its superconducting critical temperature, it has zero resistance. This enables the current to circulate in the inductor with essentially no loss after it has been generated. Then when the stored power is required, the circulating current can be delivered to the load.

Figure 8.10 Schematic diagram of a superconducting magnetic energy storage system. The superconducting coil of the inductor is typically made of an alloy of niobium and titanium and is bathed in liquid helium in a cryostat. A DC current of ~100 A circulates in the coil, producing a magnetic field of ~20 T. The stored energy is released by discharging the coil to an external circuit.

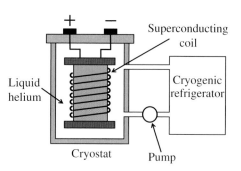

A schematic diagram of a superconducting magnetic energy storage system is shown in Figure 8.10. The inductor coil is bathed in liquid helium in a *cryostat*. The coil is usually made of a superconducting alloy of niobium and titanium. This has a critical temperature of around 9 K, which lies above the boiling point of the liquid helium (4.2 K). A DC current of ~100 A circulates in the inductor coil producing a magnetic field of ~20 T. This generates very large magnetic forces, and a robust mechanical structure is required to support the coil. Once the superconducting coil is charged, the magnetic energy, which is ~MJ can be stored indefinitely. The stored energy is released by discharging the inductor. The inductor can be charged and discharged very rapidly, delivering megawatts of power in less than a second. This ability to rapidly release stored energy makes this system very well suited to compensate for dips in the mains voltage. So these systems are used, for example, to provide very clean power in microchip manufacture. A disadvantage of this storage method is the high running costs of the liquid helium cooling system. A new generation of devices is being developed that use high-temperature superconductors. These are superconducting at the temperature of liquid nitrogen, which is much cheaper and more plentiful that liquid helium.

8.5.3   Rechargeable batteries

Rechargeable batteries are essential components of most autonomous power systems and are often used in conjunction with solar cells and small wind turbines. They are called secondary cells to differentiate them from primary cells, which are cells that consume chemicals to provide electric current and which are spent once the materials have been used up. The oldest type of rechargeable battery is the lead-acid battery, which was invented in 1859 by French physicist Gaston Planté. Despite being the oldest type, it remains the most common in use today. More recently, other rechargeable batteries such as the lithium-ion battery have been developed that are lighter and have higher energy densities.

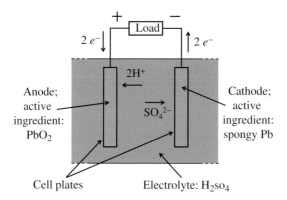

Figure 8.11 Schematic diagram of a single cell of a lead-acid battery. The cell has two plates that are made in the form of grids of typically lead-calcium alloy that support the active materials: spongy lead in the cathode and lead oxide in the anode. These plates are immersed in an electrolyte of dilute sulphuric acid, which contains hydrogen ions (H^+) and sulphate ions (SO_4^-). The ions migrate through the electrolyte, while electrons travel around the external circuit. The arrows indicate the directions of the ions and electrons during the discharge of the battery.

Lead-acid battery

A single cell of a lead-acid battery is illustrated in Figure 8.11. A battery typically consists of six cells connected in series. Each cell gives a nominal *electromotive force (emf)* or terminal voltage of about 2 V, giving a battery voltage of nominally 12 V. Each cell has two plates which are made in the form of grids, and these are typically made from lead-calcium alloy and support the active materials. The active materials are spongy lead (Pb) in the cathode and lead oxide (PbO_2) in the anode. The plates are immersed in an electrolyte of dilute sulphuric acid (H_2SO_4). This electrolyte dissociates into hydrogen ions (H^+) and sulphate ions (SO_4^{2-}) and conduction of current through the electrolyte is by the movement of these ions between the plates. Although electrons can readily pass through a conductor such as copper, free electrons cannot pass through an electrolyte. Instead, they travel through an external circuit that connects the two plates of the battery.

Discharge of the battery When the battery is discharging, SO_4^{2-} ions migrate to the cathode and H^+ ions migrate to the anode. At the cathode, the SO_4^{2-} ions react with the spongy lead and reduce it to lead sulphate:

$$Pb + SO_4^{2-} \rightarrow PbSO_4 + 2e^-.$$

At the cathode, the H^+ ions react with the lead oxide and with SO_4^{2-} ions in the electrolyte and with the electrons that travel around the external circuit from the cathode. The chemical reaction can be written as

$$PbO_2 + SO_4^{2-} + 4H^+ + 2e \rightarrow PbSO_4 + 2H_2O.$$

Again lead sulphate is formed. In the discharging reaction, water is formed and sulphuric acid is consumed, which is reflected in a reduced density of the acid. Conventional current flows through the load (in the opposite direction to the electron flow) and electrical power is delivered to it.

Charging of the battery   The battery is charged by connecting it, in opposition, to a voltage supply of greater emf. This forces a current through the cell in the opposite direction to the discharge current, so that H^+ ions are carried to the cathode and SO_4^{2-} ions to the anode. The chemical reaction at the cathode is

$$PbSO_4 + 2H^+ + 2e \rightarrow Pb + H_2SO_4$$

and at the anode is

$$PbSO_4 + 2H_2O + SO_4^{2-} - 2e \rightarrow PbO_2 + 2H_2SO_4.$$

The active materials are converted back to lead and lead oxide, water is consumed and sulphuric acid is formed. The acid therefore becomes more concentrated during the charging cycle and its density rises.

The storage capacity of a lead-acid battery is usually given in ampere-hours (Ah), where 1 Ah = 3600 C. For a motor vehicle battery the capacity is typically 30 Ah and for a terminal voltage of 12 V this gives a stored energy $(Q \times V)$ of $30 \times 3600 \times 12 \approx 1.3$ MJ. If the battery weighs 10 kg, this gives an energy density of 0.13 MJ/kg. This is a relatively small energy density compared with some other storage methods. However, a lead-acid battery has a very low internal resistance, < 0.1 ohm, which means that it can deliver very high currents and thus has a very high *power-to weight ratio*. For example, a delivered current of 50 A at 12 V gives a power density of 60 W/kg for a mass of 10 kg. This ability to deliver very large currents is a key advantage of the lead-acid battery and is put to use, for example, to start the engine of a motor vehicle. Furthermore, a lead-acid battery can undergo many hundreds of charging/discharging cycles. The round-trip efficiency of a rechargeable battery is $\sim 80\%$.

One disadvantage of a lead-acid battery is its heavy weight. This problem is reduced in the lithium-ion battery, where the plates are made from lightweight lithium and carbon. Moreover, as lithium is a highly reactive element, a relatively large amount of energy is released in the chemical reactions that occur at the plates. Consequently, lithium-ion batteries have a very high energy density compared with lead-acid batteries and so can be substantially smaller for a given application. This means that lithium-ion batteries find widespread use, especially in consumer electronics.

8.5.4   Fuel cells

A fuel cell is a device that converts chemical energy into electricity directly and does so by electrochemical processes; we can think of its operation as the opposite process to electrolysis. A fuel cell is not really a store of energy but instead requires a continuous flow of the chemical reactants to produce electrical energy. Nevertheless it is a way of harvesting the stored chemical energy in hydrogen gas which gives rise to the concept of the hydrogen economy (see Section 8.2.2). The fuel cell was invented in 1838 by the Welsh scientist William Grove. However, it was not used in a practical application until more than a century later. This was to provide both electricity and water for probes, satellites and space capsules in NASA space programmes.

There are various types of fuel cell. However, like a battery, they all consist of an anode, a cathode and an electrolyte that separates them. It is the electrolyte that distinguishes the different types of cell. Hydrogen is the most common fuel and is combined with oxygen in the fuel cell to produce electrical energy. However, other fuels that are rich in hydrogen, such as natural gas and methanol, may also be used, and *oxidising* agents other than oxygen may be used to react with the fuel. In all cases, ions pass between the anode and cathode, while electrons are drawn from the anode to the cathode through an external circuit, producing DC electricity and delivering electrical power to the load.

The principle of operation of a fuel cell, using phosphoric acid as the electrolyte, is presented schematically in Figure 8.12. This shows the anode and cathode electrodes and the electrolyte that separates them. The electrolyte is at a temperature of 150–200°C. Hydrogen gas is fed into one side of the cell and oxygen gas into the other. The anode and cathode are made of porous carbon and the porosity allows hydrogen and oxygen gases to pass into the electrolyte. Each electrode is coated with a thin layer of catalytic material, which consist of a thin layer of carbon *nanoparticles* that are coated with

Figure 8.12 Schematic diagram of a fuel cell using hydrogen as fuel and oxygen as the oxidising agent. It shows the anode and cathode electrodes and the electrolyte that separates them. Each of these electrodes is coated with a thin layer of catalytic material, which facilitates the chemical reactions occurring at each electrode. Hydrogen gas is fed into one side of the cell and oxygen gas into the other. The action of the cell is that hydrogen ions migrate from the anode to the cathode where they combine with oxygen molecules and electrons that have passed through the external circuit to form molecules of water. A voltage of ~0.7 V is developed between the anode and cathode and a DC current is delivered to the load.

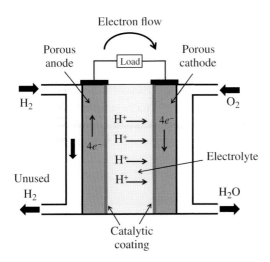

platinum. The catalytic layer lies at the interface between the electrodes and the electrolyte and its purpose is to facilitate the chemical reactions occurring at each electrode. When hydrogen molecules diffuse through the porous anode and reach the electrolyte, they ionise into hydrogen ions and electrons:

$$2H_2 \rightarrow 4H^+ + 4e^-.$$

The hydrogen ions are drawn through the electrolyte from the anode to the cathode, while the electrons travel around the external circuit. When the hydrogen ions reach the cathode, they combine with oxygen molecules and the electrons that have passed through the external circuit. The reaction is

$$4H^+ + O_2 + 4e^- \rightarrow 2H_2O.$$

The overall reaction of the cell is then

$$2H_2 + O_2 \rightarrow 2H_2O,$$

and water is the only reaction product. The potential difference that is developed between the anode and cathode is typically ~ 0.7 V, while the output current is proportional to the area of the electrodes. Typically a fuel cell delivers ~ 1 W/cm^2 of electrode area to an external load.

Fuel cells have a number of advantages. They operate silently with no moving parts and they provide continuous power as long as the chemical ingredients are fed into the cell. As the only reaction product is water, there is no pollution. Moreover, as they convert chemical energy directly into electricity, fuel cells are not subject to the thermodynamic laws that limit the efficiencies of thermal power plants. The efficiency of a fuel cell in converting the energy contained in the fuel into electricity is typically in the range 40–60%. The main disadvantage of fuel cells is their initial capital cost. In addition, the storage of hydrogen gas has technological challenges. As individual fuel cells produce relatively small voltages between the anode and cathode, they are often connected in series to obtain higher voltages. And to provide more output current they can be connected in parallel. Fuel cells are used to produce power for commercial, industrial and residential buildings and in remote or inaccessible areas. They also have much potential to power motor vehicles and several manufacturers are developing fuel-cell vehicles.

8.6   Distribution of electrical power

Energy must be delivered from its source to where it is needed. For the energy sources we have discussed, the end product is usually electricity, because electricity provides a convenient way of distributing power. The electricity is usually distributed by a network of transmission lines called a grid, as we have already mentioned. A national grid covers the whole country and indeed it may connect with several other countries. Large thermal power plants such as a gas-fired power stations or nuclear power plants may be situated a large distance away from the consumer. A grid distributes electricity from the power stations to industrial, commercial and domestic users. In addition, it connects power stations together. This facilitates balancing the power generated by these power stations with different regional demands for power. A grid also allows intermittent energy sources and energy storage systems to be connected to end-users.

Two important features of a national grid are that the electricity is transmitted at very high voltage and as alternating current (AC) rather than direct current (DC). High voltages are used to minimise joule heating losses in the power lines, while AC voltages are standard because of the ease with which they can be increased or reduced by a transformer. The AC frequency is 50 Hz in Europe and 60 Hz in the USA. However, it is worth noting that solid-state power electronic components increasingly enable DC/AC/DC conversion at large powers, so that transforming DC voltage is not as difficult as it once was.

Suppose that a power station delivers power P at an AC voltage V to a long-distance transmission line that has a resistance R. Then the current, i, flowing through the line is P/V. (Since we are dealing with AC voltages, i and V are root-mean-square values.). Some of this power will be dissipated as joule heat in the transmission line. The power dissipated is $i^2 R$ and hence the ratio of the lost power to the power leaving the power station is

$$\frac{i^2 R}{P} = \frac{PR}{V^2}. \tag{8.14}$$

This ratio decreases rapidly with V and hence it is best to deliver electrical power at high voltage, and typical values of V are 275 and 400 kV; an upper limit to the operating voltage is electrical breakdown in the air surrounding the transmission lines, which are not insulated. The transmission lines are usually made from aluminium. Its resistivity $(2.8 \times 10^{-8}\,\Omega\,\text{m})$ is higher than for copper $(1.7 \times 10^{-8}\,\Omega\,\text{m})$, but it is cheaper and lighter. There are other power losses in a high-voltage transmission line such as corona discharge, but these are usually small compared with joule heating losses.

The voltage produced by an electrical generator in a power plant is typically about 20 kV. A transformer is used to increase this to 275 kV or more for delivery down the transmission line. At the other end of the line, a transformer is used to reduce the voltage to about 30 kV. The electrical power is then further distributed to other transformers that reduce the voltage to a value suitable for industrial,

Figure 8.13 Arrangements for sending electrical power via a national grid. The AC voltage from a power station is transformed to much higher voltages to minimise power losses in the transmission line. At the other end of the line, it is transformed to lower voltages to suit industrial, commercial or domestic consumers. The transmission lines are usually made from aluminium, which has relatively low electrical resistivity and density.

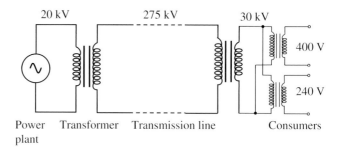

commercial or domestic purposes, e.g. 400 or 240 V. These arrangements are illustrated in Figure 8.13. As a particular example, the national grid in the UK has almost 200 large power stations connected to it. Part of the grid operates at a voltage of 400 kV over a circuit length of 11500 km, part of it operates at 275 kV over a length of 9800 km, and part operates at 132 kV over a length of 5250 km. The maximum amount of power that the grid can handle is about 80 GW and usually it is run at about 80% of its capacity.

Problems 8

8.1 Carbon combines with oxygen to form carbon dioxide in the reaction $C + O_2 \rightarrow CO_2$, releasing 394 kJ of energy per mole of carbon. Determine the energy density of coal in units of MJ/kg and kWh/kg assuming that the coal consists entirely of carbon. How much CO_2 gas is produced when 1 tonne of carbon is burnt and how much energy is produced?

8.2 It is proposed to construct a thermally insulated tank of water beneath a house to store thermal energy. The water would be heated to 50°C by solar energy during the summer and the stored thermal energy released during winter months to heat the house. Estimate the required volume of the tank. Assume that the required energy for heating the house is 24 kWh/day, that this needs to be sustained for a period of 4 months, and that the house is to be maintained at a temperature of 20°C. Neglect any heat loss from the tank to the surrounding earth. The thermal capacity of water is 4200 J/kg K.

8.3 A domestic storage heater is to deliver an average power of 1 kW for a period of 12 hours to a room at an ambient temperature of 20°C. (a) Clay bricks in the storage heater are heated to a temperature of 60°C by low-tariff electricity overnight. What is the required total volume of the bricks? The heat capacity of the bricks is 0.95 kJ/kg K and their density is 2600 kg/m³. (b) The bricks are covered by a layer of thermal insulation of thermal conductivity 0.09 W/m² K, which allows the stored heat to be delivered to the room over the 12-hour period. Assuming an average temperature difference of 20°C between the bricks and the ambient temperature, estimate the thickness of the insulating layer.

8.4   It has been suggested that a chamber for a pumped hydroelectric plant be constructed 1 km below London. Assuming that 3 kWh of stored energy is required per person and that the population of London is 8.7 million, estimate the volume of water the chamber would have to contain.

8.5   (a) Show that the moment of inertia I of a solid sphere of mass M and radius R about an axis passing through its centre is given by $I = \frac{2}{5}MR^2$.

8.6   Consider a passenger bus of total mass 5000 kg that is partly powered by a flywheel. The flywheel is a solid steel cylinder of mass 250 kg and diameter 0.75 m. At how many revolutions per minute must the flywheel spin if the stored energy is required to power the bus up a hill of height 20 m? Can a steel flywheel of these dimensions be safely operated at this rotational frequency?

8.7   (a) Show that the energy E of an inductor of self-inductance L carrying current i is given by $E = \frac{1}{2}Li^2$. (b) A superconducting inductor has a self-inductance of 1.8 H. What current circulates in the coil of the inductor when 1 MJ of energy is stored? What is a typical value of an inductor in a domestic radio?

8.8   A battery of internal resistance r is connected to a load of resistance R. Show that maximum power is delivered to the load when $R = r$.

8.9   A transmission line connects a 250 MW power station to a city 200 km away at 440 kV AC. Calculate the dimensions of the conducting wires if the dissipation of power due to joule heating must be less than 2% of the power carried. The conducting wires are manufactured from aluminium, which has an electrical resistivity of 2.8×10^{-8}ohm m.

8.10 In one kind of *district heating*, steam is transported to buildings via thermally insulated pipes. A pipe of outer diameter 100 mm carries steam at 100°C, and the pipe is insulated by a layer of material of width 20 mm and thermal conductivity 0.04 W/m^2 K. Estimate the heat loss along the pipe if the length of the pipe is 100 m and the outside temperature is 10°C.

Solutions to problems

Problems 1

1.2 Assuming six lights with a power of 100 W are used for an average of 5 hours/day, Total energy consumed $= 6 \times 100 \times 5 \times 365 = 1095$ kWh.

1.3 (a) Energy density of 36 MJ/L $= 10$ kWh/L. Hence total energy per passenger

$$= \frac{2 \times 95\,000 \times 35}{450} = 4200 \text{ kWh.}$$

(b) Total energy required

$$= \frac{48 \times 5 \times 30 \times 10}{15} = 4800 \text{ kWh.}$$

In these examples, the energy cost per person of a long-distance plane flight is comparable to the yearly energy cost of a person commuting to work by car.

1.4 Energy in chocolate bar $= 230 \times 4.2 = 966$ kJ.
Therefore vertical distance

$$= \frac{966 \times 1000}{1000 \times 9.8} \approx 100 \text{ m.}$$

Problems 2

2.1 $\lambda = 1.88 \times 10^{-12}$ m, $\nu = 1.60 \times 10^{20}$ Hz.

2.2 (a) 2.74 fm, 2.30×10^{17} kg/m³, 4.6 fm, 2.28×10^{17} kg/m³, 7.12 fm, 2.30×10^{17} kg/m³.
(c) 1.11×10^{25} C/m³, 1.03×10^{25} C/m³, 0.88×10^{25} C/m³.
Nuclear charge density reduces with A because A increases faster than Z.

2.3 $\Delta m = 1.095603$ u, $B = 1020.5$ MeV, $B/A = 8.504$ MeV.

2.4 (a) Step (1): $\Delta m = {}^4\text{He} - ({}^3\text{H} + {}^1\text{H}) = 4.002\,603\,\text{u} - (3.016\,049 + 1.007\,825)\,\text{u} = -0.021\,27\,\text{u} \equiv -19.81$ MeV The minus sign means that energy must be supplied to remove the proton.
Steps (2) and (3): -6.258 MeV, -2.224 MeV.
Adding these three values together gives the total energy required for the three steps: 28.30 MeV.
Binding energy of α particle given by $(2 \times 1.007825 + 2 \times 1.008665 - 4.002603)$ u $= 0.03038$ u $\equiv 28.30$ MeV, in agreement with previous result, as expected.
(b) 20.58 MeV; it takes less energy to remove a proton than a neutron because of the Coulomb repulsion of the other proton.

Physics of Energy Sources, First Edition. George C. King.
© 2018 John Wiley & Sons, Ltd. Published 2018 by John Wiley & Sons, Ltd.

2.5

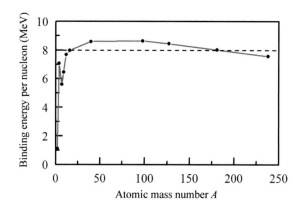

From plot, estimate $B/A = 8.5$ MeV for ^{119}Pd. $B/A = 7.57$ MeV for ^{238}u. Hence released energy $\sim 2 \times (119 \times 8.5) - (238 \times 7.57) \sim 220$ MeV.

2.6 The semi-empirical mass formula, Equation (2.28), represents a parabola of $^{A}_{Z}M$ versus Z. To find value of Z of most stable nuclide, find the minimum in $^{A}_{Z}M$ by setting $\mathrm{d}M/\mathrm{d}Z = 0$, keeping A constant and noting that $N = (A - Z)$. For $A = 121$, $\mathrm{d}M/\mathrm{d}Z = 0$ gives $Z = 51.3$ corresponding to $^{121}_{51}$Sb.

2.7 (a) 57.5 N, (b) 1.24×10^{36}, (c) work done = force $\times$ distance moved. Taking distance moved by a nucleon to be range of nuclear force, say 2 fm, then nuclear force is

$$\sim \frac{28 \times 10^{6} \times 1.6 \times 10^{-19}}{4 \times 2 \times 10^{-15}} \sim 520 \text{ N}.$$

2.8 1 g of radium-226 contains $N_{\mathrm{A}}/226 = 2.665 \times 10^{21}$ atoms, where N_{A} is Avogadro's number. $t_{1/2} = 1580$ years. Remaining amount = 191 g.

2.9

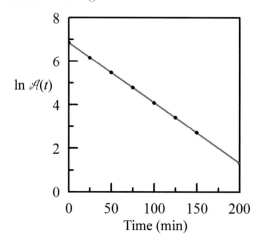

$$\text{Slope} = \frac{1.31 - 6.86}{200 - 0}, \text{ giving } \lambda = 0.0278 \text{ s}^{-1}, \text{ and } t_{1/2} = 25 \text{ min}.$$

2.10 ^{238}U has a half-life $\gg$ 24 hours and we can consider the radioactivity to be constant. Mean count $m = 0.29 \times 10 = 2.9$ counts/10 s. $P(0) = 0.055$, $P(1) = 0.160$, $P(2) = 0.231$, $P(3) = 0.224$, $P(4) = 0.162$.

Probability of getting at least one count = $[1 - P(0)] = 0.945$.

Probability of getting at least four counts = $[1 - (0.055 + 0.160 + 0.231 + 0.162)] = 0.392$.

2.11

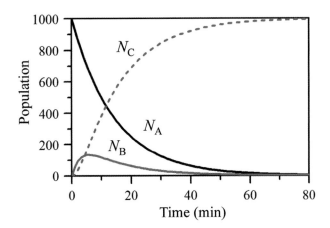

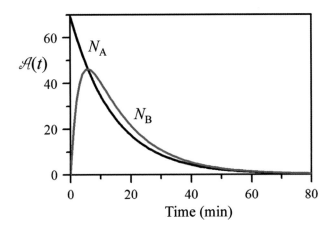

2.12 (a) Carbon has two stable isotopes, with natural abundances of 99% and 1% respectively. So ignoring the tiny amount of ^{14}C, we may say to a very good approximation that 1 g of natural carbon contains

$$\frac{1.35 \times 10^{-12} \times N_A}{12} \quad \text{atoms of } ^{14}C.$$

Activity due to ^{14}C decay $= 15.6$ decays/s.

(b) We expect 200 g of carbon to produce $250 \times 15.6 = 3900$ decays/min. Therefore the time t that elapses until the decay rate has reduced to 500 decays/min is given by

$$500 = 3900 \exp\left(-\frac{\ln 2t}{t_{1/2}}\right) \Rightarrow t = 1.7 \times 10^4 \text{ years.}$$

2.13 (a) $^{235}_{92}U_{143}$, $^{24}_{12}Mg_{12}$, $^{15}_{7}N_8$; (b) positron, alpha particle, electron.

2.14 (a)

$$\frac{1}{2}M_D V_D^2 \left[1 + \left(\frac{M_D}{m_\alpha}\right)\right] \approx \frac{1}{2}M_D V_D^2 \left(\frac{M_D}{m_\alpha}\right), \quad \text{since } M_D \gg m_a$$

$$\therefore \frac{1}{2}M_D V_D^2 \approx Q\frac{m_\alpha}{M_D} \approx Q\frac{4}{A_D}.$$

Similarly,

$$\frac{1}{2}m_\alpha v_\alpha^2 \approx Q\left(1 - \frac{4}{A_D}\right).$$

(b) 1.52×10^7 m/s, 2.76×10^5 m/s.

2.15 $^{223}_{88}\text{Ra} \rightarrow {}^{219}_{86}\text{Rn} + {}^4_2\text{He}$:

$$Q = (223.018\ 501 - 219.009\ 485 - 4.002\ 603) \times 931.5 = +5.97 \text{ MeV}.$$

$$V_0 = \frac{86 \times 2 \times (1.6 \times 10^{-19})^2}{4\pi \times 8.85 \times 10^{-12} \times (7.22 + 1.90) \times 10^{-15} \times 1.6 \times 10^{-19}} \text{eV} = 27.1 \text{ MeV}.$$

$$^{223}_{88}\text{Ra} \rightarrow {}^{209}_{82}\text{Pb} + {}^{14}_6\text{C}: \quad Q = +31.9 \text{ MeV}, \quad V_0 = 74.2 \text{ MeV}.$$

Barrier height V_0 is more than twice as large for ^{14}C emission and hence there is a much lower probability for tunnelling through barrier.

2.16 (a) Mass difference $= +0.003\ 032$ u, a positive quantity, showing that β^- decay is possible for $^{60}_{27}\text{Co}$. Indeed this nuclide is an important source of β radiation for medical applications. $E_{\max} = 2.82$ MeV. (b) β^+ decay is not allowed for a free proton because $m_p > m_n$.

2.18 Number of alpha particle decays required is

$$= \frac{15}{5.59 \times 10^6 \times 1.9 \times 10^{-19}} \text{ decays/s},$$

$$N = \frac{\mathcal{A}t_{1/2}}{\ln 2} = 6.70 \times 10^{22} \text{ atoms}.$$

$$\text{Required mass of plutonium} = \frac{238 \times 6.70 \times 10^{22}}{6.022 \times 10^{23}} = 26.5 \text{ g}.$$

Problems 3

3.1 (i) $Q = +6.84$ MeV,

(ii) $Q = [(12 + 4.002\ 603) - (15.994\ 915 + 7.115/931.5)] \times 931.5 = +0.0464$ MeV.

(iii) $Q = -0.389$ MeV. (b) 3.01605 u.

3.2 Each fission reaction produces $200 \times 10^6 \times 1.6 \times 10^{-19}$ J. Therefore we require, 3.1×10^{19} fissions/s. The mass of a ^{235}U atom is $\frac{235.0 \times 10^{-3}}{6.02 \times 10^{23}} = 39.0 \times 10^{-26}$ kg. Hence, total mass of ^{235}U required/day is $39 \times 10^{-26} \times 3.1 \times 10^{19} \times 60 \times 60 \times 24 = 1.04$ kg ≈ 1 kg.

3.3 Mass of $^{235}_{92}\text{U}_{143} = 92M_H + 143m_n - B(92,235)/c^2$; mass of $^{87}_{35}\text{Br}_{52} = 35M_H + 52m_n - B(35,87)/c^2$; mass of $^{147}_{57}\text{La}_{90} = 57M_H + 90m_n - B(57,147)/c^2$. Taking into account the incident neutron and the two neutrons produced in the reaction, the masses M_H and m_n cancel. Hence,

$$Q = \Delta mc^2 = -B(92,235) - [-B(35,87)] - [-B(57,147)].$$

For A-odd nuclei, $\delta = 0$ in SEMF.

$$B(92,235) = 15.5 \times 235 - 16.8 \times 235^{2/3} - \frac{0.72 \times 92(91)}{235^{1/3}} - \frac{23 \times (235 - 184)^2}{235} = 1772 \text{ MeV}$$

$$B(35,87) = 750 \text{ MeV}, \quad B(57,147) = 1206 \text{ MeV}.$$

Hence $Q = -1772 - (750 + 1206) = 184$ MeV.

This problem shows that the energy release is equal to the difference in the total binding energy of the particles before and after the reaction.

3.4
$$R = \frac{1700 \times 10^{-24} \, (\mathrm{cm}^2) \times 6.02 \times 10^{23} \times 40 \times 10^{-3} \times 2.5 \times 10^{13} \, (\mathrm{cm}^{-2})}{58} = 1.76 \times 10^{13} \text{ reactions/s.}$$

Therefore number of ^{59}Co nuclei are formed after 1 hour $= 1.76 \times 10^{13} \times 60 \times 60 = 6.4 \times 10^{16}$.

3.5
$$R = \frac{2000 \times 6.02 \times 10^{23} \times 0.72 \times 10^{-2}}{238.029} \times 1.0 \times 10^{13} \, (\mathrm{cm}^{-2}) \times 584 \times 10^{-2} \, (\mathrm{cm}^2)$$
$$= 2.13 \times 10^{14} \text{ fissions/s.}$$

Assuming each fission produces an average energy of 200 MeV, power $= 6.8$ kW.

3.6 (i) 25; (ii) 30; (iii) 114; (iv) 2092.

3.7 (a)
$$\lambda_{\mathrm{a}} = \frac{235}{18.95 \times 6.02 \times 10^{23} \times 1.4 \times 10^{-24}} = 14.7 \text{ cm.}$$

$$v = \sqrt{\frac{1 \times 10^6 \times 1.6 \times 10^{-19} \times 2}{1.67 \times 10^{-27}}} = 1.38 \times 10^7 \text{ ms}^{-1} \Rightarrow \tau_{\mathrm{a}} = 1.07 \times 10^{-8} \text{ s}$$

(b) Similarly, $\lambda_{\mathrm{a}} = 2.98$ cm with

$$\tau_{\mathrm{s}} = \frac{2.98 \times 10^{-2}}{1.38 \times 10^7} = 0.22 \times 10^{-8} \text{ s.}$$

Hence, in pure ^{235}U with fast neutrons, scattering happens more frequently than absorption, which might lead to fission, which is not a good situation in a reactor.
(c) For thermal energies (1/40 eV), obtain $\lambda_{\mathrm{a}} = 0.03$ cm and $\lambda_{\mathrm{s}} = 2.06$ cm, with $v = 2.19 \times 10^3$ m/s, giving $\tau_{\mathrm{a}} = 1.37 \times 10^{-7}$ s and $\tau_{\mathrm{s}} = 9.41 \times 10^{-6}$ s. Most neutrons get absorbed before they scatter, which is much better for chain reaction.

3.8 For a 3% enriched uranium,

$$\sigma_{\mathrm{f,eff}} = \frac{3}{100} 584 = 17.52 \text{ b,}$$

since $\sigma_{\mathrm{f},238} = 0$, and

$$\sigma_{\mathrm{a,eff}} = \frac{3}{100} \times 96 + \frac{97}{100} \times 2.72 = 5.52 \text{ b.}$$

Hence $\eta = 1.84$.

3.9 Considering the fission reaction $^{240}_{94}\mathrm{Pu}_{146} + \mathrm{n} \rightarrow {}^{241}_{94}\mathrm{Pu}_{147} + Q$, $Q = [(240.053\,808 + 1.008\,665)$ u $- 241.05\,6846\mathrm{u}] \times 931.494 = 5.24$ MeV. Since the activation energy for $^{240}_{94}\mathrm{Pu}_{146}$ is 6.3 MeV, thermal neutrons would not provide the extra energy required for induced fission.
For $^{239}_{94}\mathrm{Pu}_{145}$, $Q = 6.53$ MeV and thermal neutrons will induce fission.

For $^{238}_{92}\mathrm{u}_{146}$, $Q = 4.81$ MeV and thermal neutrons will not induce fission.

For $^{242}_{95}\mathrm{Am}_{147}$, $Q = 6.36$ MeV and thermal neutrons will induce fission.

3.10 (a) $R = R_0 \left(A_1^{1/3} + A_2^{1/3} \right)$ gives $R = 11.7 \times 10^{-15}$ m, and electrostatic energy $= 249$ MeV. From atomic masses, released energy $Q = 180$ MeV.

3.11 $Q = +5.0$ MeV.

3.12 463 stages.

3.13 The most probable speed v_{m} can be found by setting $\frac{\mathrm{d}P(v)}{\mathrm{d}v} = 0$, which gives $v_{\mathrm{m}} = \sqrt{\frac{2kT}{m}}$. Hence, kinetic energy for most probable speed $= kT$.

3.14 (a) (i) $Q = 3.27$ MeV, energy of neutron $= 2.45$ MeV. (ii) $Q = 4.03$ MeV, energy of proton is again 75% of total energy release.

(b)
$$\tau > \frac{12 \times 10^4}{4.03 \times 10^6 \times 5 \times 10^{-25} \times 10^{20}} \approx 600 \text{ s}.$$

3.15 (a) Distance of proton centres when they are just touching $= 2.4 \times 10^{-15}$ m. Total electrostatic energy $= 596$ keV. Hence, each proton must be accelerated through 298 kV. (b) $T = \frac{2K_{\text{mean}}}{3k} = 2.3 \times 10^9$ K.

3.16 $\Delta m = [4M_{\text{H}} - 4m_{\text{e}}] - [M_{\text{He}} - 2m_{\text{e}} + 2m_{\text{e}}] = [4M_{\text{H}} - 4m_{\text{e}}] - [M_{\text{He}}] = 0.026\,501 \text{ u} \equiv 24.69 \text{ MeV}.$
Adding the energy of the two γ rays gives a total energy of 26.7 MeV.

3.17 (a) Density $\rho =$ number density n (/m) $\times$ particle mass (kg). Assuming a mean atomic mass of 2.5 u,

$$n = \frac{200}{2.5 \times 1.66 \times 10^{-27}} = 4.8 \times 10^{28} \text{m}^{-3}$$

in uncompressed pellet. Therefore, n in compressed pellet is $4.8 \times 10^{31}/\text{m}^3$. (b) According to the Lawson criterion, $n\tau > 10^{20}$ s/m^3 for a T-D reaction. Hence, $\tau > 2 \times 10^{-12}$ s.

3.18 For fusion of ^{2}D:

$$E = \frac{1}{2} \times \frac{6.02 \times 10^{23} \times 10^3 \times 3.27 \times 10^6 \times 1.6 \times 10^{-19}}{2} \approx 8 \times 10^{13} \text{ J}.$$

For fission of ^{235}U:

$$E = \frac{6.02 \times 10^{23} \times 10^3 \times 200 \times 10^6 \times 1.6 \times 10^{-19}}{235} \approx 8 \times 10^{13} \text{ J}.$$

Energy output from both is about the same. Time $\approx 2.5 \times 10^4$ years.

3.19 Radius,

$$r = \frac{\sqrt{2mE}}{qB} = \frac{\sqrt{2 \times 20 \times 10^3 \times 1.6 \times 10^{-19} \times 2 \times 1.66 \times 10^{-27}}}{1.6 \times 10^{-19} \times 3} = 9.61 \times 10^{-3} \text{ m}.$$

Cyclotron frequency,

$$f = \frac{\text{v}}{2\pi r} = \frac{qB}{2\pi m} = \frac{1.6 \times 10^{-19} \times 3}{2\pi \times 2 \times 1.66 \times 10^{-27}} = 23 \text{ MHz}.$$

Problems 4

4.1 (a) 1.06 mm, (b) 1.93 nm.

4.2 (a) Keeping a constant, let $T_{\text{E}}^4 = \dfrac{C}{e}$, where C is a constant.
Then

$$4T_{\text{E}}^3 dT_{\text{E}} = -\frac{C}{e^2} de = \frac{eT_{\text{E}}^4}{e^2} de \quad \text{and} \quad \frac{\Delta T_{\text{E}}}{T_{\text{E}}} \approx -\frac{1}{4}\frac{\Delta e}{e}.$$

Thus a 1% change in T_{E} is produced by a 4% change in e. The minus sign indicates that the temperature goes up as the emissivity goes down.

(b) In a similar way we find

$$\frac{\Delta T_{\text{E}}}{T_{\text{E}}} \approx +\frac{1}{4}\frac{\Delta a}{a}.$$

Again, a 1% change in T_{E} is produced by a 4% change in e. Now, however, the temperature goes up as the absorption factor a increases.

4.3 $hc/\lambda kT = 4.95$, $W_e(502.5 \text{ nm}) = 8.35 \times 10^{13}$ W/m^2 m.

Then, $W_e(\lambda)d\lambda = 0.43$ MW/m^2.

Total power radiated per unit area emitted by the Sun given by $W = \sigma T^4 = 64$ MW/m^2. Hence fraction of power between 500 and 505 nm = 0.7%.

4.4 Net rate of heat loss given by $W = e\sigma A \left(T_{\text{skin}}^4 - T_{\text{room}}^4\right)$. Taking the height and the waist size of the person to be 1.7 and 0.85 m, respectively, the heat loss is 130 W. This radiation heat loss is considerably reduced by clothing, which because of its low thermal conductivity has a lower temperature.

4.5
$$W_e(\lambda) = \frac{2\pi hc^2}{\lambda^5 \left(e^{hc/\lambda kT} - 1\right)}.$$

To find $\lambda_{\max}$, set $dW_e/d\lambda = 0$, which gives

$$-2\pi hc^2 \frac{d}{d\lambda}\left[\lambda^5 \left(e^{hc/\lambda kT} - 1\right)\right] = 0.$$

This leads to

$$5e^{hc/\lambda kT} - 5 = \frac{hc}{\lambda kT}e^{hc/\lambda kT}.$$

Letting $x = hc/\lambda kT$, obtain $5 - x = 5e^{-x}$, and $x = 4.965$. Hence, $\lambda_{\max} = hc/4.965kT$.

4.6 To find total emitted power, integrate $W_e(\lambda)$ over all wavelengths:

$$\int_0^\infty W_e(\lambda)d\lambda = \int_0^\infty \frac{2\pi hc^2}{\lambda^5 \left(e^{hc/\lambda kT} - 1\right)}(\lambda)d\lambda.$$

Let $x = hc/\lambda kT$. This gives the integral

$$\left(2\pi hc^2\right)\left(\frac{kT}{hc}\right)^4 \int_0^\infty \frac{x^3}{(e^x - 1)}dx = \left(2\pi hc^2\right)\left(\frac{kT}{hc}\right)^4 \frac{(2\pi)^4}{240} = \frac{2\pi^5 k^4}{15h^3 c^3}T^4.$$

4.7
$$\rho(\lambda)d\lambda = \frac{8\pi hc}{\lambda^5 \left(e^{hc/\lambda kT} - 1\right)}d\lambda.$$

$$e^{hc/\lambda kT} = 1.21 \times 10^2;\ \rho(601) = 5.31 \times 10^5 \text{ J/m}^3 \text{ m}.$$

$$W_e(\lambda)d\lambda = \frac{1}{4}c\rho(\lambda)d\lambda = 7.97 \text{ W/m}^2.$$

Therefore, power radiated through hole in the given bandwidth = 6.26 W.

4.8 Thermal resistance $R = L/\kappa$. Hence, $R_{\text{brick}} = 0.1/0.5 = 0.2$ m^2 K/W, $R_{\text{ins}} = 0.05/0.025 = 2.0$ m^2 K/W, $R_{\text{brick}} = 0.1/0.5 = 0.2$ m^2 K/W and $R_{\text{plaster}} = 0.024/0.8 = 0.03$ m^2 K/W.

Therefore $R_{\text{eff}} = 2.43$ m^2 K/W, and

$$H = \frac{6 \times (20 - 5)}{2.43} = 37 \text{ W}.$$

(c) Temperature between first brick component and insulation is $5 + \frac{0.2}{2.43} \times 15 = 6.23$ °C, Temperature between insulation and second brick component is $5 + \frac{0.2 + 2.0}{2.43} \times 15 = 18.58$ °C, Temperature between second brick component and plaster is 19.82°C.

4.9 (b) $\ln(r_2/r_1) = 2\pi l\kappa(T_2 - T_1)/H$ gives $2r_2 = 39.3$ mm ≈ 40 mm.

4.10 (a) Because of the spherical symmetry of the material, the temperature depends only on the radial distance r and so we can use the one-dimensional form of Fourier's law:

$$H = -\kappa A \frac{dT}{dr}.$$

At equilibrium, the heat flux H passing through any spherical surface of the insulator will constant. Since the area of the shell is $4\pi r^2$, we have

$$H = -4\pi r^2 \kappa \frac{dT}{dr}.$$

Thus,

$$\int_{T_1}^{T} dT = -\frac{H}{4\pi\kappa} \int_{r_1}^{r} \frac{dr}{r^2},$$

which gives

$$T(r) = T_1 - \frac{H}{4\pi\kappa} \left(\frac{1}{r} - \frac{1}{r_1} \right).$$

(b) $H = 4\pi\kappa(T_1 - T_2)\dfrac{r_1 r_2}{(r_2 - r_1)}.$

4.11

(a) $H = \dfrac{2\pi l \kappa (T_1 - T_2)}{\ln (r_2/r_1)}.$

Letting $r_2 = r_1 + \Delta r$, $r_2/r_1 = 1 + \Delta r/r_1$.
  $\log(1 + x) \approx x$, when $x \ll 1$. Hence, for

$$\Delta r \ll r_1, \quad \ln \left(\frac{r_2}{r_1} \right) \approx \frac{\Delta r}{r_1} \approx \frac{\Delta r}{r_2},$$

giving

$$H = \frac{2\pi r_2 l \kappa (T_1 - T_2)}{\Delta r},$$

where $2\pi r_2 l$ is surface area of material and Δr is the thickness.

(b) $H = 4\pi\kappa(T_1 - T_2)\dfrac{r_1 r_2}{(r_2 - r_1)}.$

If thickness $\Delta r = r_2 - r_1$ is small, $r_2 \approx r_1$, and we obtain

$$H = 4\pi r_2^2 \kappa \frac{(T_1 - T_2)}{\Delta r}.$$

4.12   As the thickness of the crust is very small compared with the radius of the Earth, we can use the linear form of the law of conduction: Then heat flux per unit area ~ 0.07 W/m^2. This rate is negligible compared to the incident solar power.

4.13   Total collection area $\sim 24 \times 0.045 \times 2 \sim 2.16$ m^2 Delivered heat $\sim 0.85 \times 500 \times 2.16 \sim 918$ W. Using

$$\frac{dT}{dt} = \frac{1}{mC}\frac{dQ}{dt},$$

obtain $t \approx 19$ min.

4.14

(a) $\dfrac{1}{2}I\omega^2 = \dfrac{1}{2}kT,$

where

$$I = \frac{m_1 m_2}{m_1 + m_2} r^2 = 1.45 \times 10^{-46} \text{ kgm}^2$$

Hence,

$$\nu = \frac{\omega}{2\pi} = \frac{1}{2\pi}\sqrt{\frac{kT}{I}} = 8.50 \times 10^{11} \text{ Hz and } \lambda = 0.353 \text{ mm}.$$

(b) Classical harmonic oscillator frequency $\omega = \sqrt{k/\mu}$, where k is force constant.

$$\mu = \frac{m_1 m_2}{m_1 + m_2} = 1.14 \times 10^{-26} \text{ kg}$$

Hence,

$$\nu = \frac{1}{2\pi}\sqrt{\frac{k}{\mu}} = 6.5 \times 10^{13} \text{ Hz, and } \lambda = 4.6 \text{ μm}.$$

4.15 (a) For an isothermal expansion $\Delta U = 0$ and $W = Q = nRT \ln(V_2/V_1)$. Hence, $W = Q = 1729$ J.
(b) For an adiabatic expansion $\Delta Q = 0$ and $W = -\Delta U = -nC_V(T_2 - T_1)$, and $T_1 V_1^{\gamma-1} = T_2 V_2^{\gamma-1}$, where $\gamma = C_p/C_V = 1.4$ for a diatomic molecule.
 Hence,

$$W = -\Delta Q = -\frac{5}{2}R(T_2 - T_1) = \frac{5}{2}R\left[T_1 - T_1\left(\frac{V_1}{V_2}\right)^{\gamma-1}\right] = \frac{5}{2}RT_1\left[1 - \left(\frac{V_1}{V_2}\right)^{\gamma-1}\right]$$

$$= 1510 \text{ J}.$$

4.16 $\varepsilon = 30\%$, $Q_C = -1400$ J, $W = 600$ J.

4.17

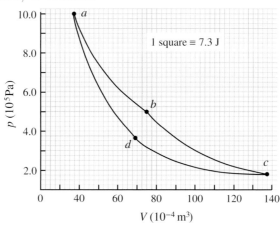

$p_a = 10.0 \times 10^5$ Pa, $V_a = 37.4 \times 10^{-4}$ m^3, $p_b = 5.0 \times 10^5$ Pa, $V_b = 74.8 \times 10^{-4}$ m^3,
$p_c = 1.82 \times 10^5$ Pa, $V_c = 137 \times 10^{-4}$ m^3, $p_d = 3.63 \times 10^5$ Pa, $V_c = 68.7 \times 10^{-4}$ m^3.

Step	Q	W	ΔU
$a \rightarrow b$	2593 J	2593 J	0
$b \rightarrow c$	0	1871 J	-1871 J
$c \rightarrow d$	-1729 J	-1729 J	0
$d \rightarrow a$	0	-1871 J	$+1871$ J
Total	864 J	864 J	0

Hence, $\varepsilon = 33\%$. The figure shows a sketch of the p–V diagram, where each small square is equivalent to 7.3 J of energy. Counting the squares within the area enclosed by the four p–V curves and counting part squares as a half square, we find the total number of squares ~ 125, giving $W \sim 900$ J, in reasonable agreement with the numerical result.

4.18 (i) 24%, (ii) 53%, (iii) 68%.

4.19 $\varepsilon = 8.3$ %, $\mathrm{d}Q_H/\mathrm{d}t = 1.2$ MW, $\mathrm{d}Q_C/\mathrm{d}t = 1.1$ MW. Using

$$\frac{\mathrm{d}Q_H}{\mathrm{d}t} = \frac{\mathrm{d}m}{\mathrm{d}t} \times C \times \Delta T,$$

obtain $\mathrm{d}m/\mathrm{d}t = 2.1 \times 10^5$ L/h. We see that OPTEC needs to pump a huge volume of water to produce a relatively modest amount of power. So far, only a few small-scale experimental OTEC power plants are in operation.

4.20 $\eta = 29.3$, $W = 0.17$ kW, and $Q_C = 4.83$ kW. Required flow of water, $\mathrm{d}m/\mathrm{d}t = 8.3 \times 10^2$ L/h. If $\eta = 5$, $W = 1.0$ kW, and $Q_C = 4.0$ kW, and required flow of water $= 6.9 \times 10^2$ L/h.

Problems 5

5.1 (a) 657, 874 and 1852 nm, respectively. (b) 1128 nm.

5.2 (a) $\lambda_{\max} = 264$ nm. (b) Number of photons/pulse

$$= \frac{5.0 \times 10^{-4} \times 226 \times 10^{-9}}{6.626 \times 10^{-34} \times 3 \times 10^8} = 5.7 \times 10^{14}.$$

Hence number of photoelectrons produced/s $= 4.3 \times 10^{14}$, which is a current of 68 μA.

5.3 The frequency ν of the light does not change, but the wavelength λ does.

$$\nu = \frac{c}{\lambda_{\mathrm{air}}} = 5.5 \times 10^{14} \text{ Hz}, \quad \lambda_{\mathrm{Si}} = \frac{\lambda_{\mathrm{air}}}{n} = \frac{550}{3.42} = 161 \text{ nm}.$$

5.4 According to the Bohr model, the electron in a hydrogen atom moves in an orbit with Bohr radius a_0 given by

$$a_0 = \frac{4\pi\varepsilon_0\hbar^2}{m_e e^2} = 0.53 \times 10^{-10} \text{ m}.$$

For an electron in the crystal we replace m_e by the effective mass m_e^* and ε_0 by ε. Hence, the Bohr radius for the loosely bound electron

$$= \frac{11.8}{0.26} \times 0.53 \times 10^{-10} = 24 \times 10^{-10} \text{ m}.$$

5.5 The radius of the electron in the impurity atom $= \frac{15.8}{0.55} \times 0.53 \times 10^{-10} = 15.2 \times 10^{-10}$ m. This gives a volume

$$= \frac{4}{3}\pi(15.2 \times 10^{-10})^3 = 1.47 \times 10^{-26} \text{ m}^3.$$

Hence, number of impurities

$$= 6.8 \times 10^{19} \text{ atoms/cm}^3.$$

72.6 g of germanium occupies a volume of 13.6 cm^3 and contains 6.022×10^{23} atoms. Hence the number density $= 4.43 \times 10^{22}$ atoms/cm^3, and maximum percentage of impurity atoms

$$\approx \frac{6.8 \times 10^{19}}{4.43 \times 10^{22}} \times 100 = 0.15\%.$$

5.6 (a)
$$n = p = n_i = 2\left(\frac{kT}{2\pi\hbar^2}\right)^{3/2}(m_e^* m_h^*)^{3/4}e^{-E_g/2kT}$$

$$= 2\left[\frac{300 \times 1.38 \times 10^{-23}}{2\pi(1.05 \times 10^{-34})^2}\right]^{3/2} \times \left[0.067 \times 0.48 \times (0.91 \times 10^{-30})^2\right]^{3/4} \times e^{-\left(\frac{1.43 \times 1.602 \times 10^{-19}}{2 \times 300 \times 1.38 \times 10^{-23}}\right)}$$

$$= 2(5.976 \times 10^{46})^{3/2} \times (2.66 \times 10^{-62})^{3/4} \times e^{-(27.7)}$$
$$= 1.8 \times 10^{12} \text{ carriers/m}^3.$$

(b) 63.55 g of copper contained Avogadro's number of atoms and occupies a volume of $\frac{63.55 \times 10^{-3}}{8.96 \times 10^3} = 7.09 \times 10^{-6}$ m^3. Hence number density

$$= \frac{6.022 \times 10^{23}}{7.09 \times 10^{-6}} = 8.5 \times 10^{28} \text{ free electrons/m}^3.$$

5.7

$$n_i = 2\left[\frac{300 \times 1.38 \times 10^{-23}}{2\pi(1.05 \times 10^{-34})^2}\right]^{3/2} \times \left[0.55 \times 0.37 \times (0.91 \times 10^{-30})^2\right]^{3/4} \times e^{-\left(\frac{0.67 \times 1.602 \times 10^{-19}}{2 \times 300 \times 1.38 \times 10^{-23}}\right)}$$

$$= 2 \times 1.46 \times 10^{70} \times 2.63 \times 10^{-46} \times e^{-12.96} = 1.81 \times 10^{19} \text{ electrons/m}^3.$$

Hence, concentration of holes in doped sample

$$= \frac{(1.81 \times 10^{19})^2}{1.4 \times 10^{23}} = 2.3 \times 10^{15}/\text{m}^3.$$

5.8 The Fermi energy of an intrinsic semiconductor lies in the middle of the band gap. Hence, $\frac{E - E_F}{kT} = 27.7$,

$$f(E) = \frac{1}{e^{27.7} + 1} = 9.33 \times 10^{-13}.$$

For $T = 310$ K:

$$f(E) = \frac{1}{e^{26.8} + 1} = 2.3 \times 10^{-12},$$

i.e. the probability more than doubles when T increases by 10 K.

5.9
$$\omega = \frac{qB}{m}.$$

Hence,

$$m_e^* = \left(\frac{1.602 \times 10^{-19} \times 1.5}{1.32 \times 10^{13} \times 9.1 \times 10^{-31}}\right) = 0.02\, m_e$$

5.10 Taking the band gap of germanium to be 0.67 eV, number of electron–hole pairs produced,

$$N = \frac{660 \times 10^3}{0.67} = 9.85 \times 10^5,$$

giving $\sqrt{N} = 9.93 \times 10^2$.

Hence energy resolution $\sim 2\sqrt{N} = 19.9 \times 10^3 \equiv 1.33$ keV. In practice, not all the energy of the γ-ray goes into producing electron–hole pairs; some of the energy goes into heating the crystal lattice. It is found that the average energy required to produce an electron–hole pair in germanium is 3 eV, so that a 660 keV γ-ray produces about 220 000 pairs.

5.11 (a) 867 nm, which lies in the infrared region of the solar spectrum. (b) Number of electrons passing through junction/s

$$= \frac{5.0 \times 10^{-3}}{1.602 \times 10^{-19}} = 3.12 \times 10^{16}.$$

Hence, number of photons emitted/s $= 3.12 \times 10^{15}$. This is a power of 0.71 mW.

5.12
$$x = \frac{1}{\mu} \ln \left[\frac{I_0}{I(x)} \right] = \frac{1}{3080} \ln \left[\frac{I_0}{0.01 \times I_0} \right] = 15 \ \mu\text{m for silicon.}$$

Similarly, $x = 1.3$ μm for gallium arsenide. (b) Number of incident photons incident on crystal surface per second $= 8.21 \times 10^{17}$. After 1 μm, this rate has reduced to 6.04×10^{17} photon/s. Hence rate of electron–hole pair generation is 2.17×10^{17} pairs/s. These are generated in a total volume of $100 \times 1.0 \times 10^{-4}$ cm^3. Hence mean rate of generation per unit volume $= 2.17 \times 10^{19}$ pairs/s cm^3.

5.13 (a)
$$E_1 = \frac{\left(1.05 \times 10^{-34}\right)^2 \times \pi^2}{2 \times 1.602 \times 10^{-19} \times 0.65 \times 9.11 \times 10^{-31} \times \left(1 \times 10^{-9}\right)^2} \left(\frac{1^2}{8^2} + \frac{0^2}{8^2} + \frac{0^2}{6^2} \right) \text{ eV}$$

$$= 0.573 \times 0.0156 = 0.9 \times 10^{-2} \text{ eV.}$$

$$E_2 = 0.573 \left(\frac{0^2}{8^2} + \frac{0^2}{8^2} + \frac{1^2}{6^2} \right) = 1.6 \times 10^{-2} \text{ eV.}$$

$$E_3 = 0.573 \left(\frac{1^2}{8^2} + \frac{1^2}{8^2} + \frac{0^2}{6^2} \right) = 1.8 \times 10^{-2} \text{ eV.}$$

(b) 1.8×10^{-4} m.

5.14 Irradiance scales as the inverse square law, $1/r^2$, but as Mars has no atmosphere we use solar irradiance above the Earth's atmosphere: 1367 W/m^2. Hence, maximum power on Mars

$$= 50 \times \frac{1367}{1000} \times \left(\frac{1.5}{2.28} \right)^2 = 30 \text{ W.}$$

5.15

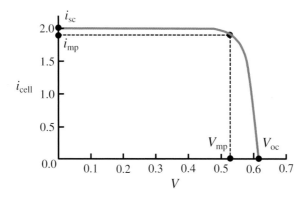

(a) $i_{sc} = 2$ A; $V_{oc} \approx \frac{kT}{e} \ln\left(\frac{i_{sc}}{i_0}\right) = 0.613$ V.

(b) $V_{mp} \approx V_{oc} - \frac{kT}{e} \ln\left(1 + \frac{eV_{oc}}{kT}\right) = 0.530$ V; $i_{mp} \approx i_{photo}\left(1 - \frac{kT}{eV_{mp}}\right) = 1.90$ A.

(c) $i_{cell} = i_{sc} - i_0\left(e^{eV/kT} - 1\right) \approx i_{sc} - i_0 e^{eV/kT} = 2.0 - \left(1.0 \times 10^{-10} \times e^{eV/kT}\right)$ A.

Although we want to see how V varies as i_{cell} increases, it's easier to plot i_{cell} against V. Rearranging the solar cell equation, we obtain

$$V = \frac{kT}{e} \ln\left(\frac{i_{sc} - i_{cell}}{i_0}\right),$$

where $e/kT = 1.602 \times 10^{-19}/1.38 \times 10^{-23} \times 300 = 38.7$ V. Fill factor $= 82\%$. To provide a nominal voltage of 12 V, 23 solar cells would be required and the optimum resistance of the load would be

$$\frac{23 \times 0.53}{1.9} = 6.4 \text{ ohms.}$$

Problems 6

6.1
$$P_0 = \frac{1}{2}\rho A_T u_0^3 = \frac{1}{2} \times 1.1 \times \pi \times (35)^2 \times (15.0)^3 = 7.1 \text{ MW.}$$

Hence, power generated by turbine $= 0.75 \times 0.59 \times 7.10 = 3.2$ MW. (i) $(1.2)^1 \times 3.2$ MW $= 3.8$ MW; (ii) $(1.2)^2 \times 3.2$ MW $= 4.6$ MW; (iii) $(1.2)^3 \times 3.2$ MW $= 5.5$ MW.

6.2 The ball takes time $t = r/v$ to travel radial distance r. In that time a point at radius r has moved distance $l = r\Omega \times t = v\Omega t^2$.

$$\text{Hence, } \frac{dl}{dt} = 2v\Omega t \text{ and } \frac{d^2 l}{dt^2} = 2v\Omega.$$

The term $2v\Omega$ is called the Coriolis acceleration.

6.3 At bathroom, we have

$$u_2 = \frac{4 \times 10^{-3}}{60 \times \pi(10 \times 10^{-3})^2} = 0.21 \text{ m/s.}$$

Hence

$$u_1 = 2.12 \times \left(\frac{20}{40}\right)^2 = 5.3 \times 10^{-2} \text{ m/s.}$$

and

$$P_1 = 2 \times 10^5 + \frac{1}{2} \times 1.0 \times 10^3 [(0.21)^2 - (5.3 \times 10^{-2})^2] + 1.0 \times 10^3 \times 9.8 \times 5.5$$

$$= 2 \times 10^5 + 21.1 + 5.39 \times 10^4 = 2.5 \times 10^5 = 2.5 \text{ atm.}$$

6.4 We have $(p_t - p_b) \approx \frac{1}{2}\rho\left(u_t^2 - u_b^2\right)$, since we can ignore the width of the wing. Hence,

$$(p_t - p_b) \approx \frac{1}{2}\rho\left(u_t + u_b\right)\left(u_t - u_b\right)$$

As $k \ll 1$, we assume that both u_t and u_b are nearly equal to u_0 and, hence,

$$(p_t - p_b) \approx \rho k u_0^2, \text{ and } F_L = A\rho k u_0^2.$$

We have

$$mg = A\rho k u_0^2 \Rightarrow u_0 = \sqrt{\frac{mg}{A\rho k}} = \sqrt{\frac{3000 \times 9.8}{40 \times 1.2 \times 0.1}} = 78.3 \text{ m/s} \approx 280 \text{ km/hour.}$$

6.6 Maximum possible value of F_A when $a = 1/2$. Then

$$F_A = \frac{1}{2}\rho u_0^2 = \frac{1}{2} \times 1.2 \times 20^2 = 240 \text{ N/m}^2.$$

If $a = 1/2$, $u_T = (1 - a)u_0 = \frac{1}{2}u_0$, and $u_f = 2u_T - u_0 = 0$, i.e. wind is stopped. When $a = 1/3$, hence $F_A = \frac{1}{2} \times 1.2 \times 20^2 \times \frac{8}{9} = 213 \text{ N/m}^2.$

6.7 (a) $u_0 = 330/6 = 55$ m/s. (b) $\Omega = v_{\text{tip}}/R = 330/50 = 6.6$ rad/s; $f = \Omega/2\pi = 1.1$ Hz.

6.8 (a) Power per turbine $= 0.8 \times 0.59 \times \frac{1}{2}\rho\pi\left(\frac{D}{2}\right)^2 u_0^3$; area per turbine $= (5D)^2$.
  Hence power per unit area of wind farm $= (8.9 \times 10^{-3})u_0^3 = 4.6 \text{ W/m}^2.$
(b) Taking typical power of nuclear reactor to be 1 GW, area of wind farm

$$= \frac{1 \times 10^9}{4.6} \text{ m}^2 = 2.2 \times 10^8 \text{ m}^2 = 220 \text{ km}^2 = 5.4 \times 10^4 \text{ acres,}$$

but note that the turbines take up less than 1% of the total area of a wind farm.

6.9

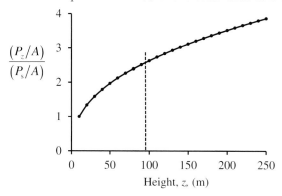

$$\left(\frac{P_z}{A}\right) = \frac{1}{2}\rho u_z^3 = \frac{1}{2}\rho\left[u_s\left(\frac{z}{h_0}\right)\right]^{0.42} = \frac{1}{2}\rho u_s^{0.42}\left(\frac{z}{h_0}\right)^{0.42}.$$

$$\left(\frac{P_s}{A}\right) = \frac{1}{2}\rho u_s^3, \text{ and hence } \left(\frac{P_z}{A}\right) = \left(\frac{P_s}{A}\right)\left(\frac{z}{10}\right)^{0.42}.$$

(P_z/A) is plotted in units of (P_s/A) against z in the figure. Because of the exponent 0.42, the power density falls faster than linearly and above say 100 m, the increase in power density is relatively small.

6.10

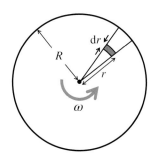

Consider a small element dr of the disk at a distance r from the centre, as shown in the figure. When the disk rotates through angle θ, the element sweeps out area $(r\theta \times dr)$. Hence emf $d\varepsilon$ across length dr of the segment is

$$d\varepsilon = -\frac{d\phi}{dt} = -rdrB\frac{d\theta}{dt} = -\omega Brdr.$$

Integrating along the radius, total emf between the centre and rim of the disk is

$$\varepsilon = -\int_0^R \omega Brdr = \frac{1}{2}\omega BR^2.$$

Note that this is a DC current generator, unlike the AC current generator in Section 6.5.

Problems 7

7.1 (a) (i) 25 mm, (ii) 42 mm, (iii) 12 Hz and (iv) 500 mm/s. Wave travels in the negative x-direction.
(b) $(dy/dt)_{\text{max}} = A\omega = 1.9$ m/s; $(d^2y/dt^2)_{\text{max}} = A\omega^2 = 140$ m/s^2.

7.2 Amplitude $= 0.10$ m, $\omega = 2\pi\nu = 20\pi$ rad/s, $k = \frac{\omega}{v} = \frac{20\pi}{40} = 0.5\pi$ m^{-1}. As displacement $= A$ at $x = 0$, the cosine function is the appropriate solution. Hence, the equation is $y = 0.10\cos(0.5\pi x - 20\pi t)$ m.

7.3 (a) $v = \sqrt{T/\rho} = \sqrt{50/0.5} = 10$ m/s; $\omega = 2\pi\nu = 2\pi \times 4.0 = 8\pi$ rad/s; $\lambda = v/\nu = 10/4 = 2.5$ m.
(b) $E_{\text{total}}/\lambda = \frac{1}{2}\mu\omega^2 A^2 = \frac{1}{2} \times 0.5 \times (8\pi)^2 \times (0.15)^2 = 3.6$ J.

$$P = \frac{1}{2}\mu\omega^2 A^2 v = 36 \text{ W}.$$

(c) The required power is: (i) four times larger; (ii) four times smaller.

7.5 The volume of water delivered to the reservoir each year $\sim 75 \times 10^6 \times 2 = 150 \times 10^6$ m^3, which has mass of 150×10^9 kg. Hence potential energy delivered by the reservoir $\sim 150 \times 10^9 \times 9.8 \times 600 \sim 9 \times 10^{14}$ J/year. This is an average power of $\sim (9 \times 10^{14})/(3.6 \times 10^6) \sim 250$ MW. Taking an efficiency of 70% for the overall efficiency, obtain ~ 175 MW of output power.

7.6 Let flow rate $= k\eta^x a^y g^z$, where k is a numerical constant. Equating dimensions on both sides: $L^3T^{-1} \equiv (MT^{-1}L^{-1})^x (L)^y (MT^{-2}L^{-2})^z$. Equating the respective indices of M, L and T on both sides, obtain: $x = -1$, $z = 1$, $y = 4$. Hence, flow rate

$$= k\frac{a^4 g}{\eta} = k\frac{pa^4}{\eta l}.$$

This is Poiseuille's formula. Further analysis of fluid flow shows that $k = \pi/8$.

7.7 Tidal force $F_T \propto R\left(M/L^3\right)$.

$$F_{T,\text{Earth}} \propto R_{\text{Earth}}\left(\frac{M_{\text{Moon}}}{L^3}\right); \quad F_{T,\text{Moon}} \propto R_{\text{Moon}}\left(\frac{M_{\text{Earth}}}{L^3}\right).$$

$$\therefore \frac{F_{T,\text{Moon}}}{F_{T,\text{Earth}}} = \frac{R_{\text{Moon}} \times M_{\text{Earth}}}{R_{\text{Earth}} \times M_{\text{Moon}}} = \frac{1.7 \times 10^3 \times 6.0 \times 10^{24}}{6.4 \times 10^3 \times 7.2 \times 10^{22}} = 22.$$

7.8

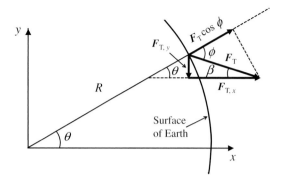

Considering the figure, we require $F_T \cos \phi = 0, \Rightarrow \phi = 90°$.

$$\theta = (90° - \beta), \text{ and } \tan \beta = \frac{F_{T,y}}{F_{T,x}} = \frac{\sin \theta}{2 \cos \theta} = \frac{1}{2} \tan \theta.$$

$$\tan \theta = \tan (90° - \beta) = \frac{1}{\tan \beta}.$$

Hence, $\tan \theta = 2/\tan \theta$, which gives $\theta = 54.7°$.

7.9

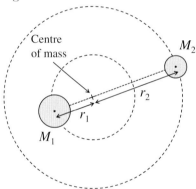

Consider two masses M_1 and M_2 that rotate about their common centre of mass under the force of gravity as illustrated in the figure. We have $M_1 r_1 = M_2 r_2$. Hence,

$$r_1 = \left(\frac{M_2}{M_1 + M_2} \right) (r_1 + r_2) ; \quad r_2 = \left(\frac{M_1}{M_1 + M_2} \right) (r_1 + r_2) .$$

(a) $r_1 = 4670$ km $= 73.2\%$ of Earth's radius; (b) $r_1 = 450$ km $= 0.00065\%$ of Sun's radius.

7.10 Starting with given equation we have

$$v_p = \left(\frac{g\lambda}{2\pi} + \frac{2\pi S}{\lambda \rho} \right)^{1/2} .$$

Then

$$\frac{\mathrm{d}v_p}{\mathrm{d}\lambda} = \frac{1}{2v_p} \left(\frac{g}{2\pi} + \frac{2\pi S}{\lambda^2 \rho} \right)$$

and setting $\mathrm{d}v_p/\mathrm{d}\lambda = 0$, find minimum phase velocity occurs at $\lambda = 2\pi(S/\rho g)^{1/2}$, which equals 1.7×10^{-2} m for given values. Using this value of λ, the minimum value of the phase velocity is 0.23×10^{-2} m/s.

7.11   (i) Wave motion dominated by surface tension. Hence, $v_p = (Sk/\rho)^{1/2}$, where $k = 200\pi$ m^{-1}. Hence

$$v_p = \left(\frac{7.2 \times 10^{-2} \times 200\pi}{1.0 \times 10^3}\right)^{1/2} = 0.21 \text{ m/s}$$

and $v_g = \frac{3}{2}v_p = 3.2$ m/s.

(ii) $v_p = (g\lambda/2\pi)^{1/2} = (9.8 \times 1.0/2\pi)^{1/2} = 1.2$ m/s. And $v_g = \frac{1}{2}v_p = 0.62$ m/s.

(iii) $v_p = (gh_0)^{1/2} = (9.8 \times 4.0)^{1/2} = 6.3$ m/s. And $v_g = v_p = 6.3$ m/s.

7.12 $P \times L = \frac{1}{2}\rho g H^2 v_g \times L = \frac{1}{2} \times 1030 \times 9.8 \times 1.0^2 \times 8.0 \times 1000 \times 10^3 = 4.0 \times 10^{10}$ kW. With conversion efficiency of 50%, power per person $= 4.0 \times 10^{10}/7 \times 10^7 \approx 600$ W.

7.13 $P_{av} = 2 \times \frac{A\rho g}{2\tau}(D_{av})^2$, as power is generated from both the incoming and outgoing tides. Then

$$P_{av} = 2 \times \frac{A\rho g}{2\tau}(D_{av})^2 = 2 \times \frac{22.5 \times 10^6 \times 1030 \times 9.8 \times 8^2}{2 \times 4.47 \times 10^4} = 320 \text{ MW}.$$

Hence power per square metre of the tide pool

$$= \frac{320 \times 10^6}{22.5 \times 10^6} \approx 14 \text{ W/m}^2.$$

7.14 Power output of an underwater turbine given by

$$P = \frac{1}{2} \times \frac{1}{2}\rho A v^3 = \frac{1}{2} \times \frac{1}{2}\pi\rho\left(\frac{D}{2}\right)^2 v^3.$$

Area per turbine $= 25D^2$. Therefore power per unit area of water farm,

$$\frac{P}{A} = \frac{1}{2} \times \frac{1}{2}\rho A v^3 = \frac{1}{2} \times \frac{1}{200}\pi\rho v^3$$

$$= \frac{1}{2} \times \frac{1}{200}\pi \times 1030 \times (1.0)^3 \approx 8 \text{ W/m}^2.$$

Problems 8

8.1   1 mole of carbon weighs 12 g. Hence, energy density

$$= \frac{1.0}{12 \times 10^{-3}} \times 394 = 33 \times 10^6 \text{ MJ/kg} \equiv 9.1 \text{ kWh/kg}.$$

These results are about 10% too high because carbon makes up about 90% of coal. We have $12\text{gC} + 32\text{gO}_2 \rightarrow 44\text{gCO}_2$. Hence, 1 tonne (1000 kg) of carbon produces $1.0 \times (44/12) = 3.7$ tonne of CO_2. Total energy produced $= 33$ GJ.

8.2   Total amount of thermal energy required is $24 \times 4 \times 30 = 2880$ kWhr $= 1.04 \times 10^{10}$ J. Hence, required mass of water is

$$\frac{1.04 \times 10^{10}}{4200 \times (50 - 20)} = 8.25 \times 10^4 \text{ kg} \equiv 82.5 \text{ m}^3.$$

If area of house is 200 m^2, the depth of the tank would be 0.4 m.

8.3   Thermal energy to be deposited $= 12$ kWh $= 12 \times 3600$ kJ. Hence, required volume

$$= \frac{12 \times 3600}{0.95 \times 40 \times 2600} = 0.44 \text{ m}^3.$$

Hence dimensions of storage heater ~ 1 m $\times$ 1.5 m $\times$ 0.3 m, with a total surface area ~ 4.5 m^2.

Using

$$\frac{Q}{t} = \kappa A \frac{\Delta T}{w},$$

thickness, w, of insulation ≈ 8 mm.

8.4   Total stored energy $= 3 \times 3.6 \times 10^6 \times 8.7 \times 10^6 = 8.7 \times 10^{13}$ J. Using $E = mgh$, find

$$m = 8.9 \times 10^9 \text{ kg} \equiv 8.9 \times 10^6 \text{ m}^3$$

of water. If the depth of the water were 30 m, the surface area of the reservoir would be 3×10^5 m^2, i.e. about 550 m $\times$ 550 m.

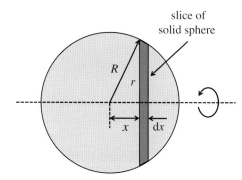

8.5   Divide sphere into thin discs of thickness $\mathrm{d}x$ (see figure). Radius r of disc $= \sqrt{R^2 - x^2}$, its volume is $\mathrm{d}V = \pi r^2 \mathrm{d}x = \pi(R^2 - r^2)\mathrm{d}x$ and its mass is $\mathrm{d}m = \rho \mathrm{d}V = \pi\rho(R^2 - r^2)\mathrm{d}x$. The moment of inertia of a disc of radius r and mass $\mathrm{d}m$ is

$$\mathrm{d}I = \frac{1}{2}r^2 \mathrm{d}m = \frac{\pi\rho}{2}(R^2 - r^2)^2 \mathrm{d}x.$$

Hence, moment of inertia of solid sphere is

$$I = \frac{\pi\rho}{2}\int_0^R (R^2 - r^2)^2 \, \mathrm{d}x = \frac{2}{5}MR^2.$$

8.6   The required stored energy $= 5000 \times 9.8 \times 20 = 9.8 \times 10^5$ J. Moment of inertia of flywheel $= \frac{1}{2} \times 250 \times 0.75^2$ kg m^2, and rotational kinetic energy of flywheel $= \frac{1}{2}\left(\frac{1}{2} \times 250 \times 0.75^2\right)\omega^2 = 35\omega^2$ J. Hence, $\omega = 167$ rad/s $\equiv 1\,600$ rpm. Energy density of flywheel $= 0.0039$ MJ/kg, which is below the limit of 0.042 MJ/kg given in Section 8.4.3.

8.7   (a) Suppose that the current, i, through the inductor is rising at rate $\mathrm{d}i/\mathrm{d}t$. Then if L is the self-inductance, the back-emf ε across it is given by $\varepsilon = L\frac{\mathrm{d}i}{\mathrm{d}t}$, and the rate P at which work is being done against the back-emf is $P = \varepsilon i = Li\frac{\mathrm{d}i}{\mathrm{d}t}$. The total work W done in bringing the current from zero to i_0 is therefore

$$W = \int P dt = \int_0^{i_0} Li\frac{\mathrm{d}i}{\mathrm{d}t}\mathrm{d}t = \frac{1}{2}Li_0^2.$$

Hence, $E = \frac{1}{2}Li^2$. (b) $1 \times 10^6 = \frac{1}{2} \times 1.8 \times i^2 = i = 1054$ A.

8.8

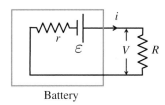

Battery

From the figure we see that voltage V across load of resistance R is $\frac{R}{R+r}\varepsilon$ and current i passing through load is $\varepsilon/R+r$. Hence, power P dissipated in load is $\frac{R}{(R+r)^2}\varepsilon^2$. Differentiating P with respect to R and equating to zero gives $r = R$.

8.9 Ratio of the lost power to the power leaving the power station is $\frac{i^2R}{P} = \frac{PR}{V^2}$.
Hence,

$$R = \frac{0.02 \times (440 \times 10^3)^2}{250 \times 10^6} = 15.5 \text{ ohms.}$$

Using $R = \frac{\rho l}{A}$, where the symbols have their usual meanings, $A = 3.6 \times 10^{-4} \text{ m}^2 = 360 \text{ mm}^2$. This could be achieved with two aluminium cables of 15 mm diameter.

8.10 We have, from Problem 4.9, heat loss

$$H = \frac{2\pi l \kappa (T_1 - T_2)}{\ln (r_2/r_1)} \text{ W.}$$

Hence, $H = 6.7$ kW.

Index

A

Acceptor atom, 212
Acceptor levels, atom, 212–213
Activation energy, 85
Adams, William Grylls, 201
Adiabatic expansion of gases, 183–185
Advanced adiabatic compressed air energy storage, 360–361
Advanced gas-cooled reactor (AGR), 106
Air-mass-ratio (AM ratio), 152
Albedo of Earth, 147–148
Alpha (α) decay, 57–62
　barrier penetration model of, 59–62
Alpha (α) particles, scattering of, 14–19, 73–74
Alpha (α) rays, 11
Alta Wind Energy Centre, 278
Aluminium, 209, 212
Angle of attack, 263
Angular distribution, of scattered α particles, 15, 17
Asymmetry term, 33
Atomic mass number, 13
Atomic mass unit, 23
Atomic nucleus, 11–12
　binding energy, 27–30
　　curve of, 30–31
　charge of, 24–27
　composition of, 12–14
　energies, 36–39
　forces, characteristics of, 35–36
　mass of, 21–24
　nuclear charge, distributions of, 19–21
　nuclear matter, distributions of, 19–21
　semi-empirical mass formula, 32–35
　size of, 14–19
Atomic number, 12–13

B

Band gap, of semiconductors, 203
Beam attenuation, 78–81
Becquerel, Antoine Henri, 11, 201
Becquerel, Edmond, 201
Becquerel (Bq), 52
Bernoulli, Daniel, 258

Bernoulli's equation, 258–263, 287
Beryllium, 101
Beta (β) decay, 57, 62–66
　electron capture, 65–66
Beta (β) rays, 11
Betz, Albert, 264
Betz criterion, 265–268
Biased p–n junction
　forward, 218–219
　reverse, 217
Binding energy
　curve of nuclides, 30–31, 72
　of nucleus, 27–30
　per nucleon, 30–31
Bioenergy, 4
Biofuels, 4
Biological energy storage, 351
Blackbody radiation, 134–135
　emissivity, 137–141
　laws of, 135–137
　photon, birth of, 141–145
　Rayleigh–Jeans formula for, 143
Bohr, Niels, 83
Bohr's model of hydrogen atom, 25–26
Boiling water reactor (BWR), 107
Bolometer, 136
Boltzmann, Ludwig, 135
Boltzmann distribution, 221
Boltzmann factor, 221
Borschberg, Andre, 242
Boyle's law, 183
Bragg, W. H., 25
Bragg, W. L., 25
Bremsstrahlung, 115
Brush, Charles, 252
Bush's wind turbine system, 252

C

Cadmium selenide (CdSe), 246
CANDU reactor, 107
Capacitors, 365–367
Carnot, Sadi, 175
Carnot cycle, 187–192

Physics of Energy Sources, First Edition. George C. King.
© 2018 John Wiley & Sons, Ltd. Published 2018 by John Wiley & Sons, Ltd.

Chain of radioactive decays, 55–57
Chain reactions, 97–101
Chapin, Darryl, 202
Characteristic X-rays, 24–25
Chemical atomic mass, 13
Chemical energy storage, 351–352
Coblentz, William C., 144
Cockcroft, John, 74
Coefficient of performance, 195, 196
Commercial nuclear reactors, 105–107
    advanced gas-cooled reactor (AGR), 106
    boiling water reactor, 107
    CANDU reactor, 107
    Magnox reactor, 106
    pressurised water reactor, 106–107
Compound nucleus, 73
Compressed air energy storage, 357–361
Consumption of energy, 1–3
    global, 2
    in UK, 2
    in USA, 2, 5
Contact potential, 215–216
Continuity equation, fluids, 257–258
Controlled fission reactions
    chain reactions, 97–101
    commercial nuclear reactors, 105–107
    control of fission reactions, 101–103
    nuclear waste, 107–109
    reactors, 103–105
Controlled thermonuclear fusion, 117–123
Control rods, 101
Convection, thermal, 169–170
Coolant, 104
Copper, 165
Coriolis, Gaspard-Gustave, 254
Coriolis force, 253–256
Coulomb barriers, 37, 60–62
    for nuclear fusion, 113
Coulomb energy, 33, 84–85
Coulomb force, 84
Coulomb repulsion, 83, 84–85, 88–89, 109–110, 129, 130
Cour, Poul la, 252
Covalent bonding, 32, 209
Cross-section for nuclear reaction, 76–82
    beam attenuation and, 78–81
    energy dependence of, 81–82
    mean free path and, 78–81
Crystalline solids, band structure of, 204–208
Curie, Marie, 11
Curie, Pierre, 11
Curie (Ci), 53
Current–voltage characteristic of p–n junction, 219–222

Cut-in/out speed, turbine blades, 274–275
Cyclotron frequency, 24, 117–118

D
Daily insolation, Sun, 154
Darcy, Henry, 288
Darcy friction factor, 288
Darcy–Weisbach equation, 288
Darrieus turbine, 271–272
Day, Richard Evans, 201
de Broglie wavelength, 20
Decay chains, 55–57
Decay laws of radioactivity, 49–57
Deep geothermal energy, 161–162
Degree of freedom, 177
Delayed neutrons, 90
Density of states function, 222–223
Deuteron, 44–45
Diffusion current, 214–215
Diffusion equation, 158
Diffusion length, 218
Dimensional analysis, in fluid mechanics, 287–288
Diode equation, 220
Diodes, 220
Direct band semiconductors, 230
Doldrums, 255–256
Donor atoms, 209–210
Donor energy levels, 211–212
Doping, 208–209
Downwind turbines, 273
Drift current, 216
Dye-sensitised solar cell, 247

E
Edinburgh Duck, 322–323
Effusion, 93–94
Einstein's equation, photoelectric effect, 202–203
Einstein's mass–energy relationship, 12, 28
Electrical energy storage, 364–365
    capacitors, 365–367
    fuel cells, 370–372
    lead-acid battery, 369–370
    rechargeable batteries, 368
    super-capacitors, 365–367
    superconducting magnetic storage, 367–368
Electrical power, distribution of, 372–374
Electricity, 4
Electromagnetic induction, 252
Electromagnetic radiation, 127, 137
Electromagnetic standing waves, 141–142
Electron capture, 65–66
Electron–hole pair, 203
Electrons, 12

Electrostatic force, 16
Emissivity, blackbody radiation, 137–141
Endothermic reaction, 75, 351–352
Energy, 1
    biofuels, 4
    consumption, 1–3
    conversion of, 6–9
    density, 300
    electricity, 4
    equivalence of, 12
    form of, 6–9
    fossil fuels, 4
    geothermal, 4, 127
    hydroelectric, 4
    importance of, 1
    mechanical, 7
    non-renewable, 5–6
    nuclear, 4
    in nuclear fission, 90–91
    in nuclear fusion, 111–112
    photovoltaic solar cells, 7–8
    renewable, 5–6
    solar, 3–4
    sources of, 3–5
    storage, 8, 349–374
    thermal, 7
    threshold, 75
    tidal, 4
    waves, 4, 319–324
    wind, 4
Energy storage, 349–350
    biological, 351
    capacity of, 350
    chemical, 351–352
    compressed air, 357–361
    density, 350
    electrical, 364–372
    flywheel, 361–364
    hydrogen, 351–352
    mechanical, 355–364
    parameters of, 350
    pumped hydroelectric, 355–357
    round-trip efficiency, 350
    systems, 8
    thermal, 352–355
    types of, 350
Equation of state of ideal gas, 175–177
Equilibrium, secular, 56–57
Equilibrium tide model, 337
Equipartition theorem, 177
Equivalence of energy, 12
Equivalence of mass, 12
Euler, Leonard, 258
Exothermic reaction, 75, 352

Extraterrestrial spectrum, 146
Extrinsic semiconductors, 208–213

F
Faraday, Michael, 252
Faraday's law, 252
Fast fission factor, 99
Fermi, Enrico, 81
Fermi–Dirac distribution, 223–226
Fermi energy, 223–225
    in p–n junction, 227–229
Fermi levels, 224, 228–229
Ferrel cell, 255
Fetch, 306
Ffestiniog Power Station, North Wales, UK, 356
Fill factors, 237
First law of thermodynamics, 177–181
Fission reactions, control of, 101–103
Fission reactors, 103–105
    coolant, 104
    efficiency of, 105
    fuel, 103–104
    moderator of, 104
    product poisoning of, 105
Flat plate water heater, 162–163
Fluids, flow of, 256–263
    Bernoulli's equation, 258–263
    continuity equation, 257–258
    laminar, 257
    steady, 256–257
    turbulent, 257
Flux Φ of beam, 76–77, 78–80
Flywheel energy storage, 361–364
Forward-biased p–n junction, 218–219
Fossil fuels, 4, 6, 7
Four-factor formula, 100
Fourier's law of conduction, 156
Fowler, William, 130
Frisch, Otto, 82
Fritts, Charles, 201
Fuel cells, 4, 370–372
Fuller, Calvin, 202
Fusion energy gain factor, 117

G
Gallium arsenide (GaAs), 230–231
Gallium indium arsenide (GaInAs), 244
Gallium indium phosphide (GaInP), 244
Gamma (γ) decay, 57, 66
Gamma (γ) rays, 11, 103
Gas centrifuge technique, for uranium enrichment, 94–95
Gas diffusion technique, for uranium enrichment, 93–94

Gases
   adiabatic expansion of, 183–185
   isothermal expansion of, 183–185
   specific heats of, 181–183
Gemasolar Solar Power Facility, 173–174
Geothermal energy, 4, 127, 159–160
   deep, 161–162
   shallow, 160
Germanium (Ge), 244
Geysers, 161
Glass cover, thermal transmission by, 170–171
Global warming potential, 155
Gluons, 35
Greenhouse effect, 155
Greybody, 138
Grove, William, 371

H
Hadley, George, 255
Hadley cell, 255
Hahn, Otto, 82
Head loss, 287–288
Heat, 180–181
Heat engines, 7, 174–196
   equation of state of ideal gas, 175–177
   and first law of thermodynamics, 177–181
   and second law of thermodynamics, 185–196
   specific heats of gases and, 181–183
   thermal efficiency of, 186
Heat pump, 195–196
Heisenberg uncertainty principle, 82
Heliostats, central receiver with, 173–174
Helium burning, 129
Hero of Alexandria, 251–252
Hoyle, Fred, 130
Huntorf compressed air storage plant, Germany, 360
Hydroelectric power, 4, 284
   flow of viscous fluid in pipe, 286–288
   hydroelectric turbines, 288–291
   plants and principles of operation, 284–286
Hydroelectric power plants, 284–286
   principle of, 286
Hydroelectric turbines, 288
   impulse, 288
   Pelton impulse turbine, 288–291
   reaction, 288
Hydrogen burning, 129
Hydrogen economy, 351
Hydrogen energy storage, 351–352
Hydrogen gas, 4

I
Ideal gas equation, 175–177
Ignition point, 115–116

Impact parameter, 15
Impulse hydroelectric turbines, 288
Indirect band semiconductors, 230
Induced nuclear fission, 86–87
Inertial confinement, 109–110
Inertial confinement fusion, 121–123
Internal energy, 177–178
Intrinsic semiconductors, 208–213. *See also* Fermi
      energy
Irradiance of solar radiation, 137–138
Isothermal expansion of gases, 183–185
Isotopes, 13–14
   natural abundances of, 13
Itaipu Dam, 285
ITER, 120–121

J
Jeans, James, 141
Joule, James, 174
Jupiter, 334–335

K
Kelvin, Lord, 175

L
Laminar flow, fluid, 257
Laser ionisation, for uranium enrichment, 95–96
Law of mass action, for semiconductors, 226
Lawson, John D., 116
Lawson criterion, for performance of nuclear fusion,
      116–117
Lead-acid battery, 369–370
   charging, 370
   discharging, 369–370
Light concentrators, 244–245
Limpet, 322
Linear attenuation coefficient, 79
Liquid-drop model, of nuclear fission, 83–86
Lithium, 111
Llyn Stwalan, 356
London Array, 278–279
Longitudinal waves, 292
Low-pass filter, 276
Lummer, Otto, 136
Lunar tides, period of, 335

M
Magic numbers, 39
Magnetic confinement, 109
   fusion, 117–121
Magnox reactor, 106
Mass defect, 28
Mass equivalence, 12
Mass number, 13

Mass spectrometer, 21–24
Mass spectroscopy, 21–24
Maxwell–Boltzmann distribution of speeds, 94
McIntosh plant, Alabama, USA, 360
Mean free path, 78–81
Mean lifetime τ, 53–54
Mean time, particle traveling, 80
Mechanical energy
    sources, 7
    storage, 355–364
Mechanical equivalence of heat, 174–175
Meitner, Lise, 82
Mixed-oxide (MOX) fuel, 108
Moderator, 104
Modes of vibration, of molecule, 155
Modulated wave, propagation of, 302–303
Monochromatic waves, 300
Moseley, Henry, 24
Moseley's law, 25
Multi-junction solar cells, 243–244

N
Nacelle, 272
National Ignition Facility (NIF), 123
Neap tides, 337
Nellis Solar Power Plant, 242
Neutron number, 13
Neutrons, 12–13
    delayed, 90
    moderation of, 91–93, 99–100
    prompt, 89
    reproduction factor, 97
    thermal, 100–101
Non-renewable energy, 5–6
n-type semiconductors, 209–212
Nuclear binding energies, 12, 27–30
    curve of, 30–31
Nuclear charge, distributions of, 19–21
    mean radius of, 20–21
    skin thickness of, 20
Nuclear energies, 4, 6, 36–39
Nuclear fission, 7, 12, 82–83
    controlled reactions, 97–109
    cross-sections for, 87–88
    energy in, 90–91
    induced, 86–87
    liquid-drop model of, 83–86
    neutrons, moderation of fast, 91–93
    process of, 71–72
    products, 88–90
    versus radiative capture, 98–99
    reactions, 88–90
    uranium enrichment, 93–97
Nuclear forces, characteristics of, 35–36

Nuclear fuels, 6, 103–104
Nuclear fusion, 12, 109–110
    Coulomb barrier for, 113
    energy in, 111–112
    inertial confinement fusion, 121–123
    magnetic confinement fusion, 117–121
    performance criteria of, 115–117
    process of, 71–72
    reaction rates, 113–114
    reactions, 110–111
    thermonuclear, controlled, 117–123
Nuclear masses, 13
Nuclear matter, distributions of, 19–21
Nuclear radiation, 57
Nuclear reactions, 73–74. See also Nuclear fission;
        Nuclear fusion
    cross-sections, 76–82
    Q-value of, 74–76
    rates, 76–82
Nuclear waste, 107–109
Nucleon number, 13
Nucleons, 13
Nucleus, 12–13
    binding energy of, 27–30
    charge of, 24–27
    compound, 73
    energy from, 71–72
    liquid drop model of, 32
    mass of, 21–24
    shell model of, 32, 38
    size, 14–19
Nuclides, 13
    binding energy curve of, 30–31, 72
    lifetime of, 53–54
    radioactive, 47–48
    Segré chart of, 48–49
    transuranic, 108

O
One-dimensional heat equation, 158
One-dimensional wave equation, 294–295
Open circuit voltage, 232
Optical window, 150
Order of magnitude, 15
Oscillating water column systems, 321–322
Overtopping devices, 320–321
Ozone–oxygen cycle, 150

P
Parabolic trough concentrator, 172–173
Particle accelerator, 74
Pauli exclusion principle, 38–39
Pearson, Gerald, 202
Pelamis energy converter, 323–324

Pelton impulse turbine, 288–290, 288–291
Penstock, 285
Phonon, 230
Phosphorus, 209–210
Photo-dissociation, 29–30
    oxygen, 149–150
Photoelectric effect
    Einstein's equation of, 202–203
    quanta of energy and, 202
Photoelectrons, 202
Photon
    absorption at p–n junction, 229–231
    birth of, 141–145
    random walk of, 132–133
Photosynthesis, 3–4, 127
Photovoltaic solar cells, 7–8
Piccard, Bertrand, 242
Planck's radiation law, 143–144
Planté, Gaston, 368
Plasma
    fusion, 109
    heating, 119
p–n junction
    biased
        forward, 218–219
        reverse, 217
    current–voltage characteristic of, 219–222
    electron and hole concentrations in semiconductor,
        222–227
    in equilibrium, 214–216
    Fermi energy in, 227–229
    photon absorption at, 229–231
Poisson probability formula, 50
Polar cell, 255
Poloidal field, 118
Positron emission tomography (PET), 65
Potential well, 36
    barrier penetration, 45–47
    deuteron, 44–45
    one-dimensional finite, 42–44
    one-dimensional infinite, 40–42
    particle, quantum mechanical description of, 39–47
Power coefficient, 267
Pressurised water reactor (PWR), 106–107
Primary energy sources, 3–4
Prinsheim, Ernst, 136
Prompt neutrons, 89
Proton–proton chain, 129, 131–132
Protons, 12–13, 37
Pseudo force. See Coriolis force
p-type semiconductors, 209, 212–213
Pumped hydroelectric energy storage, 355–357
P–V diagram, 179–180

Q
Quantum dot solar cells, 245–247
Quasi-static processes, 176
Queisser, Hans, 237
Q-value
    energy, 58
    of fusion reactions, 111
    of nuclear reaction, 74–76

R
Radiant exitance, 135
Radiative capture
    cross-section for, 99–100
    nuclear fission *versus,* 98–99
Radiative recombination process, 239
Radioactive carbon dating, 47, 63–64
Radioactivity, 11
    decay laws of, 49–57
    and nuclear stability, 47–66
    Segré chart of stable nuclides, 48–49
    α decay, 57–62
    β decay, 57, 62–66
    γ decay, 57, 66
Radioisotopes, 78
Radio window, 150
Radon, 11
Rance Tidal Power Station, 343–344
Random walk of photon, 132–133
Range, of tide, 324–325
Rated output speed, 275
Rated power output, 275
Rayleigh, Lord, 141
Rayleigh–Jeans formula for blackbody radiation,
    143
Rayleigh scattering, 150–151
Reaction hydroelectric turbines, 288
Rechargeable batteries, 368
Refrigerators, 195
Relative wind, 269
Renewable energy, 5–6
    disadvantage of, 6
    sustainability and, 6
Reservoir capacitor, 367
Resonance escape probability, 100
Reverse-biased p–n junction, 217
Reverse saturation current, 220
Rotational wind, 269
Round-trip efficiency, energy storage, 350
Rutherford, Ernest, 11
Rutherford postulated, 11
Rutherford's model of structure of atom, 14
Rutherford's scattering, 14–19, 73 74
    inelastic scattering, 73

S

Salpeter, Edwin, 130
Schrödinger equation, 40
SeaGen tidal current plant, 345
Secondary cells. *See* Rechargeable batteries
Secondary energy sources, 4
Second law of thermodynamics, 185–196
Segré chart of stable nuclides, 48–49
Semiconductors, 164–165, 204–213
   band gap of, 203
   crystalline solids, band structure of, 204–208
   direct band, 230
   electron and hole concentrations in, 222–227
   extrinsic, 208–213
   indirect band, 230
   intrinsic, 208–213
   law of mass action for, 226
   majority carriers in, 209
   minority carriers in, 209
   n-type, 209–212
   p-type, 209, 212–213
   solar cells, 229–247
Semiconductor solar cells, 229–247
   construction, 240–242
   equation, 233–235
   maximum power delivery from, 235–237
   p–n junction, photon absorption at, 229–231
   power generation by, 231–235
   Shockley–Queisser limit, 237–240
Semi-empirical mass formula, 32–35
   asymmetry term, 33
   Coulomb term, 33
   pairing term, 33–34
   surface term, 32
   volume term, 32
Semi-empirical mass formula (SEMF), 87
Shallow geothermal energy, 160
Shell model of nucleus, 32, 38
Shockley, William, 237
Shockley–Queisser limit, 237–240
   crystal lattice, dissipation of energy in, 238–239
   electron–hole recombination, 239–240
Short-circuit current, 231–232
Significant wave height, 319
Silicon, 209–210, 212, 230
Sinusoidal waves, 295–296
Siting of wind turbines, 277–279
Solar cells
   construction, 240–242
   efficiency maximisation, 243–247
   equation, 233–235
   light concentrators, 244–245
   maximum power delivery from, 235–237

multi-junction, 243–244
   quantum dot, 245–247
Solar constant, 145
Solar energy, 3–4, 127
   harvesting of, 128
Solar heaters, 162
   heat transfer processes, 165–174
      thermal conduction, 165–169
      thermal convection, 169–170
      thermal transmission by glass cover, 170–171
      vacuum tube collectors, 171
   water heaters, 162–165
Solar panels, 242
Solar power, 127–196
   blackbody radiation, 134–145
   geothermal energy, 159–162
   heat engines, 174–196
   solar heaters, 162–174
   solar radiation and interaction with earth, 145–159
   stellar fusion, 128–133
Solar radiation, 132–133
   characteristics of, 145–147
   intensity of, 137
   interaction with earth and atmosphere, 147–155
      absorption processes, 148–150
      greenhouse effect, 155
      scattering processes, 150–151
      season, latitude and daily insolation, 152–154
   irradiance of, 137–138
   penetration into ground, 155–159
Solar thermal power systems, 162, 172–174
   central receiver with heliostats, 173–174
   parabolic trough concentrator, 172–173
Solar water heaters, 127–128
   absorption of radiation, 163–165
   flat plate, 162–163
Specific heats of gases, 181–183
Spectral absorption factor, 138
Spectral emissivity, 138
Spectral power distribution, 135
Spectral radiant exitance, 135
Spring tides, 337
Standard mass of atmosphere, 152
Standing waves, 292
Star, formation and evolution, 128–131
Steady/laminar flow, fluid, 256–257
Steam reforming, 351–352
Steam turbines, 193–195
Stefan, Josef, 135
Stefan–Boltzmann law, 135, 137
Stefan's constant, 135

Stellar fusion
    solar radiation, 132–133
    star formation and evolution, 128–131
    thermonuclear fusion in Sun, 131–132
Strassman, Fritz, 82
Streamlines, 256–257
Sun–Earth system, 153–154
Super-capacitors, 365–367
Superconducting magnetic storage, 367–368

T
Tan-y-Grisan reservoir, 356
Tapered channel (Tapchan) method, 320–321
Thermal conduction, 165–169
Thermal convection, 169–170
Thermal efficiency, of heat engines, 186
Thermal energy
    sources, 7
    storage, 352–355
Thermal pollution, 6
Thermal reservoir, 183
Thermal resistance, 167–168
Thermal transmission by glass cover, 170–171
Thermal utilisation factor, 100–101
Thermographs, 134
Thermonuclear fusion, 109, 127
    in Sun, 131–132
Thermonuclear fusion, controlled, 117–123
Threshold energy, 75
Tidal current power, 344–346
Tidal energy, 4
Tidal force, 328–335
Tidal power, 324–325
    force, 328–335
    harnessing, 341–346
    lunar tides, period of, 335
    origin of, 325–328
    tidal range, variation and enhancement of, 335–341
Tidal range
    power, 342–344
    resonant enhancement of, 339
    variation and enhancement of, 335–341
Tidal wave, 338
Tip-speed ratio, 271, 274
Titanium oxide (TiO_2), 246
TOKAMAK, 118–120
Toroidal solenoid, 118
Transuranic nuclides, 108
Transverse waves, 292, 294
Travelling waves, 292
Triple-alpha process, 129–131
Tritium, 120–121
Trochoid, 308
Tsunamis, 338

Turbines, wind
  Betz criterion, 265–268
  blades
    action of, 268–270
    rotational speed of, 270–271
  design and operation, 271–277
  siting of, 277–279
  wind power extraction by, 263–271
Turbulent flow, fluid, 257

U
Ultraviolet catastrophe, 143
Uncertainty principle, quantum mechanics, 38, 82
Upwind turbines, 273
Uranium, 11
    enrichment, 93–97
    in fission cross-sections, 87–88
    in induced nuclear fission, 86–87

V
Vacuum tube collectors, 171
Varactor diode, 219
Virial theorem, 299
Viscosity, 286–287
Vitrification, 108–109
'1/v' law, 81–82
Volume flow rate, 258
von Fraunhofer, Joseph, 146

W
Walton, Ernest, 74
Water power, 283
    hydroelectric power, 284–291
    tidal power, 324–346
    wave power, 291–324
Water waves
    on deep water, 313
    energy of, 313
    physical characteristics of, 306–309
    power of, 318–319
    on shallow water, 310–313
    velocity of, 309–310
Wave energy converters, 319–320
    challenges for, 320
    Edinburgh Duck, 322–323
    oscillating water column systems, 321–322
    overtopping devices, 320–321
    Pelamis energy converter, 323–324
Wave power, 291–292
    water, 306–319
    wave energy converters, 319–324
    wave motion and, 292–306
Waves
    on deep water, 316–318

dispersion of, 300–303
energy, 4
groups, 296–300, 303–304
group velocity of, 301–302
longitudinal, 292
monochromatic, 300
motion, 292–306
one-dimensional equation,
    294–295
phase velocity of, 296–301
on shallow water, 313–316
significant height, 319
sinusoidal, 295–296
standing, 292
tidal, 338
transport energy, 296–300
transverse, 292, 294
travelling, 292
water, 306–319
Wells turbine, 321
Wheeler, John A., 83
Wien's displacement law, 136, 137
Wind energy, 4
Wind farm, 278
Windmills, 252

Wind power, 251. *See also* Wind turbines
    extraction by wind turbine, 263–271
    flow of ideal fluids, 256–263
    history of, 251–253
    origin and directions of, 253–256
Wind turbines
    Betz criterion, 265–268
    blades
        action of, 268–270
        rotational speed of, 270–271
    design and operation, 271–277
    siting of, 277–279
    wind power extraction by, 263–271
Work, 177–180

X
X-rays
    characteristic, 24–25
    emission from atom, 26–27

Y
Yaw mechanism, 273

Z
Zenith angle, 152–153